Animal–Computer Interaction: Advances and Opportunities

Animal–Computer Interaction: Advances and Opportunities

Guest Editors

Fiona French
Christopher Flynn Martin

MDPI

Basel • Beijing • Wuhan • Barcelona • Belgrade • Novi Sad • Cluj • Manchester

Guest Editors
Fiona French
School of Computing and
Digital Media
London Metropolitan
University
London
UK

Christopher Flynn Martin
Indianapolis Zoo
Indianapolis
USA

Editorial Office
MDPI AG
Grosspeteranlage 5
4052 Basel, Switzerland

This is a reprint of the Special Issue, published open access by the journal *Animals* (ISSN 2076-2615), freely accessible at: https://www.mdpi.com/journal/animals/special_issues/954F00ES71.

For citation purposes, cite each article independently as indicated on the article page online and as indicated below:

Lastname, A.A.; Lastname, B.B. Article Title. *Journal Name* **Year**, *Volume Number*, Page Range.

ISBN 978-3-7258-5181-2 (Hbk)
ISBN 978-3-7258-5182-9 (PDF)
https://doi.org/10.3390/books978-3-7258-5182-9

Cover image courtesy of Fiona French

Contents

About the Editors

Fiona French

Fiona French is an Associate Professor (Enterprise) in the School of Computing and Digital Media at London Metropolitan University. She is course leader for BSc Games Programming and her field of research is Animal–Computer Interaction. She investigates how playful technology can be used to support the welfare and enrichment of all animals, including humans, and has published widely in this field. She organises regular Zoo Jam workshops at the annual Animal–Computer Interaction conference. These are cross-disciplinary, collaborative events where people share skills, discuss species-specific design briefs, and devise exciting technology-enabled solutions that aim to enhance welfare and offer environmental enrichment for different species. She is a member of the Animal Welfare Research Group (AWRG), the Association for Study of Animal Behaviour (ASAB), the recently formed Acoustic Enrichment Interest Group (AEIG) and a founder of the Multispecies Interaction Design research group at London Metropolitan University.

Christopher Flynn Martin

Christopher Flynn Martin is the Director of Research at The Indianapolis Zoo and an Affiliate Professor at Indiana University's Luddy School of Informatics. He holds a D.Sc. and M.Sc. in Primatology from Kyoto University and a B.A. in Psychology from the University of Pennsylvania. Martin founded Zenrichment, a company that develops custom touchscreen systems and software for primate cognitive research and enrichment at zoos worldwide. His research examines great ape social cognition, communication, imitation, and strategic reasoning. At The Indianapolis Zoo, he leads a cognitive research program with 12 orangutans and 21 chimpanzees, using voluntary touchscreen tasks that provide enrichment, reveal cognitive insights, and educate visitors on species intelligence and conservation. He has authored numerous publications in the fields of primate cognition, animal enrichment, and behavioral research methods.

Preface

Animal–Computer Interaction (ACI) is a multidisciplinary field concerned with the design of technology for, with, and from the perspective of all species of animals. In some cases, interactions with technology are explicit and direct, requiring specially designed interfaces between system and user; in others, the technology may play a hidden role, monitoring and analysing an animal's behaviour or changing their environment. An emphasis on animal welfare gives rise to philosophical considerations, as well as driving new research methods and technologies.

This Special Issue shares new ideas and developments in the field of ACI, including those that advance scientific knowledge about animals, enhance connections between species, improve the stewardship of animals in human care, and articulate the design of systems that offer greater autonomy to other species. This collection of articles features research with mammals, birds, fish, reptiles, and amphibians, including wild, farmed, zoo-housed, and working contexts. The authorship represents five continents (Europe, Asia, North America, South America, and Australia) and amply demonstrates the potential for interesting and insightful collaborations between animal experts and technologists.

One of the themes addressed is the design of monitoring systems and environmental enrichment opportunities for zoo-managed and farm-dwelling species, aiming to support the expression of natural behaviours and analyse markers of well-being. Jeon et al. consider precision farming methods that use computer vision techniques to evaluate the health and welfare of pigs, while Golfidis et al. describe attempts to assess chicken emotions through analysing their vocalisations. French et al. offer a set of guidelines to support the use of technology-enhanced enrichment opportunities aimed at managed nocturnal and crepuscular species, and Vonk discusses the design of an analogue scale for zoo-housed bears so they can express their preferences.

Research in the Special Issue informs our understanding of other species' perceptive and cognitive capabilities. Examples of studies that focus on this aspect include those of Goulart and Young, who explore colour vision in zoo-housed primates, and Wood and Wood, who study how early exposure to three-dimensional geometry relates to future object recognition in farmed chickens.

Several papers focus on the deployment of machine learning techniques to identify individuals or species, monitor behavioural patterns, and analyse genetic markers. Wang et al. describe the development of a system for identifying different species of fish in blurry underwater scenarios, and Xu et al. explain their model for identifying individual endangered Amur tigers. Using a similar approach, Zhang et al. attempt to address conflict resolution between felines and humans in China through early animal detection, while Lund et al. compare the behavioural characteristics of two African elephants. Thorsrud et al. investigate the potential for genomic selection in working dogs by exploring heritable and non-heritable characteristics across a selection of breeds.

The range of topics and approaches is indicative of the wide-ranging interest in ACI across different disciplines and areas of study. Collectively, these diverse studies converge to offer innovative methods and deeper insights into animal technology use, paving the way for future efforts to design more voluntary and participatory interactions that enhance the autonomy, choice and control offered to animals in managed care. Research in this field is always challenging and ground-breaking, with each piece of work adding to our growing pool of knowledge about the others with whom we share our beautiful world.

Fiona French and Christopher Flynn Martin
Guest Editors

animals

MDPI

Article

Investigation through Animal–Computer Interaction: A Proof-of-Concept Study for the Behavioural Experimentation of Colour Vision in Zoo-Housed Primates

Vinícius Donisete Lima Rodrigues Goulart [1] and Robert John Young [2,*]

[1] Transportation Research and Environmental Modelling Laboratory—TREM, Institute of Geosciences, Universidade Federal de Minas Gerais, Belo Horizonte 31270-901, Brazil; viniciusdonisete@gmail.com
[2] School of Science, Engineering and Environment, Peel Building, University of Salford Manchester, Salford M5 4WT, UK
* Correspondence: r.j.young@salford.ac.uk

Simple Summary: Animal–computer interactions provide an opportunity for behavioural investigations. Once appropriate interfaces are established, it is possible to provide new possibilities for improving animal welfare and experimentation. Zoos are an important repository of animals and provide a great source of biological knowledge. By using accessible materials and with low levels of programming, we were able to develop a safe and reliable design for testing the sensory abilities of New World primates. The proof-of-concept for testing colour vision through a behavioural experiment resulted in engagement from the tested animals and an alternative for investigating the sensory abilities of this complex group of animals. In conclusion, we encourage the use of animal-computer interaction frameworks to enrich and develop scientific knowledge from captive animals.

Abstract: Zoos are an important repository of animals, which have a wide range of visual systems, providing excellent opportunities to investigate many comparative questions in sensory ecology. However, behavioural testing must be carried out in an animal welfare-friendly manner, which is practical for zoo staff. Here, we present a proof-of-concept study to facilitate behavioural research on the sensory ecology of captive primates. A system consisting of a tablet computer and an automated feeder connected wirelessly was developed and presented to captive primate species to evaluate interactions with and without previous training. A colour stimulus, analogous to the Ishihara test, was used to check the level of interaction with the device, supporting future studies on sensory ecology with zoo animals. Animals were able to use the system successfully and displayed signs of learning to discriminate between the visual stimuli presented. We identified no risk for small primates in their interactions with the experimental setup without the presence of keepers. The use of electronic devices should be approached with caution to prevent accidents, as a standard practice for environmental enrichment for larger animals (e.g., spider monkeys). In the long term, the system developed here will allow us to address complex comparative questions about the functions of different visual systems in captive animals (i.e., dichromatic, trichromatic, etc.).

Keywords: colour vision; zoo; animal experimentation; behavioural research; sensory ecology; animal–computer interaction; ACI; animal welfare

1. Introduction

Zoos are an important repository of animals, housing tens of thousands of different animal species around the world and millions of individuals, providing environmental education, aiding conservation efforts, and contributing to scientific research [1–3]. The use of artificial elements in exhibits has received some criticism from advocates of strictly naturalistic exhibits advocates, which may have caused reticence in the uptake of modern technology in research [4]. Zoos have a great potential for behavioural research, but for pure research, such as sensory ecology, this resource is largely under-utilised. The most obvious explanation for this is that zoos focus on conservation research, along with the fact that zoos have many other restrictions on the type of research they will allow [5]. For instance, a challenge faced by researchers conducting studies in zoos is the restriction on handling animals or modifying the enclosures, which is often necessary to achieve the best experimental design [6]. Despite the possible mischaracterization of naturalistic enclosures, environmental enrichment devices such as touchscreen computers have not affected the zoo visitors' perceptions towards animal welfare [7]. Zoo-based research often faces difficulties with small sample sizes, which can be overcome by multi-zoo studies with equipment that is easy to adapt to different enclosures. Thus, using methods less demanding to keepers' routines by automating data collection and animal–computer interaction experiments shared among zoological institutions could be the answer. Furthermore, zoo-based research should not have a negative impact on animal welfare. Given that these restrictions (i.e., animal welfare and keeper time) are red lines that zoos will not cross, researchers need to develop alternative methods for conducting their research.

Computer interfaces are centred in human abilities, such as the visual system and physical input hardware (i.e., keyboard or mouse); however, some interfaces are cross-species, such as audio, video tracking, accelerometers, and haptic sensors [8]. This leads to the development of interfaces appropriate for animal use. The animal–computer interaction (ACI) is a growing field in environmental enrichment and the animal experimentation field [9]. It can be a source of stimuli in animal enclosures, providing an enriched environment for better animal welfare [10]. Moreover, this approach can be used to experiment and test hypotheses, thereby allowing the study of animal behaviour in animal collections such as zoos [11].

Tablet computers are an affordable technology to investigate the sensory abilities of different animals: their thin profile and touch interface allow an intuitive interaction when compared to traditional computers, where mouse and keyboards comprise the main input hardware. Therefore, there is no need to translate physical movement into virtual movement [9]. The number of sensors present in commercially available tablets increases the possibility of their application to address behavioural and sensory questions. For instance, accelerometers built in the device can be used to record positional behaviour [12]; proximity sensors can trigger data loggers [13]; or the display and touch screen can be used for stimuli presentation [14,15].

One challenge imposed by using touch screens for colour stimuli presentation is that colour replication accuracy cannot be guaranteed. For instance, sensory ecology and vision research usually employs no digital compression in photos taken from cameras, colour checker cards, or colour-referenced stimuli [16,17]. Therefore, any behavioural research should take into consideration variations in colour stimuli presented on different screen types and proceed with calibration methods.

Colour discrimination in animals is related to the number of photoreceptors sensitive to different wavelengths of light [18]. Vertebrates, such as the nocturnal owl monkey (*Aotus* sp.), have one photoreceptor and are not able to distinguish colours (monochromats); most mammals that have two photoreceptors are dichromats; the majority of Old World primates are trichromats and have three photoreceptors; fishes, reptiles, and birds are tetrachromats and have the best colour vision acuity among vertebrates [19,20]. The greatest number of photoreceptors has been found in the mantis shrimp (*Neogonodactylus oerstedii*), which has 12 photoreceptors [21]. Therefore, when using an animal–computer interface display to perform behavioural research, it should be appropriate to the colour vision system of the species in terms of the colour discrimination tasks involved.

New World primates have a polymorphic colour vision system [22,23]. Within the same species, males are obligatory colour blind with a colour vision similar to a red–green colour blind human, whereas females can be either dichromats or trichromats [24–26]. Each phenotype has its own advantages: dichromats are best suited for vision in low light levels and camouflage breaking, whereas trichromats outperform dichromats in detecting ripe food sources or detecting predators in photopic light levels [27–31]. For instance, dichromat phenotypes under natural environmental conditions are notably more efficient at detecting camouflaged insects, especially in low light, suggesting an ability to break camouflage for detecting food sources and predators [32]. The trichromacy advantage conferred for ripe fruits foraging in New World primates is supported by field studies on intake rates of conspicuous coloured fruits favouring an trichromatic phenotype advantage conferring nutritional benefit on fruit foraging [33]. Further studies investigating social cooperative behaviour related to visual perception would help us understand the role of polymorphic colour vision in New World primates.

Here, we present a proof-of-concept study concerning the realisation of behavioural vision research in zoos. Thus, we investigated the use of commercial devices (i.e., off-the-shelf tablet computers) and the development of a customisable feeder to be used by keepers in zoo enclosures without causing disturbances to animal management or animal welfare.

2. Materials and Methods

2.1. Animals

A mixed group of two marmoset individuals (*Callithrix geoffroyi*), one male and one female, and three titi monkeys (*Plecturocebus cupreus*), a pair and a young male, kept in the same enclosure in Twycross Zoo, United Kingdom, were subjected to experimental sessions. The experimental sessions took place in the same enclosure where the animals were housed; the animals were free to move outdoors. There was no need for handling or capturing the animals during the study, and, therefore, this did not occur.

A single group of three variegated spider monkeys (*Ateles hybritus*), two females and one male, also housed in Twycross Zoo, were also used in the experiment. The experimental setup was placed in an animal management area, with the animals being free to move to their normal indoor enclosure or outdoors during the experimental session.

The animals were not deprived of food, and routine feeding and food enrichments were maintained during this study. The experimental sessions were performed according to the keeper's routine husbandry sessions from December 2016 to April 2017. Each session lasted 10 min and would have been terminated in case of any undesired circumstances (e.g., aggression) towards the apparatus or group members.

2.2. Stimuli

The presented stimuli comprised an image composed of circles in varying sizes and colours, comparable to the Ishihara colour blind test (i.e., pseudoisochromatic plate), which was produced using Java code in Processing v2.2.1. Each circle had a maximum diameter of 22 pixels and a minimum diameter of 8 pixels. A white background was set, and the code produced a random pattern where a static figure in PNG file format was placed randomly on a canvas of 1024 × 576 pixels. The PNG file had no background, and the outline and filling of the image were replaced by red-coloured circles. A rounded shape was used to produce a red target (Figure 1).

An Android app was created using the MIT app inventor, where the images created were shown on a tablet screen (Amazon Fire). If the target was touched, a clicker sound was emitted, and the target's position changed. If the wrong area was touched, a horn sound was played, and the target did not change its location. The tablet had a wireless connection to a feeder, which provided a raisin as a food reward if the right area in the tablet's screen was touched. A red–green colour blind human volunteer checked whether the target was visible or not. Also, a spectrophotometer, Ocean Optics USB 2000+ VIS-NIR (Halma plc, Amersham, UK), attached to a light source (LS-1 Tungsten Halogen light, Ocean Optics, Halma plc, Amersham, UK), was used to collect the relative irradiance from the tablet screen, and the colours of the target and background were compared by calculating their JND (Just Noticeable Difference) modelling trichromatic and dichromatic phenotypes. The tablet used did not have readily available colour calibration; therefore, the colours selected for developing the pseudoisochromatic image used in this study were calibrated by programming, selecting adequate hex-codes given the validation by volunteers and spectrophotometer measurements. Sensory analysis was performed using R and the package Pavo [34,35].

Figure 1. A screen capture of the colour stimuli created to investigate the interaction of captive primates with the apparatus developed to behaviourally investigate colour vision. A food reward was provided by a feeder connected via Bluetooth when the red dots were pressed. A horn was played by the tablet when the green dots were touched.

2.3. *Apparatus*

The feeder consisted of a DC motor attached to a plastic spiral that turned, pushing the reward (raisins) from a plastic container. An Arduino UNO microcontroller board (ATmega328P) was connected to the DC motor using an L298N dual H-bridge DC motor driver module. An Hm-10 Bluetooth module was also connected to the Arduino, allowing the wireless connection to the tablet. A Kindle Fire tablet with a 7-inch screen with a resolution of 1024 × 600 pixels and a rugged case was used (Table 1; Figures 2 and 3). A Ricoh theta 360 degrees camera in a protective case was used to film the experimental sessions.

Table 1. The components used to build a visual stimuli presentation device to behaviourally measure colour vision.

Component	Cost
Kindle Fire tablet computer, 7-inch 1024 × 600 screen, 313 g	GBP 35.00
Arduino Uno microcontroller board ATmega328P	GBP 17.30
DC motor-powered dispenser	GBP 5.00
Bluetooth module Hm-10	GBP 6.00
L298N dual H-bridge DC motor module	GBP 5.00
Rugged tablet case with screen protector	GBP 15.00

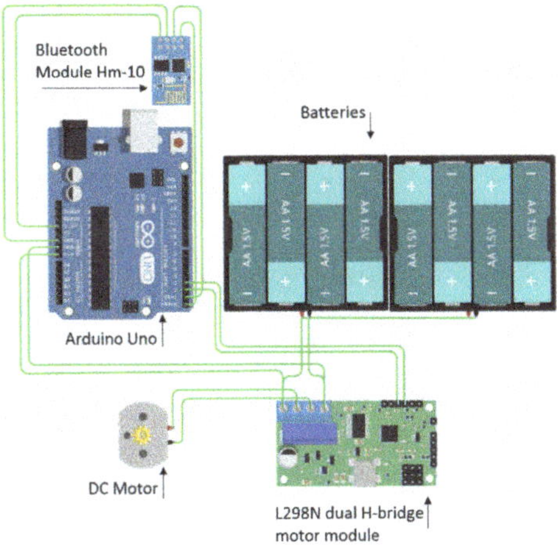

Figure 2. A schematic diagram of the automated wireless feeder used in behavioural colour vision research.

Figure 3. Experimental apparatus during the performance of behavioural tests.

2.4. Procedure

The feeder, tablet, and camera were placed inside the marmoset/titis enclosure for 10 trials. At the start of the session, a raisin was placed on top of the correct stimuli to help the association between the red target and the reward. No training was performed with the marmosets and titis.

In the spider monkey enclosure, the camera and the feeder were positioned outside the enclosure, and the keeper held the tablet, allowing the animals to touch the screen through the mesh. The training consisted of placing the raisins on top of the target during the first two minutes of the presentation, leaving the animals free to interact with the tablet after the presentation. Behavioural data were collected ad libitum, reporting behaviours expressed during interactions with the device. The time and the result of the interaction were recorded. A total of 11 trials were performed.

The data were checked for normality, and non-parametric tests were used as we found the data did not fulfil parametric requirements. A logistic regression was used to evaluate the association between the tablet and the feeder. The number of sequential interactions between the tablet and feeder occurred during the experimental session in the marmoset/titi enclosure. The frequency of training (only with the spider monkey group) and the number of correct uses was checked for correlation to the accumulative time using a Spearman's rank correlation test. All statistics were performed in the R statistical computing language [35].

3. Results

3.1. Stimuli

The colours suitable for a colour blind test are shown in Table 2 and Figure 4. By averaging the target colours and background colour from the tablet screen using their relative irradiance values, we found a colour distance of 2.8040 JND (Just Noticeable Difference) for a trichromatic phenotype, and 0.2849 JND for a dichromat viewer. Therefore, the target and background were not distinguishable in colour for dichromat phenotypes.

Table 2. Colour references of the pseudoisochromatic stimuli found to be undistinguishable by di-chromatic colour vision phenotype (i.e., a red–green colour blind individual).

Colour Category	Hex-Colour Code	CIE-Lab L	CIE-Lab A	CIE-Lab L
Green	d9CA594	66.6087	−6.3958	7.7143
Green	ACB4A5	72.3152	−5.5814	6.6789
Green	BBB946	73.3536	−13.7646	56.6906
Green	D1D6AF	84.3514	−8.2588	18.7266
Green	D7DAAA	85.7519	−8.8320	23.3500
Green	E5D57D	84.8571	−6.2664	45.3062
Red	EBA170	72.4431	22.2198	36.4731
Red	F9BB82	80.3436	15.6987	37.4752
Red	FCCD84	85.0087	7.6967	42.4568

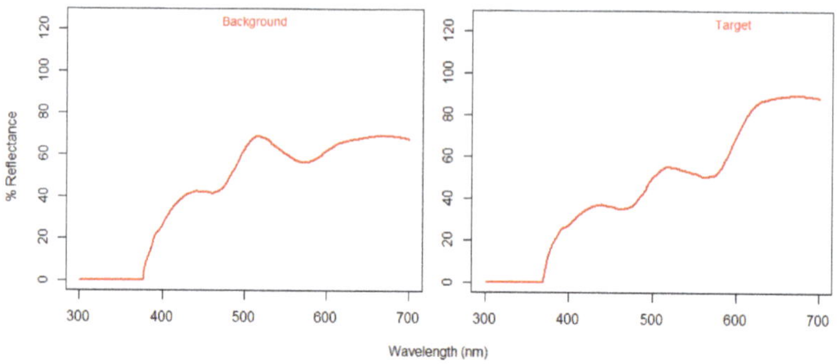

Figure 4. The relative irradiance of the pseudoisochromatic image taken from a commercial tablet (Kindle Fire).

3.2. Marmosets and Titis

A total of 82.52 min of stimuli presentation were obtained from ten experimental sessions. The device screen was sensitive to the touch of marmosets and titis (activated also by accidental touches, such as inspections and stepping on it). Interactions with the device lasted an average of 16 s with a range of 2 to 101 s. During the experimental session, device interactions accounted for 24.92% of the total time. Most of the interactions in the mixed species enclosure were performed by the marmosets (92.21%), who expelled the titis from the testing platform. We observed an association between the tablet and the feeder, with the animals inspecting the tablet and the feeder sequentially and looking to the display. From the experimental sessions, we did not find marmosets and titis using the device only to receive the rewards but also as an environmental enrichment as they were curious about activating the feeder. Likewise, the individuals associated the tablet with the feeder significantly, according to a logistic regression ($\beta \pm SE = 0.001 \pm 0.0003$; $Z = 4.131$; $p < 0.001$).

3.3. Spider Monkeys

A total of 11 experimental sessions were performed, with a total interaction time of 66.68 min. All time was used by the experimental animals interacting with the device due to a different experimental setup from titis and marmosets. We found that spider monkeys were able to learn to use the device because they were able to receive rewards from the feeder five times in a row (i.e., no incorrect interactions). This result suggests that the female individual may possess trichromatic colour vision, indicating behavioural experiments as a potential method for colour vision assessment in zoo animals. Agonistic behaviours were observed in two experimental sessions, where the male, not being able to select the right colour target, repelled the female from receiving the reward. No collaborative behaviour was observed between the males and females. The number of times that the keeper had to demonstrate and habituate the animal to use the device (i.e., training) was negatively correlated with the cumulative time of the experiment ($rs = -0.6245$; $N = 11$; $p = 0.0399$). However, we found no correlation between the cumulative time and correct use of the device ($rs = -0.2535$, $N = 11$; $p = 0.4520$). Overall, the experimental animals were participative and appeared motivated to use the experimental apparatus in the behavioural tests.

4. Discussion

The zoo staff found the device simple to use; it was easy to build and met the requirements for behavioural vision research in zoos. The touch screen of the commercial tablet with a rugged case and screen protector was reactive to the primates' touch. Even small primates (i.e., <500 g), such as marmosets, were able to interact and receive a reward. Therefore, the device is a suitable alternative to expensive scientific equipment used in some behavioural research [15]. Efforts to bridge the gap between captive and wild animal research will provide more research opportunities and support an interdisciplinary approach [36]. For instance, research on colour vision requires the knowledge of the physical properties of light, physiological aspects of colour vision, and implications of different phenotypes on the sensory ecology [37–39]. Methodologies that allow us to access the sensory abilities of animal subjects are crucial to increasing knowledge in the field.

Two major hypotheses about the importance of polymorphic colour vision for New World primates are niche divergence and mutual benefit association [40,41]. By applying similar setups, the niche divergence could be verified by changing the stimuli type from colour conspicuous to cryptic and analysing the behaviour demonstrated by the different phenotypes. The mutual benefit association hypothesis could be tested by investigating collaboration among individuals in response to a given colour vision task. The use of animal–computer interaction-based experiments with captive primates contributes substantially to the collection of behavioural data and controlling environmental variables.

The marmosets used in this study also inhabit forest fragments in cities, are frequently in contact with city dwellers, and are often hand-fed [42–44]; yet, they preserve many aspects of their natural behaviour [42]. The small sample sizes that often limit the value of research in zoos could be reduced by investigating urban animals. The portability of the setup used in this proof-of-concept study could easily be extrapolated to research in urban environments, which are more flexible regarding the access to animals and possibilities of modifying the environment. The marmosets in our study were able to associate the feeder and the tablet without assistance (training) from keepers. Thus, further studies with urban marmosets should be investigated.

One of our main goals with this study was to verify the use of zoo enclosures as an experimental area, having the keepers performing the experiment. Keepers have close contact with their animals, and their interaction is relevant in designing scientific experiments [45]. It was possible to shape the behaviour of spider monkeys with a few presentations of the experimental equipment. Nevertheless, preparation to perform the research and interest from the zoo staff are critical for success. Fortunately, Twycross Zoo is an institution interested in supporting behavioural research. Further presentations could be performed without the keeper's presence, as the experimental setup permits its use without any human assistance.

As the opportunity to manipulate zoo enclosures is limited, certain precautions must be taken to ensure the successful realisation of experiments. We found that mixed species enclosures can lead to certain species not being able to participate in the experiments. This should be considered in future studies concerning socially housed species. A critical aspect of video recording is illumination. Indoor areas should be assessed regarding their light sources, since the identification of the individual performing the behaviour is important in behavioural studies. Keepers often have a profound knowledge of each individual animal in their care, being able to recognise them without marking them. However, individual animal identification might be not possible if a 'control' researcher is used to analyse the recordings, for instance, if a double-blind experimental design is used, where the executive and analysis of experiments are performed by two different researchers independently.

Touch-sensitive tablet computer screens can have different colour reproduction from the colour selected on the computer used for programming the tablet. However, we controlled for this by selecting colours that were not visible to a dichromatic (human) colour vision phenotype. This problem is of major concern when developing colour-based tasks in which the colour stimuli may be altered on different screens. Despite this, it was possible to use a commercial tablet to generate colour targets and backgrounds that are not detectable by red–green colour blind viewers.

Unfortunately, due to time limitations, it was not possible to proceed with detailed behavioural studies of vision, but the concept proved that a computer tablet-based system can be used to behaviourally assess visual perception in zoo-housed primates.

5. Conclusions

The experimental design focused on the feasibility of animal–computer interaction-based experiments with captive animals to study sensory behaviour, which is expected to vary among the experimental subjects. We successfully tested a safe device that provided an interactive interface for studying animal perception with little impact on keepers' routines. We observed learning in spider monkeys with minimal training, and marmosets showed a significant interest in interacting with the system. Further studies can evaluate the differences in interactions between group members and infer about perception abilities.

The interest shown for interactive devices, such as touch screens and reward mechanisms, has the potential to be used as environmental enrichment in small primate enclosures. The use of simple builds with components "off-the-shelf" can be employed securely for small primate species. No aggressive behaviours, which could terminate the experimental sessions, were observed, and no damaging behaviours (e.g., biting) were observed towards the device.

Behavioural research involving the development of interfaces for interactive devices can contribute to increasing the scientific knowledge of species biology, conciliating the demands from animal management in zoological institutions, and providing the controlled environment needed for experimentation.

Author Contributions: V.D.L.R.G. and R.J.Y. conceived the study, analysed the data, and wrote the manuscript. All authors have read and agreed to the published version of the manuscript.

Funding: This research received no external funding. Vinícius Goulart was in receipt of a CAPES postgraduate scholarship Proc. 1213-13-0.

Institutional Review Board Statement: This study was conducted in accordance with the Declaration of Helsinki and approved by the University of Salford, Manchester.

Informed Consent Statement: Not applicable.

Data Availability Statement: The data are available upon request from the author.

Acknowledgments: We would like thank the Twycross zoo and the animal care team for allowing the development of this research and for encouraging, supporting, and facilitating the behavioural sessions.

Conflicts of Interest: The authors declare no conflicts of interest.

References

1. Barongi, R.; Fisken, F.A.; Parker, M.; Gusset, M. *Committing to Conservation: The World Zoo and Aquarium Conservation Strategy*; WAZA: Gland, Switzerland, 2015.
2. Patrick, P.G.; Matthews, C.E.; Ayers, D.F.; Tunnicliffe, S.D. Conservation and Education: Prominent Themes in Zoo Mission Statements. *J. Environ. Educ.* **2007**, *38*, 53–60. [CrossRef]
3. Tribe, A.; Booth, R. Assessing the Role of Zoos in Wildlife Conservation. *Hum. Dimens. Wildl.* **2003**, *8*, 65–74. [CrossRef]
4. Fernandez, E.J.; Martin, A.L. Animal Training, Environmental Enrichment, and Animal Welfare: A History of Behavior Analysis in Zoos. *J. Zool. Bot. Gard.* **2021**, *2*, 531–543. [CrossRef]
5. Kleiman, D.G. Behavior Research in Zoos: Past, Present, and Future. *Zoo Biol.* **1992**, *11*, 301–312. [CrossRef]
6. Hosey, G.R. Behavioural Research in Zoos: Academic Perspectives. *Appl. Anim. Behav. Sci.* **1997**, *51*, 199–207. [CrossRef]
7. Jacobson, S.L.; Hopper, L.M.; Shender, M.A.; Ross, S.R.; Leahy, M.; McNernie, J. Zoo Visitors' Perceptions of Chimpanzee Welfare Are Not Affected by the Provision of Artificial Environmental Enrichment Devices in a Naturalistic Exhibit. *J. Zoo Aquar. Res.* **2017**, *5*, 56–61. [CrossRef]
8. McGrath, R.E. Species-Appropriate Computer Mediated Interaction. In Proceedings of the CHI '09 Extended Abstracts on Human Factors in Computing Systems, Boston, MA, USA, 4–9 April 2009; ACM: New York, NY, USA, 2019; pp. 2529–2534.
9. Ritvo, S.E.; Allison, R.S. Designing for the Exceptional User: Nonhuman Animal-Computer Interaction (ACI). *Comput. Hum. Behav.* **2017**, *70*, 222–233. [CrossRef]
10. Wirman, H.; Zamansky, A. Toward Characterization of Playful ACI. *Interactions* **2016**, *23*, 47–51. [CrossRef]
11. Mancini, C. Animal-Computer Interaction: A Manifesto. *Interactions* **2011**, *18*, 69–73. [CrossRef]
12. Graf, P.M.; Wilson, R.P.; Qasem, L.; Hackländer, K.; Rosell, F. The Use of Acceleration to Code for Animal Behaviours; A Case Study in Free-Ranging Eurasian Beavers Castor Fiber. *PLoS ONE* **2015**, *10*, e0136751. [CrossRef]
13. Nathan, R.; Getz, W.M.; Revilla, E.; Holyoak, M.; Kadmon, R.; Saltz, D.; Smouse, P.E. A Movement Ecology Paradigm for Unifying Organismal Movement Research. *Proc. Natl. Acad. Sci. USA* **2008**, *105*, 19052–19059. [CrossRef]
14. Takemoto, A.; Miwa, M.; Koba, R.; Yamaguchi, C.; Suzuki, H.; Nakamura, K. Individual Variability in Visual Discrimination and Reversal Learning Performance in Common Marmosets. *Neurosci. Res.* **2015**, *93*, 136–143. [CrossRef] [PubMed]
15. Takemoto, A.; Izumi, A.; Miwa, M.; Nakamura, K. Development of a Compact and General-Purpose Experimental Apparatus with a Touch-Sensitive Screen for Use in Evaluating Cognitive Functions in Common Marmosets. *J. Neurosci. Methods* **2011**, *199*, 82–86. [CrossRef]
16. Melin, A.D.; Kline, D.W.; Hickey, C.M.; Fedigan, L.M. Food Search through the Eyes of a Monkey: A Functional Substitution Approach for Assessing the Ecology of Primate Color Vision. *Vis. Res.* **2013**, *86*, 87–96. [CrossRef] [PubMed]
17. Pessoa, D.M.A.; Cunha, J.F.; Tomaz, C.; Pessoa, V.F. Colour Discrimination in the Black-Tufted-Ear Marmoset (*Callithrix penicillata*): Ecological Implications. *Folia Primatol.* **2005**, *76*, 125–134. [CrossRef] [PubMed]
18. Yokoyama, S. Molecular Evolution of Vertebrate Visual Pigments. *Prog. Retin. Eye Res.* **2000**, *19*, 385–419. [CrossRef]
19. Bowmaker, J.K. Evolution of Vertebrate Visual Pigments. *Vis. Res.* **2008**, *48*, 2022–2041. [CrossRef]
20. Jacobs, G.H. Evolution of Colour Vision in Mammals. *Phil. Trans. R. Soc. B* **2009**, *364*, 2957–2967. [CrossRef] [PubMed]
21. Thoen, H.H.; How, M.J.; Chiou, T.-H.; Marshall, J. A Different Form of Color Vision in Mantis Shrimp. *Science* **2014**, *343*, 411–413. [CrossRef]
22. Hunt, D.M.; Dulai, K.S.; Cowing, J.A.; Julliot, C.; Mollon, J.D.; Bowmaker, J.K.; Li, W.-H.; Hewett-Emmett, D. Molecular Evolution of Trichromacy in Primates. *Vis. Res.* **1998**, *38*, 3299–3306. [CrossRef]
23. Jacobs, G.H. New World Monkeys and Color. *Int. J. Primatol.* **2007**, *28*, 729–759. [CrossRef]
24. De Valois, R.L.; Jacobs, G.H. Primate Color Vision: The Macaque and Squirrel Monkey Differ in Their Color Vision and in the Physiology of Their Visual Systems. *Science* **1968**, *162*, 533–540. [CrossRef] [PubMed]
25. Dulai, K.S.; Von Dornum, M.; Mollon, J.D.; Hunt, D.M. The Evolution of Trichromatic Color Vision by Opsin Gene Duplication in New World and Old World Primates. *Genome Res.* **1999**, *9*, 629–638. [CrossRef] [PubMed]
26. Jacobs, G.H.; Neitz, M.; Deegan, J.F.; Neitz, J. Trichromatic Colour Vision in New World Monkeys. *Nature* **1996**, *382*, 156–158. [CrossRef] [PubMed]
27. Dominy, N.J.; Lucas, P.W. Ecological Importance of Trichromatic Vision to Primates. *Nature* **2001**, *410*, 363–366. [CrossRef]
28. Morgan, M.J.; Adam, A.; Mollon, J.D. Dichromats Detect Colour-Camouflaged Objects That Are Not Detected by Trichromats. *Proc. R. Soc. Lond. B* **1992**, *248*, 291–295. [CrossRef]
29. Osorio, D.; Vorobyev, M. Colour Vision as an Adaptation to Frugivory in Primates. *Proc. R. Soc. B.* **1996**, *263*, 593–599. [CrossRef]
30. Pessoa, D.M.A.; Maia, R.; De Albuquerque Ajuz, R.C.; De Moraes, P.Z.P.M.R.; Spyrides, M.H.C.; Pessoa, V.F. The Adaptive Value of Primate Color Vision for Predator Detection. *Am. J. Primatol.* **2014**, *76*, 721–729. [CrossRef] [PubMed]
31. Verhulst, S.; Maes, F.W. Scotopic Vision in Colour-Blinds. *Vis. Res.* **1998**, *38*, 3387–3390. [CrossRef]

32. Melin, A.D.; Fedigan, L.M.; Hiramatsu, C.; Sendall, C.L.; Kawamura, S. Effects of Colour Vision Phenotype on Insect Capture by a Free-Ranging Population of White-Faced Capuchins, *Cebus Capucinus*. *Anim. Behav.* **2007**, *73*, 205–214. [CrossRef]
33. Melin, A.D.; Chiou, K.L.; Walco, E.R.; Bergstrom, M.L.; Kawamura, S.; Fedigan, L.M. Trichromacy Increases Fruit Intake Rates of Wild Capuchins (*Cebus Capucinus Imitator*). *Proc. Natl. Acad. Sci. USA* **2017**, *114*, 10402–10407. [CrossRef] [PubMed]
34. Maia, R.; Eliason, C.M.; Bitton, P.; Doucet, S.M.; Shawkey, M.D. Pavo: An R Package for the Analysis, Visualization and Organization of Spectral Data. *Methods Ecol. Evol.* **2013**, *4*, 906–913. [CrossRef]
35. R Core Team. *R: A Language and Environment for Statistical Computing*; R Foundation for Statistical Computing: Vienna, Austria, 2022.
36. Shettleworth, S.J. Animal Cognition and Animal Behaviour. *Anim. Behav.* **2001**, *61*, 277–286. [CrossRef]
37. Endler, J.A. The Color of Light in Forests and Its Implications. *Ecol. Monogr.* **1993**, *63*, 1–27. [CrossRef]
38. Isbell, L.A. Snakes as Agents of Evolutionary Change in Primate Brains. *J. Hum. Evol.* **2006**, *51*, 1–35. [CrossRef] [PubMed]
39. Stevens, M. *Sensory Ecology, Behaviour, and Evolution*; Oxford University Press: Oxford, UK, 2013; ISBN 978-0-19-960177-6.
40. Melin, A.D.; Fedigan, L.M.; Hiramatsu, C.; Kawamura, S. Polymorphic Color Vision in White-Faced Capuchins (*Cebus Capucinus*): Is There Foraging Niche Divergence among Phenotypes? *Behav. Ecol. Sociobiol.* **2008**, *62*, 659–670. [CrossRef]
41. Hiwatashi, T.; Okabe, Y.; Tsutsui, T.; Hiramatsu, C.; Melin, A.D.; Oota, H.; Schaffner, C.M.; Aureli, F.; Fedigan, L.M.; Innan, H.; et al. An Explicit Signature of Balancing Selection for Color-Vision Variation in New World Monkeys. *Mol. Biol. Evol.* **2010**, *27*, 453–464. [CrossRef]
42. Duarte, M.H.L.; Goulart, V.D.L.R.; Young, R.J. Designing Laboratory Marmoset Housing: What Can We Learn from Urban Marmosets? *Appl. Anim. Behav. Sci.* **2012**, *137*, 127–136. [CrossRef]
43. Goulart, V.D.L.R.; Teixeira, C.P.; Young, R.J. Analysis of Callouts Made in Relation to Wild Urban Marmosets (*Callithrix penicillata*) and Their Implications for Urban Species Management. *Eur. J. Wildl. Res.* **2010**, *56*, 641–649. [CrossRef]
44. Teixeira, B.; Hirsch, A.; Goulart, V.D.L.R.; Passos, L.; Teixeira, C.P.; James, P.; Young, R. Good Neighbours: Distribution of Black-Tufted Marmoset (*Callithrix penicillata*) in an Urban Environment. *Wildl. Res.* **2015**, *42*, 579. [CrossRef]
45. Carlstead, K. A Comparative Approach to the Study of Keeper–Animal Relationships in the Zoo. *Zoo Biol.* **2009**, *28*, 589–608. [CrossRef] [PubMed]

animals

MDPI

Article

A Serial Multi-Scale Feature Fusion and Enhancement Network for Amur Tiger Re-Identification

Nuo Xu [1], Zhibin Ma [1], Yi Xia [1], Yanqi Dong [1], Jiali Zi [1], Delong Xu [1], Fu Xu [1,2], Xiaohui Su [1,2], Haiyan Zhang [1,2] and Feixiang Chen [1,2,*]

1 School of Information Science and Technology, Beijing Forestry University, Beijing 100083, China; xu993790@bjfu.edu.cn (N.X.); mmazb@bjfu.edu.cn (Z.M.); xiayi@bjfu.edu.cn (Y.X.); yanqidong@bjfu.edu.cn (Y.D.); jializi@bjfu.edu.cn (J.Z.); xudelong@bjfu.edu.cn (D.X.); xufu@bjfu.edu.cn (F.X.); suxhui@bjfu.edu.cn (X.S.); zhyzml@bjfu.edu.cn (H.Z.)

2 Engineering Research Center for Forestry-Oriented Intelligent Information Processing, National Forestry and Grassland Administration, Beijing 100083, China

* Correspondence: bjfxchen@bjfu.edu.cn

Simple Summary: The Amur tiger is an endangered species in the world, and effective statistics on its individuals and population through re-identification will contribute to ecological diversity investigation and assessment. Due to the fact that the fur texture features of the Amur tiger contain genetic information, the main method of identifying Amur tigers is to distinguish their fur and facial features. In summary, this paper proposes a serial multi-scale feature fusion and enhancement network for Amur tiger re-identification, and designs a global inverted pyramid multi-scale feature fusion module and a local dual-domain attention feature enhancement module. We aim to enhance the learning of fine-grained features and differences in fur texture by better fusing and enhancing global and local features. Our proposed network and module have achieved excellent results on the public dataset of the ATRW.

Abstract: The Amur tiger is an important endangered species in the world, and its re-identification (re-ID) plays an important role in regional biodiversity assessment and wildlife resource statistics. This paper focuses on the task of Amur tiger re-ID based on visible light images from screenshots of surveillance videos or camera traps, aiming to solve the problem of low accuracy caused by camera perspective, noisy background noise, changes in motion posture, and deformation of Amur tiger body patterns during the re-ID process. To overcome this challenge, we propose a serial multi-scale feature fusion and enhancement re-ID network of Amur tiger for this task, in which global and local branches are constructed. Specifically, we design a global inverted pyramid multi-scale feature fusion method in the global branch to effectively fuse multi-scale global features and achieve high-level, fine-grained, and deep semantic feature preservation. We also design a local dual-domain attention feature enhancement method in the local branch, further enhancing local feature extraction and fusion by dividing local feature blocks. Based on the above model structure, we evaluated the effectiveness and feasibility of the model on the public dataset of the Amur Tiger Re-identification in the Wild (ATRW), and achieved good results on mAP, Rank-1, and Rank-5, demonstrating a certain competitiveness. In addition, since our proposed model does not require the introduction of additional expensive annotation information and does not incorporate other pre-training modules, it has important advantages such as strong transferability and simple training.

Keywords: Amur tiger; intelligent recognition; deep learning; double branch structure; feature pyramid; attention mechanism

1. Introduction

Widespread distribution, low population density, unpredictable behavior patterns, and sensitivity to interference of wildlife pose significant challenges to monitoring work for some animal species. Traditional wildlife investigation techniques mainly include manual investigation, line sampling, collar tracking, and acoustic tracking using sound recording instruments [1,2]. However, each of these methods has certain disadvantages, so scientists strive to improve them. The Amur tiger, also known as the Siberian Tiger, is one of the subspecies of tigers. The Amur tiger is mainly distributed in the northeastern region of Asia and is listed as an endangered species in the Red List of Threatened Species by the World Conservation Union. There are only just over 500 Amur tigers left in the world, so it is crucial to strengthen the protection of the Amur tiger [3]. Moreover, the survival and reproduction of species populations are closely related to regional biodiversity and ecosystem functional integrity [4]. Therefore, re-evaluating the Amur tiger and its prey resources in natural environments such as nature reserves and national parks can help to statistically analyze the situation of Amur tiger resources and provide data reference for the next step of protection work [5,6]. At present, the most commonly used method for the re-ID of wild animals is manual discrimination. After receiving professional knowledge and training, wildlife protection professionals need to screen and distinguish a large amount of image data based on the fur pattern characteristics of the abdomen, head, neck, and other parts of the Amur tiger [7,8]. To reduce errors, it is necessary for multiple people to simultaneously identify and verify the recognition results, which requires a large workload, high cost, and low efficiency.

With the continuous development and application of machine learning and deep learning technologies, machine learning algorithms and emerging deep learning models, such as Linear Clustering [9], Classification [10,11], Detection [12,13], and Generative Adversarial Networks [14], are gradually being applied to the intelligent monitoring and protection of wildlife. Research on the intelligent recognition of wildlife mainly focuses on issues such as wildlife re-ID, species classification, population counting, and attribute recognition [15]. In the process of wildlife re-ID, the application of computer vision-related technologies can greatly improve recognition efficiency and accuracy. Research in this area has gradually become popular. Currently, the main methods used are clustering algorithms based on image hotspots [9] and convolutional neural network models based on VGG [16], AlexNet [17], and ResNet [18]. These methods have been improved in optimizing feature extraction, feature fusion, and incorporating prior knowledge of pose. Zheng et al. proposed a Transformer network structure with cross-attention block (CAB) and local awareness (CATLA Transformer) [19], which captures global information of an animal body's surface and local feature differences in fur, color, texture, or face, and fuse global features and local features through CATLA Transformer. Zhang et al. proposed using texture features as global and local features for re-ID, and proposed a pyramid feature fusion model method to extract features from both local and global perspectives, effectively matching entities [20]. Li et al. proposed an Amur tiger re-ID method, which introduces precise pose parts with deep neural networks to handle the large pose variation of tigers [3]. Liu et al. proposed a Partial Pose Guided Network (PPGNet), which uses local image features based on pose data to drive the network to extract features from the original image, and applies it to an Amur tiger re-ID system based on automatic detection and Amur tiger pose estimation [21]. He et al. proposed a Multi-pose Feature Fusion Network (MPFNet), which constructs three pose modules: standing, sitting, and lying. In each module, two parallel branches are used to extract global and local features for effective feature extraction. Finally, the features are fused [22].

There are also some very advanced studies in the field of person re-identification similar to the Amur tiger re-ID. Sun et al. proposed a Part-based Convolutional Baseline (PCB) framework and an inter-block combination method with uniform partitioning to effectively extract part-level features, and by Refined Part Pooling (RPP), closer parts are allocated together to improve the within-part consistency of parts [23]. Sun et al. considered the problem of partial re-ID and proposed a Visibility-aware Part Model (VPM). Through self-supervised learning, the model perceives the features within the visible region, extracts regional features, and compares two images within their shared regions to suppress noise in unshared regions. It better extracts fine-grained features of the image and reduces image misalignment [24]. Liu et al. proposed a multi-scale Feature Enhancement (MFE) Re-ID model and a Feature Preserving Generative Adversarial Network (FPGAN). In the MFE, the semantic feature maps of the person's body are segmented, and then multi-scale feature extraction and enhancement are performed on the person's body region. In the FPGAN, the source domain is transferred to the target domain in an unsupervised manner, maximizing the preservation of personal information integrity [25].

In current research, there are issues such as the need for prior knowledge and the complexity of training large models. Although some models have been verified to have excellent average accuracy and other indicators, and the effectiveness of model improvement has been verified through ablation experiments, a large amount of reliance on prior knowledge leads to poor model transferability, requiring staff with expertise in wildlife to perform a large amount of dataset labeling and processing work in the early stages before retraining the model, which has low feasibility in practical production applications. The networks with four, six, or more branches, or which require data preprocessing through instance segmentation models before being fed into the re-ID model, are too complex and have problems such as large model size and a need for complex training. Therefore, this paper proposes a serial multi-scale feature fusion and enhancement re-ID network of Amur tigers with global inverted pyramid multi-scale feature fusion and local dual-domain attention feature enhancement for the re-ID of Amur tiger images. Combining the re-ID methods of the Amur tiger and fine-grained task properties, the Path Aggregation Network (PANet) [26] feature fusion idea is introduced. A bottom-up unidirectional feature fusion method is proposed, which uses an inverted pyramid structure for feature fusion. This helps to better integrate high-level features with large receptive fields and rich semantic information while preserving multi-scale features. We propose a local dual-domain attention feature enhancement method that is serially connected with the global branch to enhance local feature extraction and fusion. Our goal is not to go beyond the SOTA model used for re-ID, but to propose an end-to-end model that is more suitable for removing animal pose prior knowledge and other additional attribute information, and has good transferability and re-ID performance. Our core contributions are as follows:

- We integrate and propose a lightweight, efficient, end-to-end network for the re-ID task of the Amur tiger, which does not require the introduction of prior knowledge such as posture. It can be quickly and conveniently used for the re-ID task of other large mammals. The specific network innovation and design are as follows.
- In order to better extract and integrate the global information of the high-level and low-level layers of the Amur tiger, we propose a multi-scale feature fusion method of the global inverted pyramid. We introduce the ideas of Feature Pyramid Network (FPN) [27] and PANet into the global branch of the model for the task of wildlife re-ID. Improving the top-down connection method of traditional feature pyramid models will greatly compress the problem of key deep semantic information [28].
- In order to deepen the feature extraction of various parts of the Amur tiger and extract fine-grained features such as body fur texture, we introduce a serial local branch network and design an attention module and output feature fusion method in the local branch.

2. Materials and Methods

This section mainly introduces the dataset we use, the basic process of Amur tiger re-ID, and the structure and details of our proposed Amur tiger re-ID network.

2.1. Dataset

To validate the effectiveness of the proposed method and model, we conducted training, testing, and evaluation using the public dataset of the ATRW [3]. The ATRW dataset is a dataset jointly released by Shanghai Jiao Tong University and Intel Laboratories with the assistance of the World Wildlife Fund International (WWF) in 2019 for the detection, joint estimation, and re-ID tasks of the Amur tiger. The authors of this dataset collected over 8000 video clips of 92 tigers from approximately 10 zoos in China to create the ATRW dataset. After data organization and classification, the training dataset for the re-ID task contains 107 tiger entities, totaling 1887 images. The test dataset contains 47 tiger entities, with a total of 701 images. To enhance the robustness of the model and increase the number of training samples, we performed random rotation and random occlusion enhancement on some images in the training dataset, and set the image size H × W to 256 × 512 for better results [21]. We reduced the influence of external factors such as shooting angle and shooting position, and randomly divided the data into the training set Train and validation set Val in a 7:3 ratio (Table 1). The train set we use has an average of 18 training images per entity after data augmentation, with at least 9 training images for each entity (Figure 1).

Table 1. Training and Testing dataset used in the experiment.

Train Dataset	Amur Tiger Entities	Amur Tigers	Original Images
Train + Val	107	75	1552 + 35
Test Dataset	**Amur Tiger Entities**	**Amur Tigers**	**Original Images**
Query	47	42	701
Gallery	47	42	701

(a) (b) (c) (d) (e) (f)

Figure 1. (**a**–**d**) are the original images of the public dataset of the ATRW, and (**e**,**f**) are examples of data-enhanced images.

2.2. Methods

2.2.1. Serial Multi-Scale Feature Fusion and Enhancement re-ID Network of Amur Tiger

In this paper, we propose a network aimed at completing the task of Amur tiger re-ID, which is a serial multi-scale feature fusion and enhancement re-ID network of the Amur tiger (Figure 2). The network is mainly divided into two parts: a global branch and a local branch, which are combined to achieve the final effect. ResNet50 is a convolutional neural network with a depth of 50 layers, which has excellent classification performance on ImageNet [18]. Its pre-trained model was trained on ImageNet with over 1 million images. Therefore, our proposed re-ID network applies the backbone ResNet50 and removes the last down-sampling layer of ResNet50 to retain a larger scale [29]. On the basis of the backbone, a dual-branch structure is constructed based on the feature tensor obtained from ResNet50 layer 4 to achieve the localization of the position of the Amur tiger in the image and fine-grained re-ID requirement. Our proposed dual branches are connected in a serial manner, and the feature tensors obtained from the global branch output are sent to the local branch for the next step of feature extraction and fusion (the details of the global and local branches are in Sections 2.2.2 and 2.2.3). Finally, the feature tensors $F_{Global} \in R^{2048 \times 1}$ and $F_{Local} \in R^{2048 \times 1}$ for global and local branch outputs are obtained, which are concatenated and sent to the classifier layer. Through the linear layer, they are expanded to the corresponding number of categories.

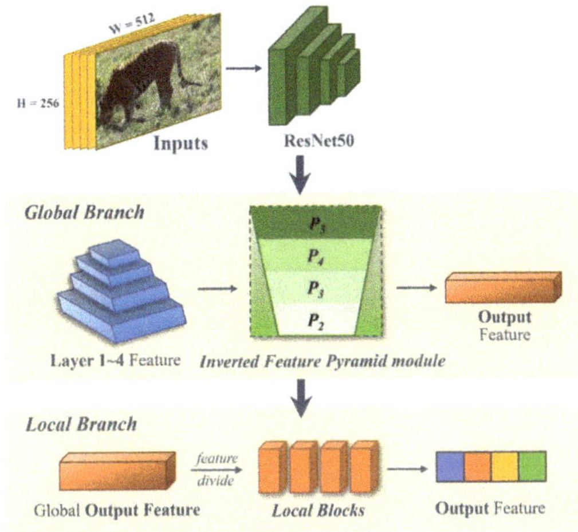

Figure 2. Our proposed Amur tiger re-ID network structure.

2.2.2. Global Inverted Pyramid Multi-Scale Feature Fusion Method

Taking inspiration from the FPN model and the PANet model, we propose an Inverted Feature Pyramid Module (IFPM) for global multi-scale feature fusion. We attempt to guide the fusion of multi-scale features from coarse to fine and from low-level to high-level, reducing the compression of deep semantic information and maximizing the retention of high-level features, in order to construct a multi-scale feature pyramid dominated by deep semantic information on the global feature branch.

As shown in the yellow part of Figure 3, we propose a reverse feature fusion path in the global branch. From bottom to top, we use the features extracted from layer 1 to layer 4 in the backbone as input feature maps of the global inverted pyramid multi-scale feature fusion module. Finally, we obtain the output feature map.

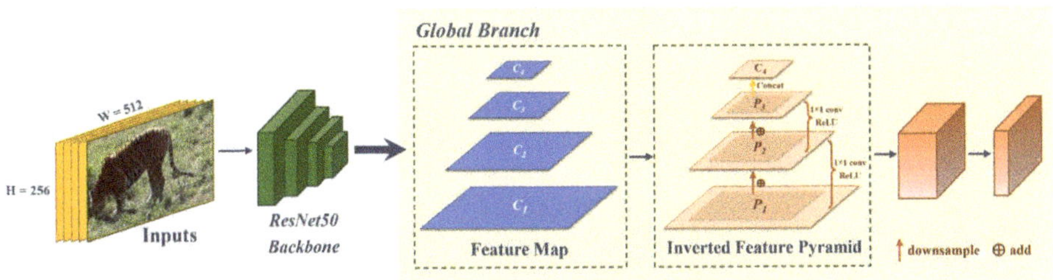

Figure 3. Principle of global inverted pyramid multi-scale feature fusion method.

Specifically, we define $C \in R^{C \times H \times W}$ to represent the features output by each layer of ResNet50, where H × W corresponds to the spatial dimensions of the feature map, and C denotes the number of channels. In the multi-scale feature fusion module of the global inverted pyramid, we designed a multi-scale feature fusion connection strategy. We utilize the features $C_1 \in R^{256 \times 32 \times 64}$, $C_2 \in R^{512 \times 16 \times 32}$, $C_3 \in R^{1024 \times 8 \times 16}$, and $C_4 \in R^{2048 \times 8 \times 16}$ extracted from layer 1 to layer 4 in ResNet50 as input features. Starting from C_1, C_1 passes through the down-sampling layer based on the H × W of C_2 to obtain P_1. After adding P_1 to C_2, the next step of feature fusion and ReLU is performed to obtain P_2. Then, based on the H × W dimension of the upper layer of the inverted pyramid, C_3 and C_4 are sequentially subjected to down-sampling, convolutional feature extraction, fusion, concatenation, and ReLU activation to complete the feature connection and fusion at each stage. By continuously mapping from low-level features to high-level features, $P_1 \in R^{512 \times 16 \times 32}$, $P_2 \in R^{1024 \times 8 \times 16}$, and $P_3 \in R^{2048 \times 8 \times 16}$ are obtained. P_3 and C_4 are connected in parallel and average pooling is performed to obtain the complete global multi-scale feature $F_{Global} \in R^{2048 \times 1 \times 1}$, maximizing the preservation of deep semantic information and better fitting the fine-grained and deep semantic features of the Amur tiger for re-ID.

2.2.3. Local Dual-Domain Attention Feature Enhancement Method

We propose a Local Attention Enhancement Module (LAEM) based on the Convolutional Block Attention Module (CBAM) [30] to enhance the feature extraction performance of multiple local blocks in local branches (Figure 4). CBAM is an attention mechanism module used to enhance the performance of convolutional neural networks, which improves the model's perception ability by introducing the mixed attention of channel attention and spatial attention (Figure 5). Channel attention helps to enhance the feature representation of different channels, while spatial attention helps to extract key information at different positions in space [30].

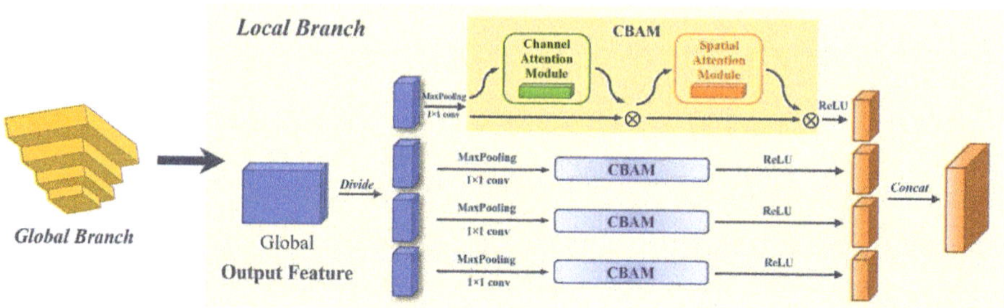

Figure 4. Principle of local dual-domain attention feature enhancement method.

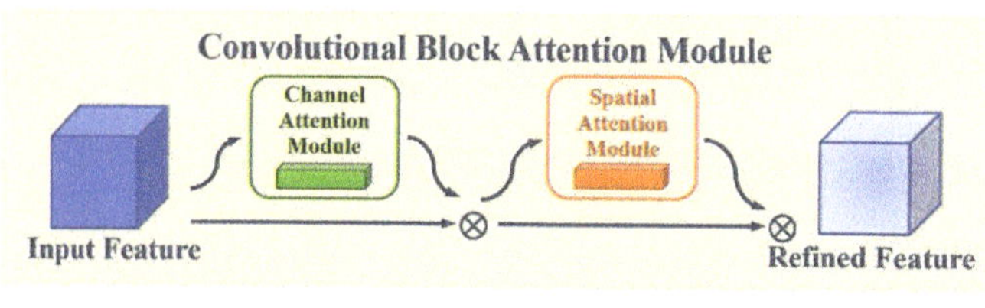

Figure 5. Principle of Convolutional Block Attention Module.

The local branch of this network is different from the global branch in fusing features at different scales. The local branch further strengthens and extracts features in different ranges through horizontal blocking, which helps to extract and optimize local details such as the texture and stripes of Amur tiger fur, and improves the accuracy of re-ID.

We divide the global feature obtained from the inverted pyramid module of the global branch into 4 blocks from left to right, each block being a local feature $block \in R^{2048 \times 8 \times 2}$ (Figure 6). In the local branch, we perform adaptive max pooling and 1×1 convolution operations on each local block feature to obtain a reduced local feature representation to 512. Then, we feed each reduced feature block into the CBAM. Channel and spatial attention feature enhancements are applied to the local block feature, dual-domain feature representation is enhanced, and important features such as local stripes after block segmentation are enhanced to obtain $\{L_1, L_2, L_3, L_4\} \in R^{512 \times 1 \times 1}$. Finally, the four feature blocks obtained through attention enhancement were activated using the activation function ReLU, and then concatenated to obtain the final feature output $F_{Local} \in R^{2048 \times 1 \times 1}$ of the local branch.

Figure 6. The method of Amur tiger feature map segmentation.

2.3. Training and Reasoning

During the training process, the training dataset of Amur tiger images is input into our proposed serial multi-scale feature fusion and enhancement re-ID network of the Amur tiger. Through global and local branches, feature extraction and classification tasks are carried out through the global inverted pyramid multi-scale feature fusion module and local dual-domain attention feature enhancement module. Finally, the classifier layer is applied for re-ID.

2.3.1. Loss Function

In probability statistics, entropy is a measure of the invariance of random variables. Cross entropy can measure the degree of similarity and difference between two distributions and is often used in image multi-classification and other problems.

We employ the Cross-Entropy Loss, a commonly used logarithmic loss function for multi-class classification problems, to regulate and optimize the training process of our network. The formula is as follows:

$$L_i = -\sum_{C=1}^{K} y_{ic} log(p_{ic}) \quad (1)$$

2.3.2. Inference

During the inference process, when wildlife conservation workers obtain a set of images from camera traps or surveillance videos, we can select any image as the Query and other images as the Gallery. Then, we input the Query and Gallery into the trained network. After feature extraction and re-ID, we can obtain the similarity ranking of all images in the Gallery compared to the Query. After descending sorting, we obtain the most similar entity image of the Amur tiger extracted by the Query in the Gallery. At this time, we can determine the probability from high to low that it is the same as the Amur tiger entity in the Query image (Figure 7).

Figure 7. The inference process for Amur tiger re-ID.

3. Results and Analysis

This section mainly introduces the experimental setup, evaluation methods, and specific experimental results.

3.1. Experimental Setup

Our proposed model is implemented in Cuda using the PyTorch framework. The computer and code environments in which we conducted the experiment were configured with PyTorch, Python 3.8, and Cuda11.3, and the entire training and testing process was conducted on a server configured with an NVIDIA GeForce RTX 3090 GPU. The training and testing sets used in the experiment are shown in Table 1. During the model training phase, we use SGD and set the momentum to 0.9, basic learning rate to 0.002, weight decay to 0.0005, and batch size to 16. The learning rate of the classifier layer is set to 0.02, and the learning rate is decreased by a factor of 10 at epoch 100. The dataset was randomly erased and underwent a total of 150 epochs of training.

3.2. Evaluation

The testing process of this task involves extracting an image from the Query dataset of the test set and calculating the Euclidean distance between this image and all images in the Gallery except for this image. The Euclidean distance calculation formula is as follows:

$$Distance = sort((A - B)\cdot(A - B)) \tag{2}$$

Sort in descending order based on the calculation results to determine whether the image extracted by Query and other images in the Gallery are entities with the same ID label:

$$F\left(l_{image}, l_{gallery}\right) \triangleq \begin{cases} 1 \ if \ l_{image} = l_{gallery} \\ 0 \ if \ l_{image} \neq l_{gallery} \end{cases} \tag{3}$$

The task of Amur tiger re-ID is similar to the sub-task of person re-ID in image retrieval, so the same testing methods and evaluation indicators, such as the CMC curve and mAP, can be used. This paper uses three indicators for re-ID evaluation: Rank-1 Accuracy, Rank-5 Accuracy, and mean average precision (mAP). Rank-1 is the probability of the first image being retrieved hitting, and Rank-5 is the probability of the first five images being retrieved hitting. Rank-1 can be explained using the formula shown in (4), and the calculation method for Rank-5 is similar to this:

$$Rank - 1 = \frac{1}{\|Q\|} \sum_{q \in Q} F\left(l^{q}_{image}, l_{gallery}\right) \tag{4}$$

mAP reflects the degree to which real images rank higher in sorting, and compared to Rank-1, Rank-5, etc., it can more comprehensively measure the effectiveness of re-ID. Therefore, this indicator is also used as the primary evaluation indicator in this paper.

3.3. *Compared with Other Advanced Methods*

3.3.1. Comparison with Improved Methods Based on ResNet50

Our proposed method has achieved good results on the public dataset of ATRW (Table 2). Compared with the comparative experimental results of the advanced person re-ID model based on ResNet50, our proposed model has achieved better results and significant improvements (Figures 8 and 9).

Table 2. Comparison with ResNet50 based method on ATRW test dataset.

Method	mAP	Rank-1	Rank-5
PCB [23]	74.5%	95.4%	98.7%
ResNet50 + Triplet Loss [29]	75.1%	92.4%	99.1%
ResNet50 + IBN [31]	75.4%	93.9%	98.4%
ResNet50 + Lifted Loss [32]	75.5%	94.2%	98.7%
ResNet50 + Circle Loss [33]	75.9%	94.2%	98.9%
Ours	78.7%	96.3%	98.9%

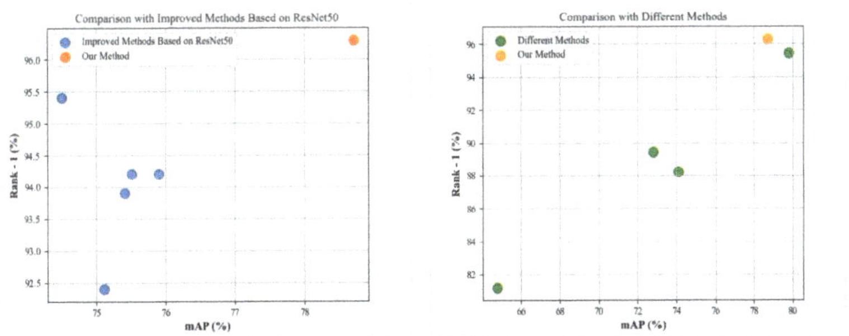

(a) (b)

Figure 8. (**a**,**b**) shows the comparison results of our proposed model experimental indicators with other advanced models.

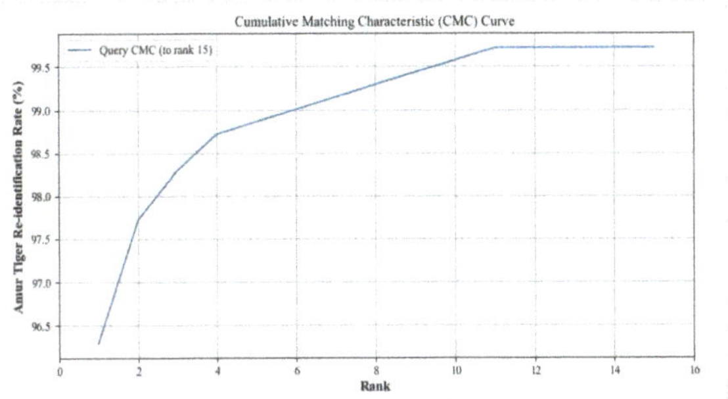

Figure 9. The CMC curve of our Amur tiger re-ID model.

3.3.2. Comparison with Different Improvement Methods

The model we propose is an end-to-end re-ID model that does not require additional prior knowledge. It has good transferability and is a significant improvement compared to Aligned-reID, which also combines global and local features. However, compared with models PPbM-a, PPbM-b, and MPFNet, which incorporate or pre-train pose estimation modules, it can still achieve good re-ID performance, with some indicators improved (Table 3). We randomly selected two Query images for Amur tiger re-ID as examples (Figure 10).

Table 3. Comparison with different methods on the ATRW test dataset.

Method	mAP	Rank-1	Rank-5
Aligned-reID [3]	64.8%	81.2%	92.4%
PPbM-a [3]	74.1%	88.2%	96.4%
PPbM-b [3]	72.8%	89.4%	95.6%
MPFNet [22]	79.8%	95.4%	98.6%
Ours	78.7%	96.3%	98.9%

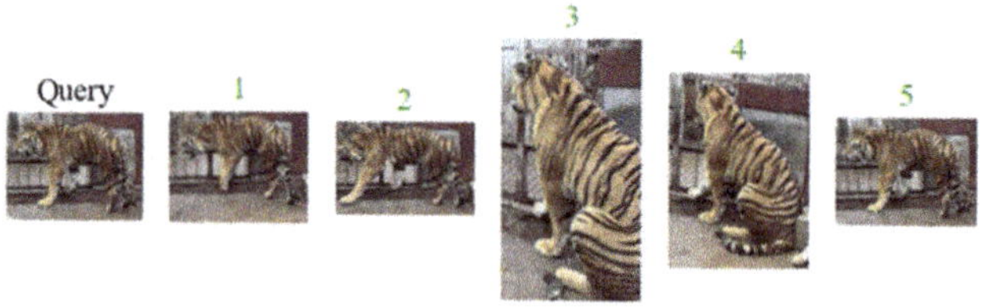

(**a**)

(**b**)

Figure 10. (**a**,**b**) are examples of the results of applying our proposed network for Amur tiger re-ID. The number above the image shows the similarity ranking result, and the green color shows correct re-ID.

3.4. Ablation Experiment

To verify the effectiveness of our proposed method and module, we conducted ablation experiments on each part. This model chooses the classic feature extraction and classification model ResNet50 as the backbone. Firstly, we experimentally verify the testing performance of only the backbone. Then, we separately verify the effectiveness of adding the proposed global inverted pyramid multi-scale feature fusion module and local feature enhancement module. The following is a detailed description and explanation.

3.4.1. Effectiveness of the Global Inverted Pyramid Multi-Scale Feature Fusion Module and the Local Dual-Domain Attention Feature Enhancement Module

Due to the fact that our proposed global inverted pyramid multi-scale feature fusion module is constructed based on the ideas of FPN and PANet, we compared the experimental results of models that only used ResNet50 and introduced FPN and PANet based on ResNet50 (Table 4). The introduction of FPN in the backbone resulted in a 7.7% decrease in mAP compared to only using the backbone, while our proposed IFPM method resulted in 76.1% mAP, 96.3% Rank-1, and 98.9% Rank-5. Compared to the model introducing FPN, mAP improved by 9.4%, while Rank-1 and Rank-5 improved by 4.6% and 1.3%, demonstrating significant advantages in these three indicators. This result proves that, as expected by our analysis, adopting deep semantic features to fuse global features in an inverted pyramid shape is more suitable for the re-ID task of the Amur tiger.

Table 4. Performing ablation experiments on the ATRW test dataset to demonstrate the effectiveness of the IFPM.

Method	mAP	Rank-1	Rank-5
ResNet50	74.4%	93.4%	98.4%
ResNet50 + FPN	66.7%	91.7%	97.6%
ResNet50 + IFPM (Ours)	76.1%	96.3%	98.9%

The experimental results of fusing IFPM and LAEM dual modules showed a 2.6% improvement in mAP compared to the model containing only IFPM modules, while the results of Rank-1 and Rank-5 remained unchanged, demonstrating the effectiveness of the LAEM module and the serial combination of IFPM and LAEM (Table 5).

Table 5. Performing ablation experiments on the ATRW test dataset to demonstrate the effectiveness of the LAME.

Method	mAP	Rank-1	Rank-5
ResNet50	74.4%	93.4%	98.4%
ResNet50 + IFPM (Ours)	76.1%	96.3%	98.9%
ResNet50 + IFPM + LAEM (Ours)	78.7%	96.3%	98.9%

3.4.2. Effectiveness of CBAM in Local Dual-Domain Attention Feature Enhancement Module

Due to the application of CBAM in our proposed local dual-domain attention feature enhancement module, we compared the experimental results of models using other robust and effective attention modules to demonstrate the optimal performance of CBAM in this module.

The experimental results of introducing Squeeze-and-Excitation Network (SENet) and Efficient Channel Attention (ECA) into our improved model, which incorporates IFPM, showed a 1.7% improvement in mAP compared to the backbone, while Rank-1 and Rank-5 increased by 1.3% and 2%, respectively (Table 6). Our proposed module incorporating CBAM improved mAP by 2.6% compared to the model incorporating SENet, while Rank-1 and Rank-5 improved by 1.6% and 0.5%, respectively. Compared to the model incorporating ECA, mAP improved by 2.6%, while Rank-1 and Rank-5 improved by 0.9% and 0.5%, respectively.

Table 6. Ablation experiments on ATRW test data set prove the progressiveness of CBAM module in the local branch.

Method	mAP	Rank-1	Rank-5
ResNet50	74.4%	93.4%	98.4%
ResNet50 + IFPM + SENet	76.1%	94.7%	98.4%
ResNet50 + IFPM + ECA	76.1%	95.4%	98.4%
ResNet50 + IFPM + LAEM (Ours)	78.7%	96.3%	98.9%

4. Discussion

The Amur tiger may have the problem of occupying a relatively large position and having a relatively complex image background in the photos captured by camera traps [34]. This is because photo shooting is triggered only when the wild animals are relatively close to the infrared camera and the infrared sensor senses their temperature [34,35]. Moreover, due to the complex forest environment, the Amur tiger has a narrow path and a larger target. Therefore, we propose a new serial multi-scale feature fusion and enhancement re-ID network of Amur tiger, which extracts and learns to input Amur tiger features in a global and local branch serial manner. We also propose a global inverted pyramid multi-scale feature fusion method and a local dual-domain attention feature enhancement method to learn Amur tiger images at multiple scales, more adaptable to this re-ID task. In the model validation stage, we applied the Amur tiger re-ID dataset of the ATRW for experimental verification. The experimental results showed that our proposed model still has good performance without introducing other prior knowledge and complex labeling, and the mAP and hit rate have been improved. In addition to the Amur tiger, our proposed network is applicable to other large quadruped animals through retraining. It can be structurally adjusted according to specific animal species and task details, without the need to introduce other prior knowledge, reducing the cost of early labeling and other inputs, and has a certain degree of universality and transferability.

In summary, since our constructed model requires vertical partitioning of extracted features in the horizontal direction, it is effective in identifying large quadruped mammals that are mostly identified by body surfaces, such as snow leopard re-ID and leopard species classification. However, we have not yet conducted further validation and model fine-tuning on many other large quadruped animal datasets, and if we apply datasets of upright animals such as monkeys, there may be issues with poor performance. Because our proposed method requires local partitioning in the horizontal direction, and for such animals, the key complete features may be segmented, resulting in the inability to learn important information. Currently, however, a suitable dataset for comparative experiments remains unavailable. This is the limitation and problem that this paper aims to address, and further in-depth research is still needed. In the future, we will strive to create datasets and complete research and comparative experiments on model transfer. In addition, this paper is conducted on a public dataset where each entity has an average of 14.5 and at least 8 training images prior to data augmentation. However, in real life, there may be small and uneven sample sizes in the dataset we obtain in the wild or in surveillance videos, which are also issues that we need to address in the future.

5. Conclusions

The re-ID and counting of Amur tigers play an important role in studying and analyzing the entity quantity and distribution of various populations, biodiversity, and individual tracking of wild animals. Therefore, improving the mAP, Rank-1, and Rank-5 of Amur tiger re-ID has enormous ecological significance and value. In order to improve the accuracy and efficiency of the Amur tiger re-ID, we propose a serial multi-scale feature fusion and enhancement re-ID network of the Amur tiger. A global inverted pyramid multi-scale feature fusion method and a local dual-domain attention feature enhancement method were designed for this network. Our proposed network and method have the advantages of no prior knowledge, high efficiency, and end-to-end functionality. Through experiments on the public dataset ATRW of Amur tiger and comparative experiments with other methods, it has been proven that this method has good performance and will help improve the re-ID performance of Amur tiger.

Author Contributions: Conceptualization, N.X., Y.D. and F.X.; Data curation, Z.M., D.X. and F.X.; Formal analysis, Z.M. and J.Z.; Funding acquisition, F.C.; Investigation, D.X. and X.S.; Methodology, N.X., J.Z. and H.Z.; Project administration, F.C.; Resources, Y.X.; Software, N.X., Y.X. and X.S.; Supervision, F.C.; Validation, Y.D. and H.Z.; Visualization, N.X.; Writing—original draft, N.X. All authors have read and agreed to the published version of the manuscript.

Funding: This research was funded by the Outstanding Youth Team Project of Central Universities: QNTD202308; the National Key R&D Program of China: 2022YFF1302700; and the Emergency Open Competition Project of the National Forestry and Grassland Administration: 202303.

Institutional Review Board Statement: Not applicable.

Informed Consent Statement: Not applicable.

Data Availability Statement: The data presented in this study are available on request from the corresponding author.

Conflicts of Interest: The authors declare no conflicts of interest.

References

1. Noad, M.J.; Cato, D.H.; Stokes, M.D. Acoustic tracking of humpback whales: Measuring interactions with the acoustic environment. In Proceedings of the Acoustics 2004, Gold Goast, Australia, 3–5 November 2024; pp. 353–358.
2. Andreychev, A.; Kuznetsov, V.; Lapshin, A.; Alpeev, M. Activity of the Russian desman *Desmana moschata* (*Talpidae, Insectivora*) in its burrow. *Therya* **2020**, *11*, 161–167. [CrossRef]

3. Li, S.; Li, J.; Tang, H.; Qian, R.; Lin, W. ATRW: A Benchmark for Amur Tiger Re-Identification in the Wild. In Proceedings of the 28th ACM International Conference on Multimedia, Seattle, WA, USA, 12–16 October 2020; pp. 2590–2598.
4. Jiang, Y.; Tian, J.; Zhao, J.; Tang, X. The connotation and assessment framework of national park ecosystem integrity: A case study of the Amur Tiger and Leopard National Park. *Biodivers. Sci.* **2021**, *29*, 1279–1287. [CrossRef]
5. Chen, J.; Nasendelger; Sun, Q.; Zhang, L.; Tang, J.; Lang, J.; Liu, T.; Liu, K.; Xiao, W.; Bao, W. Amur Tiger and Prey in Jilin Hunchun National Nature Reserve, China. *Chin. J. Zool.* **2011**, *46*, 46–52.
6. Zhang, C.; Zhang, M. Population Status and Dynamic Trends of Amur Tigers Prey in Eastern Wandashan Mountain, Heilongjiang Province. *Acta Ecol. Sin.* **2011**, *31*, 6481–6487.
7. Hiby, L.; Lovell, P.; Patil, N.; Kumar, N.S.; Gopalaswamy, A.M.; Karanth, K.U. A Tiger Cannot Change Its Stripes: Using a Three-Dimensional Model to Match Images of Living Tigers and Tiger Skins. *Biol. Lett.* **2009**, *5*, 383–386. [CrossRef] [PubMed]
8. Zhang, P. Study on Northeast Tiger Skin Texture Extraction and Recognition Based on BP Network. Master's Thesis, Northeast Forestry University, Harbin, China, June 2008.
9. Crall, J.P.; Stewart, C.V.; Berger-Wolf, T.Y.; Rubenstein, D.I.; Sundaresan, S.R. HotSpotter—Patterned Species Instance Recognition. In Proceedings of the 2013 IEEE Workshop on Applications of Computer Vision (WACV), Clearwater Beach, FL, USA, 15–17 January 2013; pp. 230–237.
10. Curran, B.; Nekooei, S.M.; Chen, G. Accurate New Zealand wildlife image classification-deep learning approach. In Proceedings of the Australasian Joint Conference on Artificial Intelligence, Sydney, NSW, Australia, 2 February 2022; pp. 632–644.
11. Binta Islam, S.; Valles, D.; Hibbitts, T.J.; Ryberg, W.A.; Walkup, D.K.; Forstner, M.R.J. Animal Species Recognition with Deep Convolutional Neural Networks from Ecological Camera Trap Images. *Animals* **2023**, *13*, 1526. [CrossRef] [PubMed]
12. Wang, L.; Ding, R.; Zhai, Y.; Zhang, Q.; Tang, W.; Zheng, N.; Hua, G. Giant Panda Identification. *IEEE Trans. Image Process.* **2021**, *30*, 2837–2849. [CrossRef] [PubMed]
13. Ghosh, S.B.; Muddalkar, K.; Mishra, B.; Garg, D. Amur tiger Detection for Wildlife Monitoring and Security. In Proceedings of the Advanced Computing: 10th International Conference, IACC 2020, Panaji, Goa, India, 5–6 December 2020; Part II 10. pp. 19–29.
14. Zhang, Q.; Yi, X.; Guo, J.; Tang, Y.; Feng, T.; Liu, R. A Few-Shot Rare Wildlife Image Classification Method Based on Style Migration Data Augmentation. *Ecol. Inform.* **2023**, *77*, 102237. [CrossRef]
15. Meng, D.; Li, T.; Li, H.; Zhang, M.; Tan, K.; Huang, Z.; Li, N.; Wu, R.; Li, X.; Chen, B.; et al. A Method for Automatic Identification and Separation of Wildlife Images Using Ensemble Learning. *Ecol. Inform.* **2023**, *77*, 102262. [CrossRef]
16. Simonyan, K.; Zisserman, A. Very Deep Convolutional Networks for Large-Scale Image Recognition. *arXiv* **2014**, arXiv:1409.1556.
17. Krizhevsky, A.; Sutskever, I.; Hinton, G.E. ImageNet Classification with Deep Convolutional Neural Networks. *Commun. ACM* **2017**, *60*, 84–90. [CrossRef]
18. He, K.; Zhang, X.; Ren, S.; Sun, J. Deep Residual Learning for Image Recognition. In Proceedings of the 2016 IEEE Conference on Computer Vision and Pattern Recognition (CVPR), Las Vegas, NV, USA, 27–30 June 2016; pp. 770–778.
19. Zheng, Z.; Zhao, Y.; Li, A.; Yu, Q. Wild Terrestrial Animal Re-Identification Based on an Improved Locally Aware Transformer with a Cross-Attention Mechanism. *Animals* **2022**, *12*, 3503. [CrossRef] [PubMed]
20. Zhang, K.; Xin, Y.; Shi, C.; Xie, Z.; Ren, Z. A Pyramidal Feature Fusion Model on Swimming Crab Portunus Trituberculatus Re-Identification. *Front. Mar. Sci.* **2022**, *9*, 845112. [CrossRef]
21. Liu, C.; Zhang, R.; Guo, L. Part-Pose Guided Amur Tiger Re-Identification. In Proceedings of the 2019 IEEE/CVF International Conference on Computer Vision Workshop (ICCVW), Seoul, Republic of Korea, 27–28 October 2019; pp. 315–322.
22. He, Z.; Qian, J.; Yan, D.; Wang, C.; Xin, Y. Animal Re-Identification Algorithm for Posture Diversity. In Proceedings of the ICASSP 2023—2023 IEEE International Conference on Acoustics, Speech and Signal Processing (ICASSP), Rhodes Island, Greece, 4 June 2023; pp. 1–5.
23. Sun, Y.; Zheng, L.; Yang, Y.; Tian, Q.; Wang, S. Beyond Part Models: Person Retrieval with Refined Part Pooling (and A Strong Convolutional Baseline). In *Computer Vision—ECCV 2018, Munich, Germany, 8–14 September 2018*; Ferrari, V., Hebert, M., Sminchisescu, C., Weiss, Y., Eds.; Lecture Notes in Computer Science; Springer International Publishing: Cham, Switzerland, 2018; Volume 11208, pp. 501–518.
24. Sun, Y.; Xu, Q.; Li, Y.; Zhang, C.; Li, Y.; Wang, S.; Sun, J. Perceive Where to Focus: Learning Visibility-Aware Part-Level Features for Partial Person Re-Identification. In Proceedings of the 2019 IEEE/CVF Conference on Computer Vision and Pattern Recognition (CVPR), Long Beach, CA, USA, 15–20 June 2019; pp. 393–402.
25. Liu, X.; Tan, H.; Tong, X.; Cao, J.; Zhou, J. Feature Preserving GAN and Multi-Scale Feature Enhancement for Domain Adaption Person Re-Identification. *Neurocomputing* **2019**, *364*, 108–118. [CrossRef]
26. Liu, S.; Qi, L.; Qin, H.; Shi, J.; Jia, J. Path Aggregation Network for Instance Segmentation. In Proceedings of the 2018 IEEE Conference on Computer Vision and Pattern Recognition (CVPR), Salt Lake City, UT, USA, 18–23 June 2018; pp. 8759–8768.
27. Lin, T.-Y.; Dollar, P.; Girshick, R.; He, K.; Hariharan, B.; Belongie, S. Feature Pyramid Networks for Object Detection. In Proceedings of the 2017 IEEE Conference on Computer Vision and Pattern Recognition (CVPR), Honolulu, HI, USA, 21–26 July 2017; pp. 936–944.
28. Li, Z.; Lang, C.; Liew, J.; Hou, Q.; Li, Y.; Feng, J. Cross-Layer Feature Pyramid Network for Salient Object Detection. *IEEE Trans. Image Process.* **2021**, *30*, 4587–4598. [CrossRef] [PubMed]
29. Schroff, F.; Kalenichenko, D.; Philbin, J. FaceNet: A Unified Embedding for Face Recognition and Clustering. In Proceedings of the 2015 IEEE Conference on Computer Vision and Pattern Recognition (CVPR), Boston, MA, USA, 7–12 June 2015; pp. 815–823.

30. Woo, S.; Park, J.; Lee, J.-Y.; Kweon, I.S. CBAM: Convolutional Block Attention Module. In *Computer Vision—ECCV 2018, Munich, Germany, 8–14 September 2018*; Ferrari, V., Hebert, M., Sminchisescu, C., Weiss, Y., Eds.; Lecture Notes in Computer Science; Springer International Publishing: Cham, Switzerland, 2018; Volume 11211, pp. 3–19.
31. Pan, X.; Luo, P.; Shi, J.; Tang, X. Two at Once: Enhancing Learning and Generalization Capacities via IBN-Net. In *Computer Vision—ECCV 2018, Munich, Germany, 8–14 September 2018*; Ferrari, V., Hebert, M., Sminchisescu, C., Weiss, Y., Eds.; Lecture Notes in Computer Science; Springer International Publishing: Cham, Switzerland, 2018; Volume 11208, pp. 484–500.
32. Song, H.O.; Xiang, Y.; Jegelka, S.; Savarese, S. Deep Metric Learning via Lifted Structured Feature Embedding. In Proceedings of the 2016 IEEE Conference on Computer Vision and Pattern Recognition (CVPR), Las Vegas, NV, USA, 27–30 June 2016; pp. 4004–4012.
33. Sun, Y.; Cheng, C.; Zhang, Y.; Zhang, C.; Zheng, L.; Wang, Z.; Wei, Y. Circle Loss: A Unified Perspective of Pair Similarity Optimization. In Proceedings of the 2020 IEEE/CVF Conference on Computer Vision and Pattern Recognition (CVPR), Seattle, WA, USA, 13–19 June 2020; pp. 6397–6406.
34. Gomez Villa, A.; Salazar, A.; Vargas, F. Towards Automatic Wild Animal Monitoring: Identification of Animal Species in Camera-Trap Images Using Very Deep Convolutional Neural Networks. *Ecol. Inform.* **2017**, *41*, 24–32. [CrossRef]
35. Zwerts, J.A.; Stephenson, P.J.; Maisels, F.; Rowcliffe, M.; Astaras, C.; Jansen, P.A.; Van Der Waarde, J.; Sterck, L.E.H.M.; Verweij, P.A.; Bruce, T.; et al. Methods for Wildlife Monitoring in Tropical Forests: Comparing Human Observations, Camera Traps, and Passive Acoustic Sensors. *Conservat. Sci. Prac.* **2021**, *3*, e568. [CrossRef]

animals MDPI

Article

The Development of Object Recognition Requires Experience with the Surface Features of Objects

Justin Newell Wood [1,*] and Samantha Marie Waters Wood [2]

1 Departments of Informatics, Cognitive Science, Neuroscience, Center for Integrated Study of Animal Behavior, Indiana University, Bloomington, IN 47408, USA
2 Departments of Informatics and Neuroscience, Indiana University, Bloomington, IN 47408, USA; sw113@iu.edu
* Correspondence: woodjn@iu.edu

Simple Summary: Understanding how brains work requires understanding the role of experience in the development of core mental abilities. Here, we show that the development of one core mental ability—object recognition—requires visual experience with the surface features of objects. We found that when newborn chicks were raised with objects containing surface features, the chicks learned to recognize objects across familiar and novel viewpoints. However, when chicks were raised with line drawings of those same objects, the chicks failed to develop object recognition. These findings shed light on the role of experience in early visual development and suggest that certain kinds of experiences are especially important for the development of core mental abilities.

Abstract: What role does visual experience play in the development of object recognition? Prior controlled-rearing studies suggest that newborn animals require slow and smooth visual experiences to develop object recognition. Here, we examined whether the development of object recognition also requires experience with the surface features of objects. We raised newborn chicks in automated controlled-rearing chambers that contained a single virtual object, then tested their ability to recognize that object from familiar and novel viewpoints. When chicks were reared with an object that had surface features, the chicks developed view-invariant object recognition. In contrast, when chicks were reared with a line drawing of an object, the chicks failed to develop object recognition. The chicks reared with line drawings performed at chance level, despite acquiring over 100 h of visual experience with the object. These results indicate that the development of object recognition requires experience with the surface features of objects.

Keywords: controlled rearing; object recognition; newborn; development; line drawing; chick

1. Introduction

Mature animals have powerful object recognition abilities. For example, after just a brief glimpse of an object, humans can recognize that object across substantial variation in the retinal images produced by the object, due to changes in viewpoint, size, illumination, and so forth (reviewed in [1]). However, the origins of object recognition are still not well understood. What role does early visual experience play in the development of object recognition? Does the development of object recognition require a specific type of visual experience with objects?

Human infants are not well suited for addressing these questions because they cannot be raised in strictly controlled environments from birth. In contrast, controlled-rearing studies of newborn animals can directly probe the role of experience in development. By systematically manipulating the visual experiences provided to newborn animals and measuring the effects of those manipulations on behavioral and neural development, controlled-rearing studies can isolate the specific experiences that drive the development of object recognition.

Prior controlled-rearing studies with newborn chicks have revealed two types of experiences that are necessary for the development of object perception: slow and smooth experiences with objects [2–5]. When newborn chicks were reared with virtual objects that changed slowly and smoothly over time (akin to natural objects), the chicks successfully developed object recognition, including the ability to recognize objects across novel viewpoints, backgrounds, and motion speeds. Conversely, when chicks were reared with objects that moved too quickly or non-smoothly, the chicks failed to develop object recognition. Without slow and smooth visual experiences, newborn chicks develop inaccurate object representations. Here, we extend these findings by examining whether the development of object recognition also requires experience with the surface features of objects. The term "surface features" refers to the features (e.g., color, texture, and shading) of the surfaces between the boundaries of an object.

There are mixed perspectives on the importance of surface features in object recognition. On one hand, a large number of studies have shown that human adults can readily recognize objects depicted in line drawings, which lack surface features such as color, texture, and shading (e.g., [6–9]). This ability to perceive and understand line drawings emerges early in development. For instance, infants begin showing enhanced attention to lines that depict corners and edges in the first year of life [10], and young children use lines to define the boundaries of objects in their first attempts to depict the world [11]. Humans have also used line drawings to capture scenes since prehistoric times [12,13]. Furthermore, many nonhuman animals can understand line drawings. Chimpanzees can recognize objects presented in line drawings [14,15] and pigeons can recognize line drawings of objects that are rotated in depth, even after exposure to just a single depth orientation [16]. Even insects appear to use line representation to some extent in biomimicry [17]. Together, these studies indicate that the ability to understand line drawings emerges early in development and is shared with a wide range of animals.

On the other hand, many studies provide evidence that surface features play an important role in object recognition (e.g., [18–23]). Human adults can recognize an object faster when the light source remains in the same location compared to when the light source moves [21], and surface features can affect the speed and accuracy of object recognition and object naming [24]. During their first months of life, human infants also rely on the motion of surface features to build object representations (reviewed in [25]).

In all of the studies cited above, the subjects had acquired months to years of visual experience with real-world objects before they were tested. Thus, these studies do not reveal whether newborn brains can understand line drawings at the onset of vision or whether the development of this ability requires experience with natural visual objects. The present study distinguishes between these possibilities by testing whether newborn chicks can recognize objects presented in line drawings at the onset of vision, in the absence of prior visual experience with natural objects. Specifically, we contrasted the object recognition performance of newborn chicks reared with line drawings of objects versus realistic objects with surface features. The three experiments presented here allow for a direct test of the importance of surface features in the development of object recognition.

Across three experiments, newborn chicks were reared in strictly controlled environments that contained no objects other than the virtual objects projected on the display walls (input phase). For one group of chicks, the virtual object(s) contained surfaces (Surface Feature Condition), whereas for the other group, the virtual object(s) was a line drawing animation that lacked surfaces (Line Drawing Condition). In the test phase, we then used a two-alternative forced-choice procedure to measure whether the chicks could recognize their imprinted object across familiar and novel viewpoints. If chicks can recognize objects presented in line drawings, then their performance should be high in both conditions. Conversely, if the development of object recognition requires visual experience with the surface features of objects, then the chicks should develop more accurate object recognition abilities in the Surface Feature Condition than the Line Drawing Condition.

In the first experiment, each chick was reared with a single virtual object (either with or without surfaces) that moved through a limited 60° viewpoint range. In the second experiment, we verified the results of Experiment 1 under different rearing and testing conditions by rearing each chick with a single virtual object (either with or without surfaces) that moved through a 360° viewpoint range. Finally, in the third experiment, we tested whether our findings would extend to new 2D object shapes. Across all three experiments, the chicks successfully recognized objects with surface features, but failed to recognize line drawings of those same objects.

Using Automated Controlled Rearing to Study the Origins of Object Recognition

To examine the role of surface features in the development of object recognition, we used an automated controlled-rearing method [26]. We used controlled-rearing chambers to eliminate any exposure to real-world objects. There are two benefits to using automated methods to probe the origins of visual intelligence. First, automation allows large amounts of precise behavioral data to be collected from each subject. In the present study, each chick's behavior was recorded continuously (24/7) for up to two weeks, providing precise measurements of their object recognition performance. Second, since computers (rather than researchers) present the stimuli and code the behavior, automation eliminates the possibility of experimenter error and bias [27].

We used newborn chicks as an animal model because they are an ideal model system for studying the origins of object recognition [28]. First, newborn chicks can be raised in strictly controlled environments immediately after hatching (e.g., environments containing no real-world objects). As a result, it is possible to control and manipulate all of the chicks' visual object experiences from the onset of vision. Second, chicks imprint to objects seen in the first few days of life and will attempt to reunite with those objects when separated [29]. This imprinting behavior emerges spontaneously and provides a reliable behavioral assay for measuring chicks' object recognition abilities. Third, newborn chicks develop high-level object recognition. For example, newborn chicks can solve the visual binding problem, building integrated object representations with bound color–shape features [30]. Chicks can also parse objects from complex backgrounds [31], build view-invariant object representations [26,32], and recognize objects rapidly, within a fraction of a second [33].

Finally, studies of chicks can also inform human development because birds and mammals process sensory input using homologous cortical circuits with similar connectivity patterns [34–37]. The cortical circuits are organized differently in birds and mammals (nuclear vs. layered organization), but the circuits share similar cell morphology, connectivity patterns of input and output neurons, gene expression, and function [34,38–40]. Architecturally, avian and mammalian brains share the same large-scale organizational principles. Specifically, their brains are modular, small-world networks. These networks are organized into a connective core of hub nodes that includes visual, auditory, limbic, prefrontal, premotor, and hippocampal structures [41]. The similarities between avian and mammalian brains suggest that controlled-rearing studies of newborn chicks can elucidate both avian and mammalian intelligence.

2. Experiment 1

In the input phase Experiment 1, newborn chicks were reared with a single virtual object moving through a limited 60° viewpoint range. In the test phase, we examined whether the chicks could recognize that object across 12 different viewpoint ranges. The chicks were either raised and tested with line drawings or with objects containing surface features. The text describing the methods is partly adapted from [26]. The data from the baseline (surface feature) conditions in Experiments 1, 2, and 3 were published previously in [4,26,42]. In the present study, we directly contrasted chicks reared with line drawings of objects versus objects with surface features.

2.1. Materials and Methods

2.1.1. Subjects

Twenty-three domestic chicks of unknown sex were tested. We tested 11 subjects in the Surface Feature Condition [26] and 12 subjects in the Line Drawing Condition. No subjects were excluded from the analyses. The eggs were obtained from a local distributor and incubated in darkness in an OVA-Easy incubator (Brinsea Products Inc., Titusville, FL, USA). After hatching, we moved the chicks from the incubation room with the aid of night vision goggles. Each chick was placed, singularly, in a controlled-rearing chamber.

This research was approved by The University of Southern California Institutional Animal Care and Use Committee.

2.1.2. Controlled-Rearing Chambers

The controlled-rearing chambers (66 cm length × 42 cm width × 69 cm height) were constructed from white, high-density plastic. Each chamber contained no real-world (solid, bounded) objects (Figure 1A). We presented object stimuli to the chicks by projecting animations of virtual objects on two display walls situated on opposite sides of the chamber. The display walls were 19″ liquid crystal display (LCD) monitors with 1440 × 900 pixel resolution. We provided food and water in transparent troughs in the ground (66 cm length × 2.5 cm width × 2.7 cm height). We fed the chicks grain because grain does not behave like an object (i.e., a heap of grain does not maintain a solid, bounded shape). The floors were wire mesh and supported 2.7 cm off the ground by transparent beams.

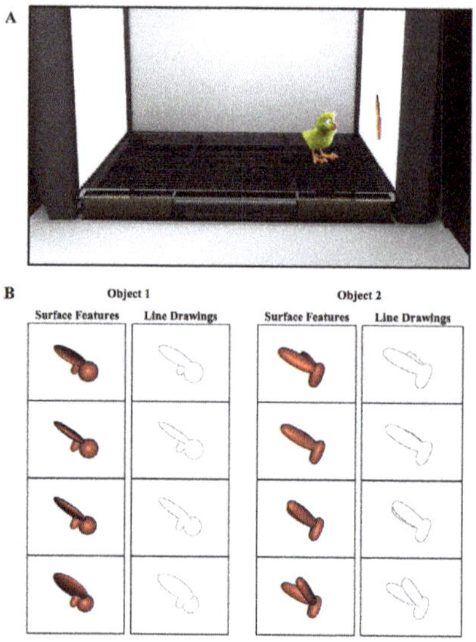

Figure 1. (**A**) Illustration of a controlled-rearing chamber. The chambers contained no real-world objects. To present object stimuli to the chicks, virtual objects were projected on two display walls situated on opposite sides of the chamber. During the input phase (1st week of life), newborn chicks were exposed to a single virtual object either with surface features or without surface features (line drawing). (**B**) Sample images of the virtual objects.

We embedded micro-cameras in the ceilings of the chambers to record all of the chicks' behavior (9 samples/s, 24 h/day, 7 days/week). We used automated image-based tracking software (EthoVision XT, version 7, Noldus Information Technology, Leesburg, VA, USA) to track their behavior throughout the experiment. This automated data collection approach allowed us to collect 168 trials from each chick. In total, 7728 h of video footage (14 days × 24 h/day × 23 subjects) were collected for Experiment 1.

2.1.3. Procedure

In the first week of life (input phase), newborn chicks were reared in controlled-rearing chambers that contained a single virtual object. On average, the object measured 8 cm (length) × 7 cm (height) and was displayed on a uniform white background. Eleven of the chicks were imprinted to Object 1 (with Object 2 serving as the unfamiliar object), and twelve of the chicks were imprinted to Object 2 (with Object 1 serving as the unfamiliar object). The objects were modeled after those used in previous studies that tested for invariant object recognition in adult rats [43].

The object moved continuously (24 frames/s), rotating through a 60° viewpoint range about a vertical axis passing through its centroid (Figure 1B). The object only moved along this 60° trajectory; the chicks never observed the object from any other viewpoint in the input phase. The object switched display walls every 2 h (following a 1 min period of darkness), appearing for an equal amount of time on the left and right display wall. In the Surface Feature Condition, the imprinted object had realistic surface features, whereas in the Line Drawing Condition, the imprinted object was a line drawing animation of the object (Figure 1B, see Movie S1 for animations).

In the second week of life (test phase), the chicks received 168 test trials (24 test trials per day). During the test trials, the imprinted object was shown on one screen and an unfamiliar object was shown on the other screen. We expected the chicks to spend a greater proportion of time in proximity to the object that they perceived to be their imprinted object.

For all test trials, the unfamiliar object was presented from the same viewpoint range as the imprinted object shown during the input phase. The unfamiliar object had a similar size, color, motion speed, and motion trajectory as the imprinted object from the input phase. Consequently, for all of the novel viewpoint ranges, the unfamiliar object was more similar to the imprinting stimulus (from a pixel-wise perspective) than the imprinted object was to the imprinting stimulus (for details, see [26]). To recognize their imprinted object, the chicks needed to generalize across large, novel, and complex changes in the object's appearance on the retina.

The chicks were tested across 12 viewpoint ranges (11 novel, 1 familiar). Each viewpoint range was tested twice per day. The test trials lasted 20 min and were separated from one another by 40 min rest periods. During the rest periods, the animation from the input phase appeared on one display wall and a white screen appeared on the other display wall. The 12 viewpoint ranges were tested 14 times each within randomized blocks over the course of the test phase. Figure 2A illustrates how the objects were presented across the display walls during the input phase and test phase. In the Surface Feature Condition, the chicks were tested with objects containing surface features, whereas in the Line Drawing Condition, the chicks were tested with line drawings of the objects. In both conditions, the test objects moved continuously through a 60° viewpoint range.

2.2. Results

To analyze the chicks' behavior, the image-based tracking software scored the chick as being in proximity to an object when the chick occupied a 22 × 42 cm zone next to the object. Then, we computed the number of test trials in which chicks preferred their imprinted object over the unfamiliar object. The chick was rated to have preferred their imprinted object on a trial if their object preference score was greater than 50%. The object preference score was calculated with the formula Object Preference Score = Time by Imprinted Object/(Time by Imprinted Object + Time by Unfamiliar Object).

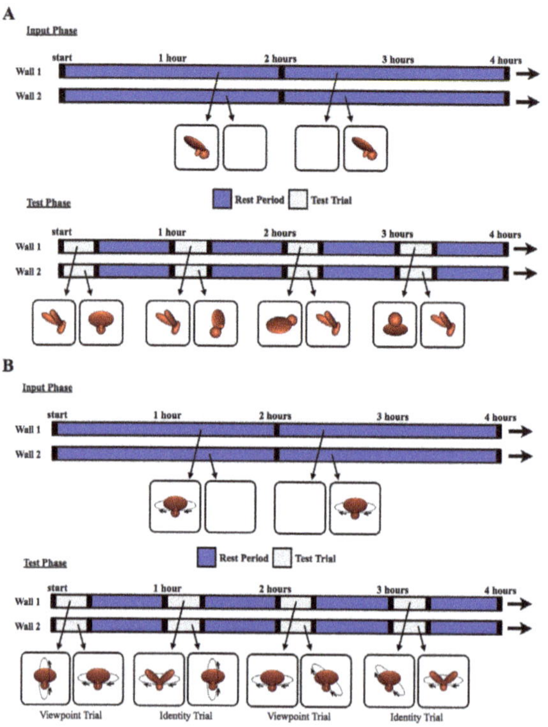

Figure 2. The experimental procedure. The schematics illustrate how the objects were presented for sample 4 h periods during (**A**) Experiment 1 and (**B**) Experiment 2. During the input phase, chicks were exposed to a single virtual object moving through a 60° (Experiment 1) or 360° (Experiment 2) viewpoint range. The object appeared on one wall at a time (indicated by blue segments on the timeline), switching walls every 2 h, after a 1 min period of darkness (black segments). During the test trials, two virtual objects were shown simultaneously, one on each wall, for 20 min per hour (gray segments). The illustrations below the timeline are examples of paired test objects displayed in four of the test trials. Each test trial was followed by a 40 min rest period (blue segments). During the rest periods, the animation from the input phase was shown on one wall, and the other wall was blank. This figure shows the stimuli from the Surface Feature Condition. In the Line Drawing Condition, the chicks were raised and tested with line drawings rather than objects with surface features.

The results are depicted in Figure 3. For each viewpoint range, we computed the percentage of time the chick spent with the imprinted object versus the unfamiliar object. Recognition performance exceeded chance level in the Surface Feature Condition ($t(10) = 9.75, p < 10^{-5}$, Cohen's $d = 2.94$), but did not exceed chance level in the Line Drawing Condition ($t(11) = 1.53, p = 0.15$, Cohen's $d = 0.44$).

To test whether performance differed between the Surface Feature Condition and the Line Drawing Condition on each of the test viewpoint ranges, we first used SPSS to compute a repeated-measures ANOVA. We used an ANOVA because we ultimately wanted to run *t*-tests on every viewpoint range, which requires running an overall ANOVA first to control the Type I error rate. We used repeated measures because the viewpoint range varied within subjects. The repeated-measures ANOVA with the viewpoint range as a within-subjects factor and condition (surface feature vs. line drawing) as a between-subjects factor revealed a significant main effect of the viewpoint range ($F(6.94, 145.81) = 2.73, p = 0.01, \eta_p^2 = 0.12$) and condition ($F(1,21) = 52.13, p < 0.001, \eta_p^2 = 0.71$). The interaction was also significant ($F(6.94, 145.81) = 2.70, p = 0.01, \eta_p^2 = 0.11$). Recognition performance was significantly higher in the Surface Feature Condition than the Line Drawing Condition, both in terms of overall recognition performance ($t(21) = 7.22$, $p < 10^{-6}$, Cohen's $d = 2.99$; Figure 3A) and for each of the 12 viewpoint ranges (all $ps < 0.05$, Figure 3B).

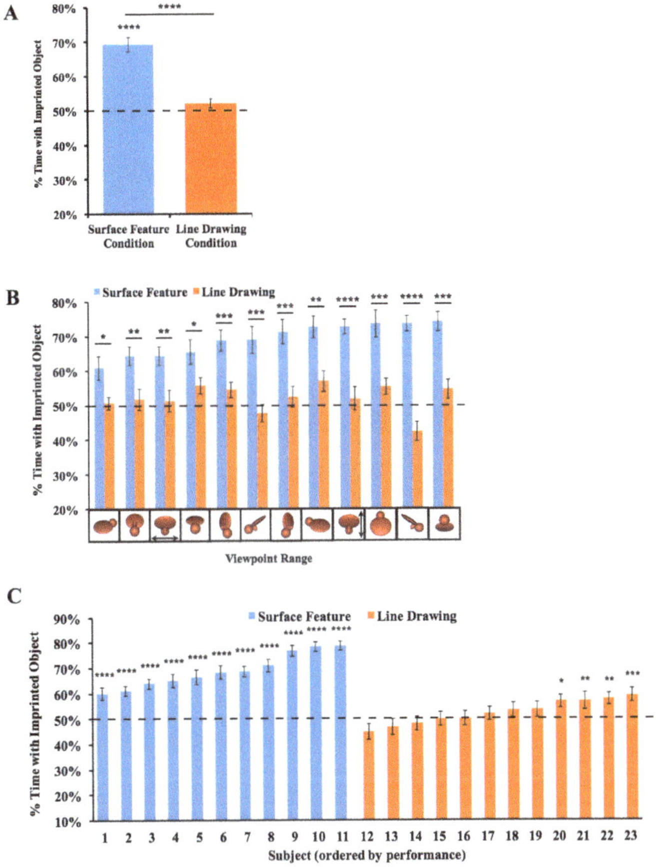

Figure 3. Results from Experiment 1. (**A**) Overall object recognition performance across the test phase. (**B**) Recognition performance on each of the 12 viewpoint ranges. (**C**) Recognition performance of each individual subject. The graphs show the percentage of time spent with the imprinted object versus unfamiliar object. The dashed lines indicate chance performance. Error bars denote ±1 standard error. Asterisks denote statistical significance: * $p < 0.05$; ** $p < 0.01$; *** $p < 0.001$; **** $p < 0.0001$ (two-tailed t-tests).

We also examined performance for each of the two imprinted objects. When the chicks were imprinted to Object 1 (see Figure 1 for reference), performance exceeded chance level in the Surface Feature Condition ($t(4) = 6.54$, $p = 0.003$, Cohen's $d = 2.92$), but not in the Line Drawing Condition ($t(5) = 1.35$, $p = 0.24$, Cohen's $d = -0.55$). Performance was also significantly higher in the Surface Feature Condition than the Line Drawing Condition for Object 1 ($t(9) = 7.11$, $p = 0.00006$, Cohen's $d = 4.13$). When the chicks were imprinted to Object 2, performance exceeded chance level in both the Surface Feature Condition ($t(5) = 7.75$, $p = 0.001$, Cohen's $d = 3.16$) and the Line Drawing Condition ($t(5) = 5.23$, $p = 0.003$, Cohen's $d = 2.13$), although performance was significantly higher in the Surface Feature Condition than the Line Drawing Condition ($t(10) = 4.55$, $p = 0.001$, Cohen's $d = 2.63$). In general, newborn chicks developed superior object recognition abilities when reared with objects containing surface features versus line drawings.

Since over 100 test trials were collected from each chick, we could also measure each chick's object recognition performance with high precision. As shown in Figure 3C, all chicks in the Surface Feature Condition successfully created view-invariant object representations (all $ps < 0.0001$). Conversely, only four of the twelve chicks in the Line Drawing Condition performed above chance level in the task, and those four subjects performed much worse than the subjects in the Surface Feature Condition.

2.3. Discussion

In Experiment 1, newborn chicks developed enhanced object recognition performance when reared with objects containing surface features versus line drawings. Overall, the chicks reared with the line drawings performed at chance level, despite acquiring over 100 h of visual experience with the line drawings during the input phase. Thus, the development of object recognition in newborn chicks requires visual experience with the surface features of objects.

To verify this conclusion under different testing conditions, we performed a second experiment with two key changes. First, rather than presenting the object from a 60° viewpoint range, the object moved through a 360° viewpoint range. As a result, the chicks were exposed to six times as many unique views of the object during the input phase. Second, we measured each chick's object recognition abilities with Identity Trials and Viewpoint Trials (Figure 2B). The Identity Trials tested whether the chicks built object representations that were selective for object identity and tolerant to changes in viewpoint. The Viewpoint Trials tested whether the chicks built object representations that were selective for familiar viewpoints. The Identity Trials tested the chicks' view-invariant object recognition abilities, whereas the Viewpoint Trials tested whether the chicks could use an image-based matching strategy to recognize their imprinted object.

3. Experiment 2

3.1. Materials and Methods

The text describing the methods is partly adapted from [4]. The methods were identical to those used in Experiment 1, except in the following ways. First, 20 different subjects were tested. Ten chicks were tested in the Surface Feature Condition [4] and ten chicks were tested in the Line Drawing Condition. No subjects were excluded from the analyses. Second, the imprinted object completed a 360° rotation every 15 s around a frontoparallel vertical axis (see SI Movie 2 for animations). Third, the chicks were tested with Viewpoint Trials and Identity Trials. In the Viewpoint Trials, one display wall showed familiar viewpoints of the imprinted object (rotation around the familiar axis), whereas the other display wall showed novel viewpoints of the imprinted object (rotation around a novel axis, Figure 2B). If the chicks created object representations that were selective for familiar viewpoints, then they should have preferred the imprinted object rotating around the familiar axis over the novel axis. In the Identity Trials, one display wall showed the imprinted object rotating around a novel axis, whereas the other display wall showed a novel object rotating around the familiar axis (Figure 2B). Thus, to recognize their imprinted object in Identity Trials, the chicks needed to build view-invariant representations that were selective for object identity and tolerant to viewpoint changes.

The chicks received 24 test trials per day (168 test trials in total). Figure 2B shows how the objects were presented on the display walls during the input phase and test phase. In total, 6720 h of video footage (14 days $\times$ 24 h/day $\times$ 20 subjects) were collected for Experiment 2.

3.2. Results and Discussion

The results are shown in Figure 4. An ANOVA with the within-subjects factor of the Trial Type (Viewpoint Trials vs. Identity Trials) and the between-subjects factor of the condition (surface feature vs. line drawing) revealed a significant main effect of the condition ($F(1,18) = 48.55$, $p < 0.001$, $\eta_p^2 = 0.73$), reflecting higher performance in the Surface Feature Condition. The ANOVA also showed a significant main effect of the Trial Type ($F(1,18) = 14.01$, $p = 0.001$, $\eta_p^2 = 0.44$), reflecting higher performance in the Identity Trials. The interaction was not significant ($F(1,18) = 1.22$, $p = 0.29$, $\eta_p^2 = 0.06$). In the Surface Feature Condition, performance was above chance level in the Identity Trials (one-sample *t*-test, $t(9) = 7.84$, $p < 0.001$, Cohen's $d = 2.48$), but not in the Viewpoint Trials ($t(9) = 1.41$, $p = 0.19$, Cohen's $d = 0.45$). In the Line Drawing Condition, performance did not exceed chance level in the Identity Trials ($t(9) = 0.70$, $p = 0.50$, Cohen's $d = 0.22$) or the Viewpoint Trials ($t(9) = 2.22$, $p = 0.053$, Cohen's $d = 0.70$). Thus, when chicks were reared with an object containing surface features, the chicks built object representations that were highly sensitive to identity features. When chicks were reared with line drawings, they did not show evidence for sensitivity to identity or viewpoint features.

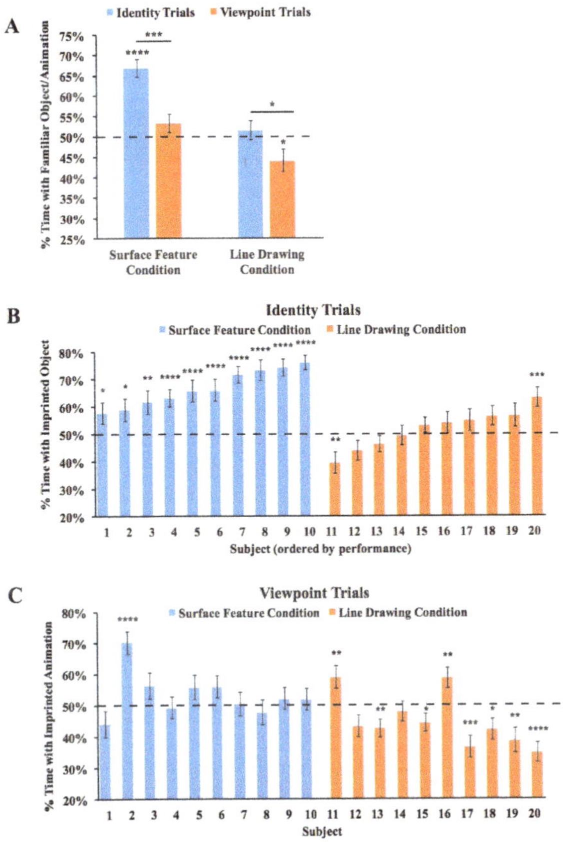

Figure 4. Results from Experiment 2. (**A**) Overall object recognition performance across the test phase. (**B**) Recognition performance of each individual subject in the Identity Trials. (**C**) Recognition performance of each individual subject in the Viewpoint Trials. The dashed lines indicate chance performance. Error bars denote ±1 standard error. Asterisks denote statistical significance: * $p < 0.05$; ** $p < 0.01$; *** $p < 0.001$; **** $p < 0.0001$ (two-tailed *t*-tests).

We also examined performance for each of the two imprinted objects. We repeated the ANOVA above, but with the addition of the object as a main effect. The ANOVA revealed the same significant effects as before (significant main effects of condition and Trial Type), and the main effect of the object was not significant, nor were the interactions (all $ps > 0.3$). When the chicks were imprinted to Object 1, performance exceeded chance level in the Identity Trials in the Surface Feature Condition ($t(3) = 4.03$, $p = 0.03$, Cohen's $d = 2.02$), but not in the Line Drawing Condition ($t(5) = 0.16$, $p = 0.88$, Cohen's $d = 0.06$). In the Identity Trials, performance was significantly higher in the Surface Feature Condition than the Line Drawing Condition ($t(8) = 3.98$, $p = 0.004$, Cohen's $d = 2.41$). Similarly, when the chicks were imprinted to Object 2, performance exceeded chance level in the Identity Trials in the Surface Feature Condition ($t(5) = 6.44$, $p = 0.001$, Cohen's $d = 2.63$), but not in the Line Drawing Condition ($t(3) = 0.67$, $p = 0.55$, Cohen's $d = 0.33$). Again, in the Identity Trials, performance was significantly higher in the Surface Feature Condition than the Line Drawing Condition ($t(8) = 2.68$, $p = 0.03$, Cohen's $d = 1.64$). In the Viewpoint Trials, performance did not exceed chance level when the chicks were imprinted to Object 1 or Object 2 in the Surface Feature Condition or the Line Drawing Condition ($ps > 0.15$).

As shown in Figure 4B, all of the chicks in the Surface Feature Condition exceeded chance level in the Identity Trials (two chicks, $p < 0.05$; one chick, $p < 0.01$; seven chicks, $p < 0.0001$). In contrast, recognition performance was low for all of the chicks in the Line Drawing Condition. Only one chick exceeded chance level in the Identity Trials, while one other chick performed significantly below chance level. As in Experiment 1, newborn chicks developed superior object recognition abilities when reared with objects containing surface features versus line drawings.

3.3. *Measuring the Strength of the Imprinting Response in Experiments 1 and 2*

To test the strength of the chicks' imprinting response, we performed two additional analyses. First, we examined the proportion of time that the chicks spent in proximity to their imprinted object during the input phase. As shown in Figure 5A, the chicks in both Experiments 1 and 2 spent the majority of their time in proximity to the imprinted object during the input phase (Experiment 1 subjects imprinted to surface feature objects: $t(10) = 40.47$, $p < 10^{-11}$, $d = 12.20$; Experiment 1 subjects imprinted to line drawings: $t(11) = 6.27$, $p = 0.00006$, $d = 1.81$; Experiment 2 subjects imprinted to surface feature objects: $t(9) = 20.55$, $p < 10^{-8}$, $d = 6.50$; Experiment 2 subjects imprinted to line drawings: $t(9) = 10.84$, $p = 0.000002$, $d = 3.43$). Thus, the chicks successfully imprinted to both the line drawings and the objects with surface features. In both experiments, however, the imprinting response was stronger in the Surface Feature Condition than the Line Drawing Condition (Experiment 1: $t(13.16) = 6.22$, $p = 0.00003$, $d = 2.55$; Experiment 2: $t(18) = 2.15$, $p = 0.05$, $d = 0.96$), suggesting that the chicks imprinted less strongly to the line drawings than to the objects with surface features.

Second, we examined the proportion of time the chicks spent with their imprinted object during the rest periods. During the rest periods, the imprinted object was presented on one display wall while the other display wall was blank. The rest periods therefore provided a measure of the strength of the chick's attachment to the imprinted object during the test phase. As shown in Figure 5B, the chicks in both experiments spent the majority of their time in proximity to the imprinted object during the rest periods (Experiment 1 subjects imprinted to surface feature objects: $t(10) = 24.91$, $p < 10^{-9}$, $d = 7.51$; Experiment 1 subjects imprinted to line drawings: $t(11) = 5.35$, $p = 0.0002$, $d = 1.54$; Experiment 2 subjects imprinted to surface feature objects: $t(9) = 30.14$, $p < 10^{-9}$, $d = 9.53$; Experiment 2 subjects imprinted to line drawings: $t(9) = 7.76$, $p = 0.00003$, $d = 2.45$). However, the imprinting response was stronger in the Surface Feature Condition than the Line Drawing Condition (Experiment 1: $t(21) = 5.91$, $p = 0.000007$, Cohen's $d = 2.50$; Experiment 2: $t(11.30) = 2.89$, $p = 0.01$, Cohen's $d = 1.29$), providing additional evidence that the chicks imprinted less strongly to the line drawings than to the objects with surface features. Importantly, this reduction in the strength of the imprinting response cannot fully explain the low recognition performance because even the chicks that imprinted strongly to the line drawings still built inaccurate object representations (Figure 5C). Together, these analyses suggest that when chicks are reared with line drawings versus objects with surface features, the chicks develop an impaired imprinting response and build less accurate object representations.

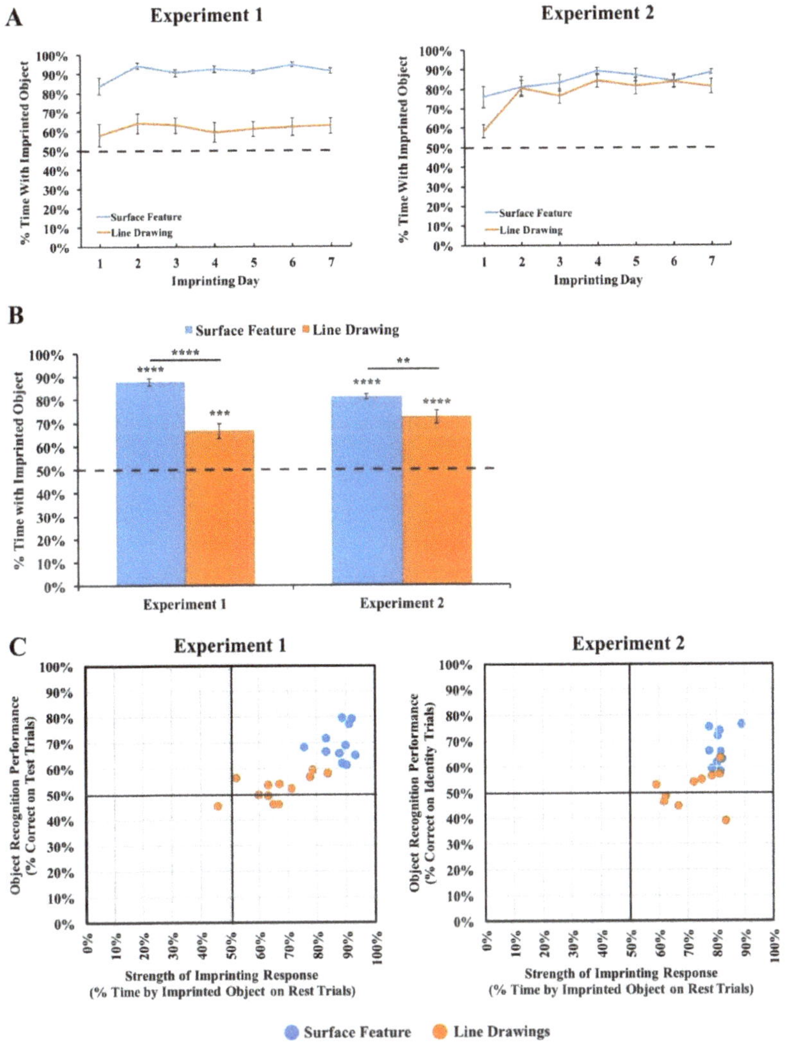

Figure 5. (**A**) Strength of the imprinting response in Experiments 1 and 2 during the input phase. (**B**) Strength of the imprinting response in Experiments 1 and 2 during the rest periods of the test phase. The dashed lines indicate chance performance. Error bars denote ±1 standard error. Asterisks denote statistical significance: ** $p < 0.01$; *** $p < 0.001$; **** $p < 0.0001$ (two-tailed t-tests). The chicks successfully imprinted to both the line drawings and the objects with surface features, and this effect was stronger for the objects with surface features. (**C**) Comparison of the chicks' object recognition performance and the strength of their imprinting response. The chicks developed enhanced object recognition performance when reared with objects with surface features compared to line drawings, even when the chicks imprinted to the objects at similar strengths.

4. Experiment 3

Experiments 1 and 2 indicate that the development of object recognition requires experience with the surface features of objects. However, there are three limitations to these results. First, the line drawings and the objects with surface features differed in several respects, including their color, contrast, hue homogeneity, and complexity. Thus, it is unclear which particular features caused the observed differences in recognition performance across the conditions. Indeed, there is extensive evidence that color is one of the most distinctive features encoded in imprinting (e.g., [30,44–48]), which raises the possibility that color differences may have influenced performance across the conditions. Second, Experiments 1 and 2 tested chicks' recognition performance with the same two objects, so it is unclear whether these results generalize to other objects. Third, the imprinting response was less strong in the Line Drawing Condition than the Surface Feature Condition, potentially because of the color differences across objects. Given that recognition performance in this task is directly constrained by the strength of the imprinting response, the weaker imprinting response in the Line Drawing Condition likely produced lower recognition performance in the test trials.

To provide a more direct comparison of chicks' recognition performance across conditions, we performed a third experiment in which the objects in the Surface Feature and Line Drawing Conditions were the same color (red). We also ensured that the objects did not have shadows and regions with different luminance values (as in Experiments 1 and 2), by using two-dimensional objects, rather than three-dimensional objects. The only difference between the conditions was whether the objects did, or did not, have surface features (Figure 6A).

For the Surface Feature Condition, we used the data previously reported in [42]. In that paper, we tested whether newborn chicks can encode the transitional probabilities (TPs) between shapes in a sequence. During the input phase, the chicks were reared with an imprinting sequence consisting of a stream of four shapes, and the order of the shapes was defined by the TPs within and between shape pairs. During the test phase, we presented two types of test trials. In the shape recognition trials, one monitor showed a sequence of familiar shapes, and the opposite monitor showed a sequence of novel shapes. In the TP trials, both monitors showed the familiar shapes, but we manipulated the TPs between shapes. One monitor showed a familiar TP sequence, in which the TPs between shapes matched the imprinting sequence, and the opposite monitor showed a novel TP sequence, in which the TPs between shapes did not match the imprinting sequence. In the original study, we found that the chicks successfully distinguished between the sequences in the shape recognition trials, but failed to distinguish between the sequences in the TP trials. Here, we repeated this experiment with one crucial change: rather than presenting sequences of shapes with surface features, we presented sequences of red line drawing shapes (Figure 6A).

4.1. Materials and Methods

For a detailed description of the methods, see [42]. In the present study, we used a similar design to the original study, except that the chicks were imprinted and tested with red line drawings, rather than red objects with surface features. As in the original study, we tested the chicks with both shape recognition and TP test trials. We tested 12 subjects in the Surface Feature Condition [42] and 10 subjects in the Line Drawing Condition. No subjects were excluded from the analyses.

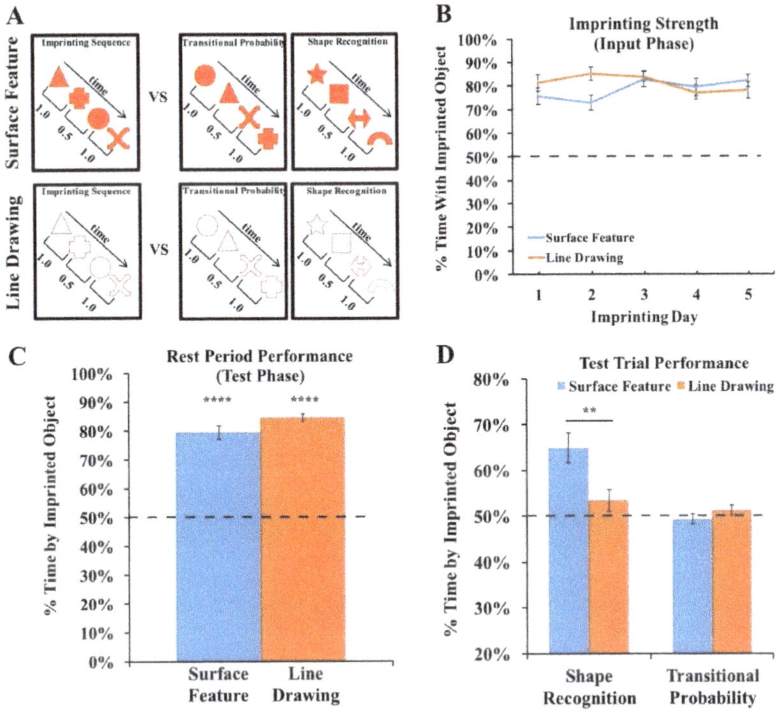

Figure 6. (**A**) Experiment 3 method. During the input phase, an imprinting sequence defined by the transitional probabilities (TPs) within and between shape pairs appeared on one display wall at a time. The imprinting sequence either contained shapes with surface features or line drawings of shapes. During the test phase, we presented chicks with two-alternative forced-choice tasks. In each test trial, one display wall showed the imprinting sequence, and the other display wall showed either the same shapes as in the imprinting sequence but in novel orders (TP trials) or a sequence of novel shapes (shape recognition trials). (**B**) Strength of the imprinting response in Experiment 3 during the input phase. (**C**) Strength of the imprinting response in Experiment 3 during the rest periods of the test phase. The chicks imprinted equally strongly across the Surface Feature and Line Drawing Conditions. (**D**) Recognition performance in the Shape Recognition and Transitional Probability Conditions. The chicks showed superior object recognition performance in the Surface Feature Condition compared to the Line Drawing Condition. The dashed lines indicate chance performance. Error bars denote ±1 standard error. Asterisks denote statistical significance: ** $p < 0.01$; **** $p < 0.0001$ (two-tailed t-tests).

4.2. Results and Discussion

We first examined the strength of the imprinting response to ensure that the chicks imprinted equally strongly across the two conditions. As shown in Figure 6B, the chicks in both conditions spent the majority of their time in proximity to the imprinted object during the input phase (one-sample t-tests, Surface Feature Condition: $t(11) = 12.35$, $p < 10^{-7}$, Cohen's $d = 3.57$; Line Drawing Condition: $t(11) = 18.77$, $p < 10^{-8}$, Cohen's $d = 5.42$). Unlike in Experiments 1 and 2, the chicks did not show a stronger imprinting response in the Surface Feature Condition than the Line Drawing Condition (independent samples t-test, $t(22) = 0.7$, $p = 0.94$, Cohen's $d = 0.03$). Similarly, as shown in Figure 6C, the chicks in both conditions spent the majority of their time in proximity to the imprinted object during the rest periods (one-sample t-tests, Surface Feature Condition: $t(11) = 13.06$, $p < 10^{-7}$, Cohen's $d = 3.77$; Line Drawing Condition: $t(9) = 28.59$, $p < 10^{-10}$, Cohen's $d = 9.04$). Unlike in Experiments 1 and 2, the chicks did not show a stronger imprinting response during the rest periods in the Surface Feature Condition than the Line Drawing Condition (independent samples *t*-test, $t(20) = 1.69$, $p = 0.11$, Cohen's $d = 0.75$). Together, these analyses show that the chicks did not imprint more strongly in the Surface Feature Condition, allowing for a more direct comparison of recognition performance across the conditions.

The chicks' recognition performance is shown in Figure 6D. An ANOVA with the within-subjects factor of the Trial Type (shape recognition vs. TP trials) and the between-subjects factor of the condition (surface feature vs. line drawing) revealed a significant main effect of the Trial Type ($F(1,20) = 15.95$, $p = 0.001$, $\eta_p^2 = 0.44$), a significant main effect of the condition ($F(1,20) = 4.38$, $p = 0.049$, $\eta_p^2 = 0.18$), and a significant interaction between the Trial Type and condition ($F(1,20) = 9.16$, $p = 0.007$, $\eta_p^2 = 0.31$).

When reared with a sequence of shapes containing surface features, the chicks could reliably distinguish between familiar shapes and novel shapes (one-sample *t*-test, $t(11) = 4.67$, $p = 0.0007$, Cohen's $d = 1.35$). In contrast, when reared with a sequence of line drawing shapes, the chicks failed to distinguish between familiar shapes and novel shapes (one-sample *t*-test, $t(9) = 1.30$, $p = 0.23$, Cohen's $d = 0.41$). As in the original study [42], the chicks in both conditions failed to distinguish between the sequences based on the TPs between shapes (Surface Feature Condition: $t(11) = 0.56$, $p = 0.59$, Cohen's $d = -0.16$; Line Drawing Condition: $t(9) = 1.04$, $p = 0.33$, Cohen's $d = 0.33$).

On the individual-subject level, 8 of the 12 chicks in the Surface Feature Condition showed a statistically significant preference for the familiar shapes (7 chicks: $p < 0.0001$; 1 chick: $p < 0.05$). In contrast, in the Line Drawing Condition, only one chick exceeded chance level in the shape recognition trials, while one other chick performed significantly below chance level. As in Experiments 1 and 2, newborn chicks developed superior object recognition performance when reared with objects containing surface features versus line drawings.

5. General Discussion

A deep understanding of object recognition requires understanding the role of visual experience in development. Here, we reveal a set of conditions under which object recognition fails to develop in newborn animals: when a newborn's visual experience with objects consists solely of line drawings. When newborn chicks were reared with objects containing surface features, the chicks developed robust view-invariant object recognition. In contrast, when chicks were reared with line drawings of objects, the chicks failed to develop object recognition. Notably, the chicks reared with the line drawings performed at chance level, despite acquiring over 100 h of experience with the objects. Thus, the development of object recognition requires visual experience with the surface features of objects.

Interestingly, the chicks reared with line drawings failed to build accurate object representations despite being raised in environments that contained some surface features. The walls and floor of the chamber contained surface features, as did the heaps of grain consumed during feeding. Nevertheless, when the *objects* in the chicks' visual environment lacked surface features, the chicks failed to build accurate representations. This finding suggests that experience with surface features per se is not sufficient for the development of object recognition; rather, newborn visual systems need experience with the surface features of objects.

These results add to a growing body of work mapping out the conditions under which object recognition does, and does not, emerge in newborn animals. For instance, studies with newborn chicks have revealed two constraints on the development of object recognition. First, there is a "slowness constraint" on newborn vision: object recognition emerges when newborn chicks are reared with slowly moving objects, but not quickly moving objects [4]. When chicks are reared with quickly moving objects, their object representations become distorted in the direction of object motion and fail to generalize to novel viewpoints and rotation speeds. Second, there is a "smoothness constraint" on newborn vision: object recognition emerges when newborn chicks are reared with temporally smooth objects, but not temporally non-smooth objects [3,49]. When chicks are reared with temporally non-smooth objects, their object representations are less selective for object identity. The present study extends this literature by demonstrating that experience with slow and smooth objects is not sufficient for the development of object recognition. The line drawings in Experiments 1–3 moved slowly and smoothly over time, but the chicks nevertheless failed to develop object recognition. Together, these findings indicate that the development of object recognition requires experience with naturalistic objects: objects that have surface features and move slowly and smoothly over time.

More generally, these results build on a large body of research using animal models to examine the mechanisms of object recognition and early visual learning. For decades, newborn chicks have been used to characterize the effects of visual experience on the brain (e.g., [50–52]) and to isolate the neural mechanisms that underlie imprinting (e.g., [53,54]). Studies of chicks have also revealed predispositions that might shape early visual learning (e.g., [55–57]). Another important animal model for studying early visual learning is rodents. Studies of rats provide converging evidence that high-level vision is not unique to primates. Like newborn chicks, rats can recognize objects across novel viewpoints (e.g., [43]). Normal visual development in rats also requires experience with a slow and smooth visual world [58], suggesting that avian and mammalian brains are subject to common developmental constraints. Specifically, when newborn rats were reared with frame-scrambled versions of natural movies (which preserved the natural spatial statistics but resulted in quickly changing, temporally unstructured input), the rats developed fewer complex cells in the primary visual cortex, the cells showed abnormally fast response dynamics, and the cells were less likely to support stable decoding of stimulus orientation. Thus, depriving newborn animals of slowly changing visual experiences disrupts normal visual development in both birds and mammals, potentially reflecting a shared cortex-like canonical circuit found across taxa [37].

Limitations of This Study and Directions for Future Research

While these results contribute to our understanding of early visual development, there are limitations to this work that will require additional future research. First, these chicks were reared with either one 3D object (Experiments 1 and 2) or four 2D objects (Experiment 3). It is therefore possible that chicks could develop object recognition in a visual world consisting solely of line drawings if there were more line drawings in the environment and/or if those line drawings were more complex (e.g., line drawings containing polyhedral shapes that included L, Y, and T junctures between lines). Future studies could distinguish between these possibilities by rearing chicks in more complex "line drawing worlds".

Second, we used the chicks' preference for their imprinted object as a measure of object recognition performance. While successful performance provides evidence for object recognition (e.g., in the Surface Feature Conditions), the absence of a preference (e.g., in the Line Drawing Conditions) does not necessarily provide evidence for a lack of object recognition. For instance, it is possible that the chicks in the Line Drawing Conditions perceived the test objects as different but grouped them in the same object category. Of course, this alternative explanation must then explain why the presence of surface features would lead chicks to categorize objects differently from one another, whereas line drawings would lead chicks to categorize objects together. It would be interesting for future studies to test chicks using alternative methods (e.g., reinforcement learning) to explore whether the present findings generalize to other object recognition tasks.

Third, these results do not reveal why surface features are necessary for the development of object recognition. Why do newborn chicks fail to understand line drawings, when mature animals (including birds) can readily recognize objects presented in line drawings? One possibility is that the mechanisms underlying object recognition require patterned input from natural visual objects in order to develop a receptive field structure that efficiently recovers edges and lines. Specifically, in natural visual environments, the edges of objects and surfaces are typically marked by discrete changes in surface attributes, and mature visual systems contain neurons tuned to the orientation of these contours, responding to edges and lines [59,60]. To develop these orientation-tuned detectors, newborn brains may require visual experience of objects with surface features.

Moreover, line drawings are impoverished compared to real objects, and the surface features that appear on real objects may provide valuable information for building accurate representations of an object's three-dimensional shape. For example, the surface features on the objects used in Experiments 1 and 2 had gradients of luminance that moved as the object rotated, creating flow field cues for three-dimensional shape. Accordingly, it is a possibility that experience with realistic objects is necessary to develop the ability to recognize objects in line drawings. Future controlled-rearing experiments could test this hypothesis directly by examining whether experience with realistic objects allows for the development of line drawing understanding.

Ultimately, a deep mechanistic understanding of the role of experience in the development of object recognition will require task-performing computational models that can simulate the complex interactions between newborn brains and the visual environment. The present results should be valuable for this enterprise because they provide precise descriptions of how specific visual inputs relate to specific object recognition outputs in a newborn model system. These input–output patterns can serve as benchmarks for measuring the accuracy of computational models (e.g., [61–64]). Specifically, to explain the development of object recognition, a computational model would need to produce two patterns. First, the model should successfully develop view-invariant object recognition when trained with realistic objects that move slowly and smoothly over time. Second, the model should fail to develop object recognition when trained solely with line drawings.

6. Conclusions

The present study provides evidence that the development of object recognition requires experience with the surface features of objects. Newborn chicks develop enhanced object recognition performance when reared with objects containing surface features compared to line drawings. This study sheds light on how a fundamental ability emerges in newborn animals and provides precise input–output patterns for measuring the accuracy of task-performing computational models of visual development. Furthermore, these insights into the role of experience in object recognition may offer valuable parallels to understanding and potentially enhancing early visual development in human infants.

Supplementary Materials: The following supporting information can be downloaded at: https://www.mdpi.com/article/10.3390/ani14020284/s1, Movie S1: Experiment 1 Imprinting Stimuli; Movie S2: Experiment 2 Imprinting Stimuli. Data S1: Chick Performance Data for Experiments 1–3.

Author Contributions: Conceptualization, J.N.W.; methodology, J.N.W.; formal analysis, S.M.W.W.; resources, J.N.W.; data curation, J.N.W.; writing—original draft preparation, J.N.W.; writing—review and editing, S.M.W.W. and J.N.W.; visualization, S.M.W.W. and J.N.W.; project administration, J.N.W.; funding acquisition, J.N.W. All authors have read and agreed to the published version of the manuscript.

Funding: This research was funded by NSF CAREER Grant BCS-1351892 and a James S. McDonnell Foundation Understanding Human Cognition Scholar Award. This research was conducted at the University of Southern California.

Institutional Review Board Statement: The animal study protocol was approved by the Institutional Review Board (or Ethics Committee) of University of Southern California (protocol code #11316, approved 1 February 2016).

Data Availability Statement: Data are contained within the article and Supplementary Materials.

Conflicts of Interest: The authors declare no conflicts of interest.

References

1. DiCarlo, J.J.; Zoccolan, D.; Rust, N.C. How does the brain solve visual object recognition? *Neuron* **2012**, *73*, 415–434. [CrossRef] [PubMed]
2. Prasad, A.; Wood, S.M.; Wood, J.N. Using automated controlled rearing to explore the origins of object permanence. *Dev. Sci.* **2019**, *22*, e12796. [CrossRef] [PubMed]
3. Wood, J.N. A Smoothness constraint on the development of object recognition. *Cognition* **2016**, *153*, 140–145. [CrossRef]
4. Wood, J.N.; Wood, S.M.W. The development of newborn object recognition in fast and slow visual worlds. *Proc. R. Soc. B* **2016**, *283*, 20160166. [CrossRef]
5. Wood, J.N.; Wood, S.M. The development of invariant object recognition requires visual experience with temporally smooth objects. *Cogn. Sci.* **2018**, *42*, 1391–1406. [CrossRef]
6. Biederman, I. Recognition-by-components: A theory of human image understanding. *Psychol. Rev.* **1987**, *94*, 115. [CrossRef] [PubMed]
7. Biederman, I.; Ju, G. Surface versus edge-based determinants of visual recognition. *Cogn. Psychol.* **1988**, *20*, 38–64. [CrossRef]
8. Ishai, A.; Ungerleider, L.G.; Martin, A.; Haxby, J.V. The representation of objects in the human occipital and temporal cortex. *J. Cogn. Neurosci.* **2000**, *12*, 35–51. [CrossRef]
9. Walther, D.B.; Chai, B.; Caddigan, E.; Beck, D.M.; Fei-Fei, L. Simple line drawings suffice for functional MRI decoding of natural scene categories. *Proc. Natl. Acad. Sci. USA* **2011**, *108*, 9661–9666. [CrossRef]
10. Yonas, A.; Arterberry, M.E. Infants perceive spatial structure specified by line junctions. *Perception* **1994**, *23*, 1427–1435. [CrossRef]
11. Goodnow, J. *Children Drawing*; Harvard University Press: Cambridge, MA, USA, 1977; ISBN 0-674-49214-5.
12. Clottes, J. Chauvet Cave (ca. 30,000 BC). In *Heilbrunn Timeline of Art History*; The Metropolitan Museum of Art: New York, NY, USA, 2000; Volume 2010.
13. Kennedy, J.M.; Ross, A.S. Outline picture perception by the Songe of Papua. *Perception* **1975**, *4*, 391–406. [CrossRef]
14. Itakura, S. Recognition of line-drawing representations by a chimpanzee (Pan troglodytes). *J. Gen. Psychol.* **1994**, *121*, 189–197. [CrossRef]
15. Tanaka, M. Recognition of pictorial representations by chimpanzees (Pan troglodytes). *Anim. Cogn.* **2007**, *10*, 169–179. [CrossRef]
16. Wasserman, E.A.; Gagliardi, J.L.; Cook, B.R.; Kirkpatrick-Steger, K.; Astley, S.L.; Biederman, I. The pigeon's recognition of drawings of depth-rotated stimuli. *J. Exp. Psychol. Anim. Behav. Process.* **1996**, *22*, 205. [CrossRef]
17. Sayim, B.; Cavanagh, P. What line drawings reveal about the visual brain. *Front. Hum. Neurosci.* **2011**, *5*, 118. [CrossRef]
18. Hayward, W.G.; Williams, P. Viewpoint dependence and object discriminability. *Psychol. Sci.* **2000**, *11*, 7–12. [CrossRef] [PubMed]
19. Naor-Raz, G.; Tarr, M.J.; Kersten, D. Is color an intrinsic property of object representation? *Perception* **2003**, *32*, 667–680. [CrossRef] [PubMed]
20. Rossion, B.; Pourtois, G. Revisiting Snodgrass and Vanderwart's object pictorial set: The role of surface detail in basic-level object recognition. *Perception* **2004**, *33*, 217–236. [CrossRef] [PubMed]
21. Tarr, M.J.; Kersten, D.; Bülthoff, H.H. Why the visual recognition system might encode the effects of illumination. *Vis. Res.* **1998**, *38*, 2259–2275. [CrossRef]
22. Vuong, Q.C.; Peissig, J.J.; Harrison, M.C.; Tarr, M.J. The role of surface pigmentation for recognition revealed by contrast reversal in faces and Greebles. *Vis. Res.* **2005**, *45*, 1213–1223. [CrossRef] [PubMed]
23. Wurm, L.H.; Legge, G.E.; Isenberg, L.M.; Luebker, A. Color improves object recognition in normal and low vision. *J. Exp. Psychol. Hum. Percept. Perform.* **1993**, *19*, 899. [CrossRef] [PubMed]
24. Price, C.J.; Humphreys, G.W. The effects of surface detail on object categorization and naming. *Q. J. Exp. Psychol.* **1989**, *41*, 797–828. [CrossRef] [PubMed]
25. Spelke, E.S. Principles of object perception. *Cogn. Sci.* **1990**, *14*, 29–56. [CrossRef]
26. Wood, J.N. Newborn chickens generate invariant object representations at the onset of visual object experience. *Proc. Natl. Acad. Sci. USA* **2013**, *110*, 14000–14005. [CrossRef] [PubMed]
27. Wood, S.M.; Wood, J.N. Using automation to combat the replication crisis: A case study from controlled-rearing studies of newborn chicks. *Infant Behav. Dev.* **2019**, *57*, 101329. [CrossRef]
28. Wood, S.M.W.; Wood, J.N. A chicken model for studying the emergence of invariant object recognition. *Front. Neural Circuits* **2015**, *9*, 7. [CrossRef] [PubMed]
29. Horn, G. Pathways of the past: The imprint of memory. *Nat. Rev. Neurosci.* **2004**, *5*, 108–120. [CrossRef]
30. Wood, J.N. Newly hatched chicks solve the visual binding problem. *Psychol. Sci.* **2014**, *25*, 1475–1481. [CrossRef] [PubMed]
31. Wood, S.M.; Wood, J.N. One-shot object parsing in newborn chicks. *J. Exp. Psychol. Gen.* **2021**, *150*, 2408. [CrossRef] [PubMed]
32. Wood, J.N. Characterizing the information content of a newly hatched chick's first visual object representation. *Dev. Sci.* **2015**, *18*, 194–205. [CrossRef]
33. Wood, J.N.; Wood, S.M. Measuring the speed of newborn object recognition in controlled visual worlds. *Dev. Sci.* **2017**, *20*, e12470. [CrossRef] [PubMed]
34. Jarvis, E.D.; Güntürkün, O.; Bruce, L.; Csillag, A.; Karten, H.; Kuenzel, W.; Medina, L.; Paxinos, G.; Perkel, D.J.; Shimizu, T. Avian brains and a new understanding of vertebrate brain evolution. *Nat. Rev. Neurosci.* **2005**, *6*, 151–159. [CrossRef] [PubMed]
35. Güntürkün, O.; Bugnyar, T. Cognition without cortex. *Trends Cogn. Sci.* **2016**, *20*, 291–303. [CrossRef] [PubMed]
36. Karten, H.J. Neocortical evolution: Neuronal circuits arise independently of lamination. *Curr. Biol.* **2013**, *23*, R12–R15. [CrossRef]

37. Stacho, M.; Herold, C.; Rook, N.; Wagner, H.; Axer, M.; Amunts, K.; Güntürkün, O. A cortex-like canonical circuit in the avian forebrain. *Science* **2020**, *369*, eabc5534. [CrossRef] [PubMed]
38. Dugas-Ford, J.; Rowell, J.J.; Ragsdale, C.W. Cell-type homologies and the origins of the neocortex. *Proc. Natl. Acad. Sci. USA* **2012**, *109*, 16974–16979. [CrossRef] [PubMed]
39. Wang, Y.; Brzozowska-Prechtl, A.; Karten, H.J. Laminar and columnar auditory cortex in avian brain. *Proc. Natl. Acad. Sci. USA* **2010**, *107*, 12676–12681. [CrossRef] [PubMed]
40. Calabrese, A.; Woolley, S.M. coding principles of the canonical cortical microcircuit in the avian brain. *Proc. Natl. Acad. Sci. USA* **2015**, *112*, 3517–3522. [CrossRef] [PubMed]
41. Shanahan, M.; Bingman, V.P.; Shimizu, T.; Wild, M.; Güntürkün, O. Large-scale network organization in the avian forebrain: A connectivity matrix and theoretical analysis. *Front. Comput. Neurosci.* **2013**, *7*, 89. [CrossRef]
42. Wood, S.M.; Johnson, S.P.; Wood, J.N. Automated study challenges the existence of a foundational statistical-learning ability in newborn chicks. *Psychol. Sci.* **2019**, *30*, 1592–1602. [CrossRef]
43. Zoccolan, D.; Oertelt, N.; DiCarlo, J.J.; Cox, D.D. A rodent model for the study of invariant visual object recognition. *Proc. Natl. Acad. Sci. USA* **2009**, *106*, 8748–8753. [CrossRef]
44. Bateson, P.; Jaeckel, J.B. Chicks' preferences for familiar and novel conspicuous objects after different periods of exposure. *Anim. Behav.* **1976**, *24*, 386–390. [CrossRef]
45. Ham, A.D.; Osorio, D. Colour preferences and colour vision in poultry chicks. *Proc. R. Soc. B Biol. Sci.* **2007**, *274*, 1941–1948. [CrossRef] [PubMed]
46. Johnson, M.; Bolhuis, J.; Horn, G. Interaction between acquired preferences and developing predispositions during imprinting. *Anim. Behav.* **1985**, *33*, 1000–1006. [CrossRef]
47. Miura, M.; Nishi, D.; Matsushima, T. Combined predisposed preferences for colour and biological motion make robust development of social attachment through imprinting. *Anim. Cogn.* **2020**, *23*, 169–188. [CrossRef] [PubMed]
48. Nakamori, T.; Maekawa, F.; Sato, K.; Tanaka, K.; Ohki-Hamazaki, H. Neural basis of imprinting behavior in chicks. *Dev. Growth Differ.* **2013**, *55*, 198–206. [CrossRef]
49. Wood, J.N.; Prasad, A.; Goldman, J.G.; Wood, S.M. Enhanced learning of natural visual sequences in newborn chicks. *Anim. Cogn.* **2016**, *19*, 835–845. [CrossRef]
50. Bateson, P.; Horn, G.; Rose, S. Effects of early experience on regional incorporation of precursors into RNA and protein in the chick brain. *Brain Res.* **1972**, *39*, 449–465. [CrossRef]
51. Horn, G.; McCabe, B.; Bateson, P. An autoradiographic study of the chick brain after imprinting. *Brain Res.* **1979**, *168*, 361–373. [CrossRef]
52. Horn, G. Review lecture: Neural mechanisms of learning: An analysis of imprinting in the domestic chick. *Proc. R. Soc. Lond. Ser. B. Biol. Sci.* **1981**, *213*, 101–137.
53. McCabe, B.; Horn, G.; Bateson, P. Effects of restricted lesions of the chick forebrain on the acquisition of filial preferences during imprinting. *Brain Res.* **1981**, *205*, 29–37. [CrossRef]
54. McCabe, B.; Cipolla-Neto, J.; Horn, G.; Bateson, P. Amnesic effects of bilateral lesions placed in the hyperstriatum ventrale of the chick after imprinting. *Exp. Brain Res.* **1982**, *48*, 13–21. [CrossRef]
55. Horn, G.; McCabe, B.J. Predispositions and preferences. Effects on imprinting of lesions to the chick brain. *Anim. Behav.* **1984**, *32*, 288–292. [CrossRef]
56. Versace, E.; Martinho-Truswell, A.; Kacelnik, A.; Vallortigara, G. Priors in animal and artificial intelligence: Where does learning begin? *Trends Cogn. Sci.* **2018**, *22*, 963–965. [CrossRef]
57. Wood, J.N. Spontaneous preference for slowly moving objects in visually naïve animals. *Open Mind* **2017**, *1*, 111–122. [CrossRef]
58. Matteucci, G.; Zoccolan, D. Unsupervised experience with temporal continuity of the visual environment is causally involved in the development of V1 complex cells. *Sci. Adv.* **2020**, *6*, eaba3742. [CrossRef] [PubMed]
59. Hubel, D.H.; Wiesel, T.N. Receptive fields, binocular interaction and functional architecture in the cat's visual cortex. *J. Physiol.* **1962**, *160*, 106. [CrossRef]
60. Hubel, D.H.; Wiesel, T.N. Receptive fields and functional architecture of monkey striate cortex. *J. Physiol.* **1968**, *195*, 215–243. [CrossRef]
61. Lee, D.; Gujarathi, P.; Wood, J.N. Controlled-rearing studies of newborn chicks and deep neural networks. *arXiv* **2021**, arXiv:2112.06106.
62. McGraw, J.; Lee, D.; Wood, J. Parallel development of social preferences in fish and machines. *arXiv* **2023**, arXiv:2305.11137.
63. Pandey, L.; Wood, S.M.; Wood, J.N. Are vision transformers more data hungry than newborn visual systems? *arXiv* **2023**, arXiv:2312.02843.
64. Pak, D.; Lee, D.; Wood, S.M.; Wood, J.N. A newborn embodied turing test for view-invariant object recognition. *arXiv* **2023**, arXiv:2306.05582.

animals

MDPI

Article

DiffusionFR: Species Recognition of Fish in Blurry Scenarios via Diffusion and Attention

Guoying Wang [1,†], Bing Shi [1,†], Xiaomei Yi [1], Peng Wu [1], Linjun Kong [2,*] and Lufeng Mo [1,3,*]

1 College of Mathematics and Computer Science, Zhejiang A&F University, Hangzhou 311300, China; wgy@zafu.edu.cn (G.W.); 2021611011034@stu.zafu.edu.cn (B.S.); yxm@zafu.edu.cn (X.Y.); wp@zafu.edu.cn (P.W.)

2 Office of Information Technology, Zhejiang University of Finance & Economics, Hangzhou 310018, China

3 Information and Education Technology Center, Zhejiang A&F University, Hangzhou 311300, China

* Correspondence: klj@zufe.edu.cn (L.K.); molufeng@zafu.edu.cn (L.M.)

† These authors contributed equally to this work.

Simple Summary: Blurry scenarios often affect the clarity of fish images, posing significant challenges to deep learning models in terms of the accurate recognition of fish species. A method based on deep learning with a diffusion model and an attention mechanism, DiffusionFR, is proposed herein to improve the accuracy of fish species recognition in blurry scenarios caused by light reflections and water ripple noise. Using a self-constructed dataset, BlurryFish, extensive experiments were conducted and the results showed that the proposed two-stage diffusion network model can restore the clarity of blurry fish images to some extent and the proposed learnable attention module is effective in improving the accuracy of fish species recognition.

Abstract: Blurry scenarios, such as light reflections and water ripples, often affect the clarity and signal-to-noise ratio of fish images, posing significant challenges for traditional deep learning models in accurately recognizing fish species. Firstly, deep learning models rely on a large amount of labeled data. However, it is often difficult to label data in blurry scenarios. Secondly, existing deep learning models need to be more effective for the processing of bad, blurry, and otherwise inadequate images, which is an essential reason for their low recognition rate. A method based on the diffusion model and attention mechanism for fish image recognition in blurry scenarios, DiffusionFR, is proposed to solve these problems and improve the performance of species recognition of fish images in blurry scenarios. This paper presents the selection and application of this correcting technique. In the method, DiffusionFR, a two-stage diffusion network model, TSD, is designed to deblur bad, blurry, and otherwise inadequate fish scene pictures to restore clarity, and a learnable attention module, LAM, is intended to improve the accuracy of fish recognition. In addition, a new dataset of fish images in blurry scenarios, BlurryFish, was constructed and used to validate the effectiveness of DiffusionFR, combining bad, blurry, and otherwise inadequate images from the publicly available dataset Fish4Knowledge. The experimental results demonstrate that DiffusionFR achieves outstanding performance on various datasets. On the original dataset, DiffusionFR achieved the highest training accuracy of 97.55%, as well as a Top-1 accuracy test score of 92.02% and a Top-5 accuracy test score of 95.17%. Furthermore, on nine datasets with light reflection noise, the mean values of training accuracy reached a peak at 96.50%, while the mean values of the Top-1 accuracy test and Top-5 accuracy test were at their highest at 90.96% and 94.12%, respectively. Similarly, on three datasets with water ripple noise, the mean values of training accuracy reached a peak at 95.00%, while the mean values of the Top-1 accuracy test and Top-5 accuracy test were at their highest at 89.54% and 92.73%, respectively. These results demonstrate that the method showcases superior accuracy and enhanced robustness in handling original datasets and datasets with light reflection and water ripple noise.

Keywords: blurry scenarios; fish recognition; deep learning; diffusion models

1. Introduction

Fish are vital for humans as a protein source and for maintaining marine biodiversity [1]. However, they face challenges like overfishing, habitat destruction, and climate change.

Recognition of fish species benefits animal welfare, ecological protection, and wildlife support. Fish image recognition helps researchers understand fish behavior and improve habitats. It also aids in accurate population counting and the monitoring [2] of wild fish populations. Additionally, it enables rapid recognition of fish at customs and in markets, preventing the illegal trade of endangered species.

In blurry marine scenarios, fish species recognition is challenging, requiring accurate methods [3]. This contributes to surveys, population analyses, and the sustainable utilization of fish as a biological resource.

Underwater cameras are commonly used for fish surveys [4]. Unlike other methods, they minimize ecosystem impact and allow continuous recording of fish activity. However, limitations include a restricted field of view and factors like water turbidity, lighting conditions, and flow magnitude that affect image quality and recognition accuracy.

Previous research mainly focused on high-resolution fish recognition [5]. However, practical scenarios often feature blurry images due to water quality [6], relative movement [7] between the shooting device and the fish, water ripples [8], and light reflection [9]. This poses significant challenges for fish recognition in real-life situations.

In order to overcome the challenges mentioned above, a method of fish image recognition in blurry scenarios based on the diffusion model and attention [10–12] mechanism, DiffusionFR, is proposed herein. DiffusionFR offers a comprehensive set of technical solutions for fish recognition in blurry scenarios. It shows the selection and application of this correcting technique.

The main contribution list of this paper is summarized as follows:

(1) A two-stage diffusion model for fish recognition in blurry scenarios, TSD, was designed to maximize the removal of bad, blurry, and otherwise inadequate effects in fish images.

(2) A learnable attention module, LAM, was designed to ensure that the semantic features learned at the end of the network can distinguish fish for fine-grained recognition.

(3) A method for fish image recognition in blurry scenarios that synthesizes TSD and LAM, DiffusionFR, was proposed to present a complete solution for fish image recognition in blurry scenarios, and the selection and application of this correcting technique are presented herein.

(4) A dataset of fish images in blurry scenarios, BlurryFish, was constructed and used to validate the effectiveness of DiffusionFR, and integrated the bad, blurry, and otherwise inadequate images from the publicly available dataset Fish4Knowledge.

The structure of this paper is as follows. In Section 2, we review the relevant works on fish species recognition. Section 3 provides a detailed explanation of the key concepts and methodology used in this study. This includes the main ideas behind the method, DiffusionFR, the two-stage diffusion model (TSD), the learnable attention module (LAM), the modified ResNet as the recognition network, the dataset, and the experimental design. Moving on to Section 4, we present the treatment and analysis of the experimental findings. In Section 5, we thoroughly discuss the implications and significance of the results. Finally, in Section 6, we summarize the essential findings and draw conclusions based on the research conducted in this paper.

2. Background

Previous studies on fish species recognition commonly used different deep neural network architectures or employed layered and phased strategies.

Numerous studies on fish recognition have utilized various deep neural networks, such as CNN, Tripmix-Net, DAMNet, MobileNetv3, and VGG16. Villon et al. [13] employed CNN to enhance the accuracy of coral reef fish recognition by using rule-based techniques. They achieved a model accuracy of 94.9%, surpassing manual accuracy. Similarly, Villon et al. [14] used a convolutional neural network to analyze images from social media, providing support in monitoring rare megafauna species. Li et al. [15] proposed Tripmix-Net, a fish image classification model that incorporates multiscale network fusion. Qu et al. [16] introduced DAMNet, a deep neural network with a dual-attention mechanism for aquatic biological image classification. However, due to the incorporation of the dual-attention mechanism, the DAMNet model may exhibit a relatively higher level of complexity. Meanwhile, Alaba et al. [17] developed a model using the MobileNetv3-large and VGG16 backbone networks for fish detection. However, their method still encounters certain challenges, such as dealing with low-light conditions, noise, and the limitations posed by low-resolution images.

A hierarchical and phased approach to fish target recognition refers to dividing the recognition process into multiple phases and levels. Liang et al. [18] divided the recognition process into multiple stages to enhance accuracy and robustness. However, their method suffers from a high number of parameters and computational complexity, which can make the training process extremely time-consuming. Similarly, Ben et al. [5] proposed a hierarchical CNN classification method for automatic fish recognition in underwater environments.

In blurry scenarios [19], intelligent fish image recognition technology aims to improve image clarity using image processing techniques. These techniques include image denoising, image enhancement, and image alignment. Image denoising [20] reduces noise in the image using filters. Image enhancement [21] improves clarity through techniques like histogram equalization. Image alignment [22] addresses image blurring through registration. Neural heuristic video systems [23] analyze video frames automatically using heuristic algorithms, extending image analysis to video analysis. The bilinear pooling with poisoning detection (BPPD) module [24] utilizes bilinear pooling of convolutional neural networks. This algorithm combines data from two networks through bilinear pooling to achieve improved classification accuracy. Intelligent fish image recognition technology utilizes the diffusion model to deblur images. This model enhances image quality, recovers lost information, and improves feature extraction. As a result, it provides better inputs for subsequent image recognition tasks, significantly improving the accuracy of fish image recognition in blurry scenarios [25].

3. Materials and Methods

3.1. Main Ideas

Figure 1 presents the framework of the method based on the diffusion model and attention mechanism for fish image recognition in blurry scenarios, DiffusionFR. The framework visually illustrates the selection and application of the correction technique.

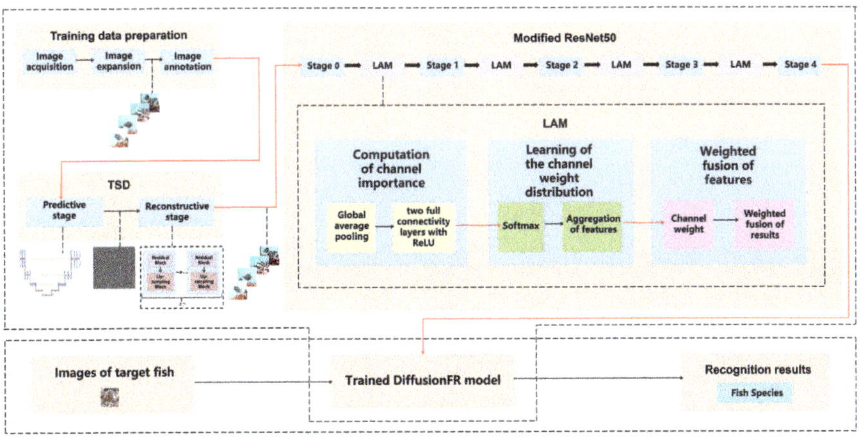

Figure 1. The structure for DiffusionFR, consisting of TSD, LAM, and the modified ResNet50.

The main ideas behind DiffusionFR can be summarized as follows:

(1) Two-stage diffusion (TSD): This model consists of two stages—the predictive stage and the reconstructive stage. In the predictive stage, a U-Net structure generates feature probability maps for the bad, blurry, and otherwise inadequate fish images. Each pixel in the maps represents the probability of belonging to a specific class of fish image. In the reconstructive stage, four identical modules comprising a residual block and an up-sampling block are employed to convert the feature probability maps into clear fish images.

(2) Learnable attention module (LAM): The attention mechanism in DiffusionFR comprises three processes—the computation of channel importance, the learning of channel weight distribution, and the weighted fusion of features. The computation of channel importance involves global average pooling and two fully connected layers with ReLU activation. The learning of channel weight distribution includes SoftMax and the aggregation of features. Finally, the weighted fusion of features incorporates the channel weight and performs a weighted fusion of the results.

(3) Modifying ResNet as the recognition network: In DiffusionFR, the ResNet feature extraction network is modified by adding the LAM between each pair of adjacent stages. This modification aims to minimize the loss of accuracy, train a more precise recognition model, and enhance recognition accuracy for fish images in blurry scenarios.

3.2. *Two-Stage Diffusion (TSD)*

Recently, deep neural network-based diffusion models [26–28] have become popular for image denoising and super-resolution. These models utilize the capabilities of deep learning to learn image features and predict image evolution. As a result, they can quickly and efficiently denoise and enhance images.

The proposed TSD method in this study consists of two stages: a predictive stage and a reconstructive stage. The predictive stage detects fish image features in blurry images, while the reconstructive stage analyzes and processes diffusion data to address errors or deficiencies in the model. This stage significantly enhances the model's accuracy and reliability. The entire TSD process is visually depicted in Figure 2.

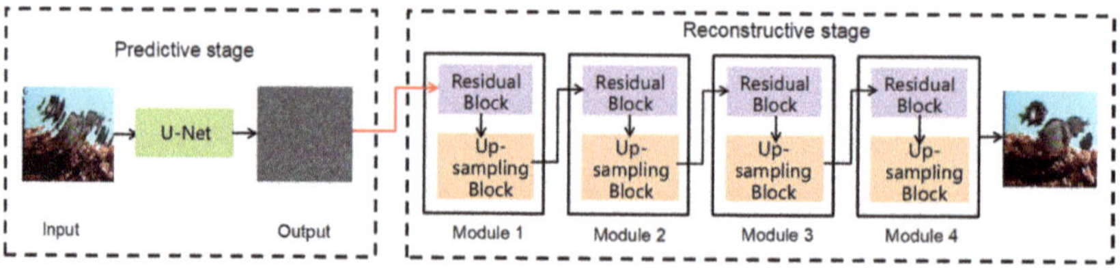

Figure 2. Structure of two-stage diffusion (TSD), including a predictive stage and a reconstructive stage.

The predictive stage of the proposed method takes the fish image as an input and generates a probability map, which represents the likelihood of each pixel belonging to a specific fish species category, as an output. This probability map provides valuable insights into the model's classification probabilities for different fish species.

To achieve this, the predictive stage utilizes the U-Net architecture [29], as shown in Figure 3. U-Net consists of symmetrical encoders and decoders. The encoder extracts image features using convolution and pooling operations, encoding the input image into a low-dimensional tensor. The decoder then reconstructs the encoder's output into an image of the same dimensions as the input, with each pixel containing a probability value for the target category. To address information loss, U-Net incorporates jump connections that connect the feature maps of the encoder and decoder. It consists of four 2D convolutional layers and four maximum pooling layers, enabling the model to handle fish images of varying sizes and shapes within blurry images.

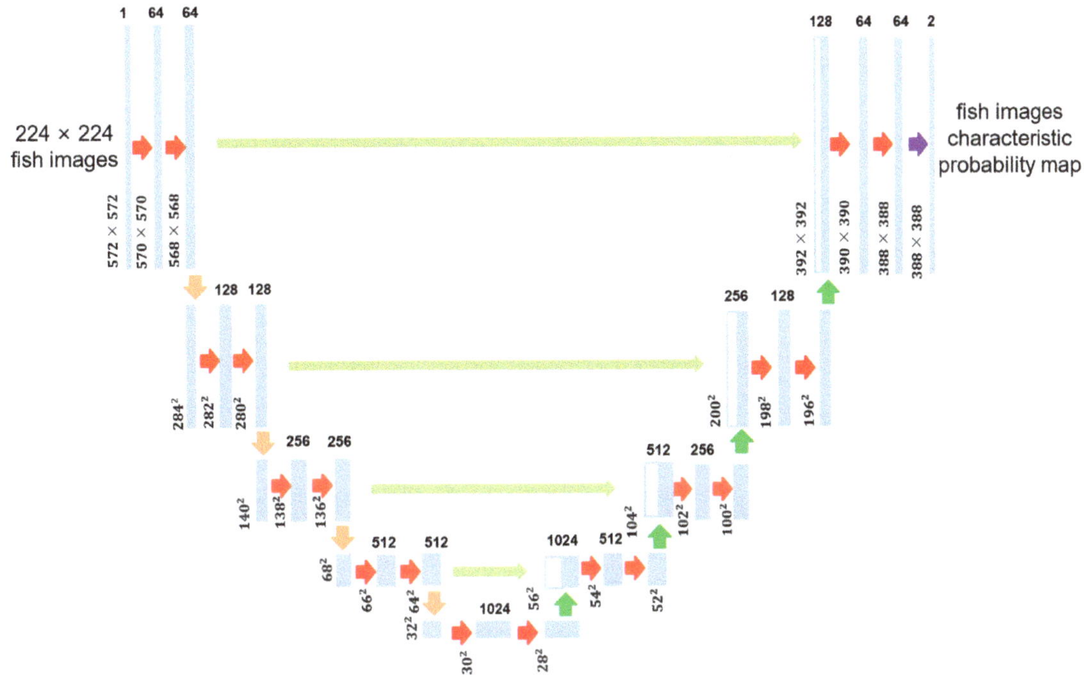

Figure 3. U-Net architecture, consisting of an encoder and a decoder.

The restoration stage is responsible for generating the final restored image. It utilizes both thc output image from the prediction stage and the input image.

The reconstructive stage is composed of four modules. Each module consists of a Residual Block [30] and Up-Sampling Block [31]. The Residual Block addresses issues of gradient vanishing and explosion during deep neural network training, as shown in Figure 4. It includes two convolutional layers and a jump connection, where the input is added directly to the output to form residuals. This helps the network capture the mapping relationship between inputs and outputs, improving the model's performance and robustness. During model training, special attention is given to the error generated by the Up-Sampling Block in the network, as shown in Equation (1).

$$q(x_s|x_t, x_0) = N\left(x_s \middle| \frac{1}{{g_{t0}}^2}\left(f_{s0}{g_{ts}}^2x_0 + f_{ts}{g_{s0}}^2x_t\right), \frac{{g_{s0}}^2{g_{ts}}^2}{{g_{t0}}^2}I\right) \tag{1}$$

where $q(x_s|x_t, x_0)$ denotes the conditional probability distribution of x_s; given conditions x_t and x_0, the mean part of this is a series of linear combination terms including f_{s0}, g_{ts}, f_{ts}, and g_{s0}. In addition, f_{ts} is a ratio indicating the relative scale that maps the input variable t to the input variable s, as shown in Equation (2), and g_{ts} is computed from the scale parameters of the input variables t and s and is used to adjust the propagation process of the error, as shown in Equation (3).

$$f_{ts} = \frac{f(t)}{f(s)} \tag{2}$$

$$g_{ts} = \sqrt{g(t)^2 - {f_{ts}}^2g(s)^2} \tag{3}$$

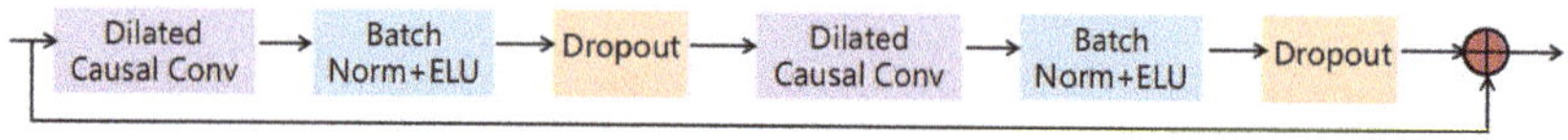

Figure 4. Residual Block, consisting of Dilated Causal Conv, BatchNorm, ELU, and Dropout.

Equation (4) describes the gradual process of recovering the image from noise, as illustrated.

$$x_t = \sqrt{\overline{\alpha}_t}x_0 + \overline{Z}_{t-1} = \sqrt{\overline{\alpha}_t}x_0 + \sqrt{1-\overline{\alpha}_t}Z, Z \sim N(0, I) \quad (4)$$

where x_t denotes the image recovered at moment t, obtained by a linear combination of the initial image x_0 and the previous recovery result $\overline{Z}_{t-1}$. This linear combination uses a scaling factor $\sqrt{\overline{\alpha}_t}$ to adjust the contribution of the initial image and the previous recovery result. Meanwhile, the noise term Z is generated through a Gaussian distribution $N(0, I)$.

Equation (5) represents the inverse diffusion process from the recovered image x_t back to the previously recovered result x_{t-1}.

$$q(x_{t-1}|x_t, x_0) = N\left(x_{t-1}; \frac{1}{\sqrt{\alpha_t}}x_t - \frac{\beta_t}{\sqrt{\alpha_{t(1-\overline{\alpha}_t)}}}Z, \frac{1-\overline{\alpha}_{t-1}}{1-\overline{\alpha}_t}\beta_t\right), Z \sim N(0, I) \quad (5)$$

where the conditional probability distribution $q(x_{t-1}|x_t, x_0)$ represents the conditional probability distribution of x_{t-1}; given conditions x_t and x_0, this conditional probability distribution is represented by a Gaussian distribution where the mean part contains a linear combination of x_t and the noise term Z.

TSD implements batch normalization techniques and dropout layers to enhance the stability, convergence, and generalization of the model.

3.3. Learnable Attention Module (LAM)

In this paper, we propose LAM, which is based on the channel attention mechanism [32] (CAM) and depicted in Figure 5. Unlike CAM, LAM assigns weights to channels by learning the importance of features.

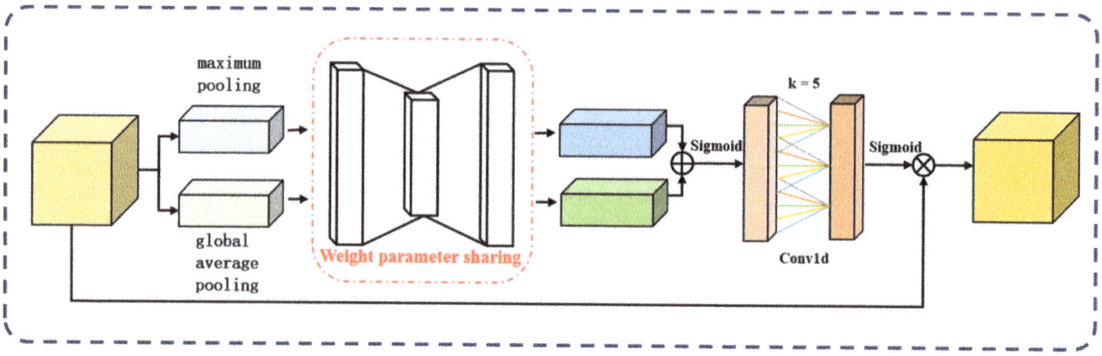

Figure 5. The framework structure of the learnable attention module (LAM).

In DiffusionFR, the LAM consists of three steps. These steps include the computation of channel importance, learning of channel weight distribution, and weighted fusion of features. These steps are illustrated in Figure 1.

The first step is the computation [33] of channel importance. First, the global pooling values for each channel in the feature map F are extracted by a global average pooling or maximum pooling operation to obtain a C-dimensional vector Z. Then, Z is processed using a network architecture containing two fully connected layers and a ReLU activation function to generate a C-dimensional weight assignment vector k, which stores the weight assignments for each channel, as shown in Equation (6).

$$k = \varphi(C) = \left| \frac{\log_2(C)}{\gamma} + \frac{b}{\gamma} \right|_{odd} \tag{6}$$

The second step involves the learning [34] of the channel weight distribution. This distribution determines the significance of each feature map channel. To compute the channel weights, we use the softmax function to map the values in the weight vector between 0 and 1. This ensures that the sum of all weights equals 1, representing the weight of each channel. To gather global information about the channels, we apply a global average pooling operation to the features, which is represented by Equation (7).

$$y = \frac{1}{H \times W} \sum_{a}^{H} \sum_{b}^{W} x_i(a, b) \tag{7}$$

In the formula, x_i represents the ith feature map of input size H × W, and y represents the global feature. In the softmax function, each feature vector element is mapped to a value between 0 and 1. With this mapping, the model can determine how much each channel contributes relative to the overall feature map.

The third step involves the weighted fusion [35] of features. Each channel in the feature map is weighted and fused based on their assigned weights. Firstly, weights are assigned to each channel and applied to their corresponding features. Then, the features of all channels are proportionally weighted and fused to generate a feature map adjusted by the attention mechanism. By incorporating the LAM, the network can dynamically adjust the contribution of each channel, improving its robustness and generalization ability. This attention mechanism enables the network to disregard irrelevant information (weights close to 0) and prioritize important features essential for successful task completion.

3.4. Modifying ResNet as the Recognition Network

DiffusionFR selected ResNet50 as the base network after comparing ResNet34 [36], ResNet50 [37], and ResNet101 [38].

ResNet50 has a layered architecture that enables it to learn hierarchical representations of input data. Lower layers capture low-level features, while higher layers capture intricate patterns and relationships. The pooling layer reduces spatial dimensionality, improving computational efficiency and translation invariance while mitigating overfitting. By leveraging ResNet50's transfer learning, this provides a solid foundation for the probabilistic graph generation task.

However, in images with complex backgrounds or noise, ResNet50 may unintentionally focus on less relevant regions, impacting model performance. To address this, an attention mechanism is introduced to dynamically adjust feature map weights based on different parts of the input data. This helps prioritize crucial features, enhancing accuracy and generalization capabilities. Therefore, the DiffusionFR approach modifies ResNet50 by incorporating LAM into the network. LAM is added between each pair of adjacent stages, as shown in Figure 6.

3.5. Dataset

This paper introduces BlurryFish, a fish image dataset created by integrating blurred images from the publicly available dataset Fish4Knowledge. The construction process involved the following steps:

(1) Data Collection

The datasets used in this paper are from three sources. The first source is the publicly available dataset Fish4Knowledge, consisting of realistically shot images. The second source is a field-photographed dataset that prioritizes challenging scenarios like low-light conditions and inclement weather to ensure representative fish images. The third source is fish images from Internet, which we organized and classified. The dataset comprises 25 fish species, and Figure 7 displays these species and example images.

(2) Data Cleaning

The collected fish images underwent a cleaning process to ensure their quality and reliability. This involved eliminating invalid samples and duplicate samples.

(3) Dataset Partition

The dataset was divided into three sets for the experiments: the training set, the validation set, and the test set. This division follows the leave-out method [39] and maintains an 8:1:1 ratio. The goal was to ensure that all sets included pictures of the same fish species, as well as similar scenarios and angles.

(4) Data Enhancement

The collected dataset has an interclass balance problem [40] due to the varying number of pictures for each fish species. This can result in lower recognition accuracy for less common fish species if the dataset is directly used for training. To address this issue, standard data enhancement methods were employed, including panning, cropping, rotating, mirroring, flipping, and brightness adjustment. These operations generated additional image samples, enhancing the model's robustness, generalization ability, and recognition accuracy for smaller fish species. Table 1 shows that the initial BlurryFish dataset contained 2754 bad, blurry, and otherwise inadequate fish images. However, after applying data enhancement techniques, the dataset expanded to 35,802 bad, blurry, and otherwise inadequate fish images, as shown in Table 2.

(5) Data Annotation

To create valuable training and testing sets from the image dataset, each image in the fish image dataset was labeled with associated fish species data. We utilized the graphical interface labeling software, LabelImg (v 1.8.5), to annotate the fish images and generate XML files. Although DiffusionFR does not impose any restrictions on the resolution and other parameters of the dataset images, we uniformly converted the dataset to RGB images with a resolution of 224 × 224. These images were then stored in the PASCAL VOC data format.

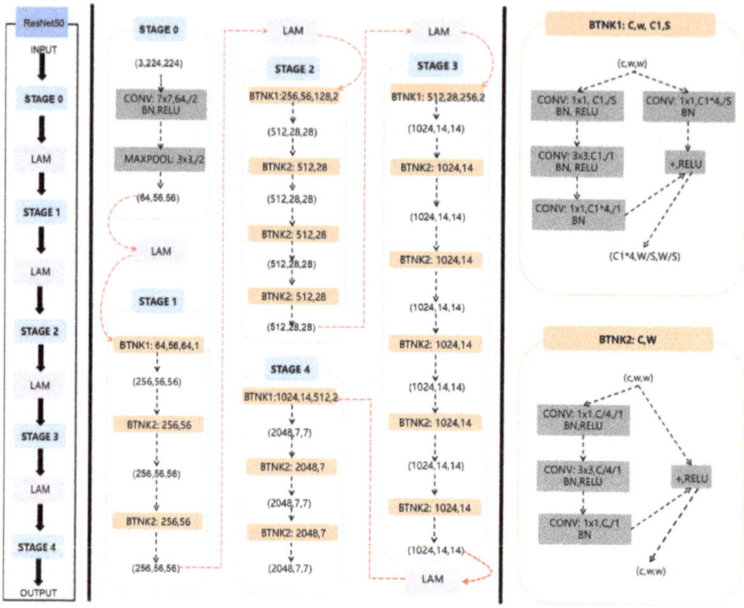

Figure 6. Structure of the modified ResNet50 with the LAM added between every two neighboring stages.

Figure 7. Fish species in the dataset. (**a**) Dascyllus reticulatus; (**b**) Neoniphon sammara; (**c**) Abudefduf vaigiensis; (**d**) Canthigaster valentini; (**e**) Pomacentrus moluccensis; (**f**) Zebrasoma scopas; (**g**) Hemigymnus melapterus; (**h**) Lutjanus fulvus; (**i**) Scolopsis bilineata; (**j**) Scaridae; (**k**) Pempheris vanicolensis; (**l**) Plectroglyphidodon dickii; (**m**) Zanclus cornutus; (**n**) Neoglyphidodon nigroris; (**o**) Balistapus undulatus; (**p**) Siganus fuscescens; (**q**) Chromis chrysura; (**r**) Amphiprion clarkii; (**s**) Chaetodon lunulatus; (**t**) Chaetodon trifascialis; (**u**) Myripristis kuntee; (**v**) Acanthurus nigrofuscus; (**w**) Hemigymnus fasciatus; (**x**) Abactochromis labrosus; and (**y**) Abalistes stellaris.

Table 1. Number of fish images in the BlurryFish dataset before data enhancement.

ID	Name of Fish	Training Set	Validation Set	Test Set	Total
1	*Dascyllus reticulatus*	91	12	12	115
2	*Neoniphon sammara*	84	11	11	106
3	*Abudefduf vaigiensis*	85	10	10	105
4	*Canthigaster valentini*	88	11	11	110
5	*Pomacentrus moluccensis*	94	12	12	118
6	*Zebrasoma scopas*	85	11	11	107
7	*Hemigymnus melapterus*	84	10	10	104
8	*Lutjanus fulvus*	83	10	10	103
9	*Scolopsis bilineata*	86	11	11	108
10	*Scaridae*	92	11	11	114

Table 1. *Cont.*

ID	Name of Fish	Training Set	Validation Set	Test Set	Total
11	*Pempheris vanicolensis*	82	10	10	102
12	*Plectroglyphidodon dickii*	85	11	11	107
13	*Zanclus cornutus*	95	12	12	119
14	*Neoglyphidodon nigroris*	85	10	10	105
15	*Balistapus undulatus*	89	11	11	111
16	*Siganus fuscescens*	92	11	11	114
17	*Chromis chrysura*	92	12	12	116
18	*Amphiprion clarkii*	84	11	11	106
19	*Chaetodon lunulatus*	91	12	12	115
20	*Chaetodon trifascialis*	95	12	12	119
21	*Myripristis kuntee*	90	11	11	112
22	*Acanthurus nigrofuscus*	87	11	11	109
23	*Hemigymnus fasciatus*	82	10	10	102
24	*Abactochromis labrosus*	89	11	11	111
25	*Abalistes stellaris*	92	12	12	116
	Total	2202	276	276	2754

Table 2. Number of fish images in the BlurryFish dataset after data enhancement.

ID	Name of Fish	Training Set	Validation Set	Test Set	Total
1	*Dascyllus reticulatus*	1195	150	150	1495
2	*Neoniphon sammara*	1102	138	138	1378
3	*Abudefduf vaigiensis*	1093	136	136	1365
4	*Canthigaster valentini*	1144	143	143	1430
5	*Pomacentrus moluccensis*	1228	153	153	1534
6	*Zebrasoma scopas*	1113	139	139	1391
7	*Hemigymnus melapterus*	1082	135	135	1352
8	*Lutjanus fulvus*	1071	134	134	1339
9	*Scolopsis bilineata*	1124	140	140	1404
10	*Scaridae*	1186	148	148	1482
11	*Pempheris vanicolensis*	1060	133	133	1326
12	*Plectroglyphidodon dickii*	1113	139	139	1391
13	*Zanclus cornutus*	1237	155	155	1547
14	*Neoglyphidodon nigroris*	1093	136	136	1365
15	*Balistapus undulatus*	1155	144	144	1443
16	*Siganus fuscescens*	1186	148	148	1482
17	*Chromis chrysura*	1206	151	151	1508
18	*Amphiprion clarkia*	1102	138	138	1378
19	*Chaetodon lunulatus*	1195	150	150	1495
20	*Chaetodon trifascialis*	1237	155	155	1547
21	*Myripristis kuntee*	1164	146	146	1456
22	*Acanthurus nigrofuscus*	1133	142	142	1417
23	*Hemigymnus fasciatus*	1060	133	133	1326
24	*Abactochromis labrosus*	1155	144	144	1443
25	*Abalistes stellaris*	1206	151	151	1508
	Total	28,640	3581	3581	35,802

3.6. Experimental Design

3.6.1. Experimental Environment Configuration

PyTorch, a deep learning framework, was employed to evaluate DiffusionFR. The specific experimental software and hardware configurations are detailed in Table 3.

Table 3. Experimental software and hardware configurations.

Item	Detail
GPU	NVIDIA GeForce RTX 3060
CPU	12th Gen Intel(R) Core(TM) i5-12400 2.50 GHz
RAM	16.0 GB
Operating system	Windows 11 64-bit
CUDA	CUDA 11.6
Python	Python 3.7.15

3.6.2. Evaluation Indicators

To evaluate the model's performance in classifying fish images in blurry scenarios, accuracy and Top-k accuracy were used as evaluation metrics. The experimental data was processed using Python code and analyzed using Excel software (12.1.0.16250).

(1) Accuracy

Accuracy is a metric that measures the proportion of correctly predicted samples compared to the total number of instances. It is calculated using Equation (8).

$$\text{Accuracy} = \frac{\text{TN} + \text{TP}}{\text{TN} + \text{FP} + \text{TP} + \text{FN}} \times 100\% \tag{8}$$

where TN denotes true negative, TP denotes true positive, FP denotes false positive, and FN denotes false negative.

(2) Top-k Accuracy

The top-k accuracy rate measures the proportion of samples where at least one of the top-k predictions matches the true label, compared to the total number of samples. In this study, we use Top-1 accuracy and Top-5 accuracy as model criteria. Equation (9) demonstrates the calculation of the top-k accuracy.

$$\begin{aligned} \text{Top} - \text{k Accuracy} \\ &= \text{number of samples correctly predicted} \\ &\quad /\text{the total number of samples} \times 100\% \end{aligned} \tag{9}$$

Here, k can be any positive integer, and it is common to have top − 1 and top − 5 accuracy rates, which indicate the accuracy in the highest confidence prediction and the first five highest confidence predictions, respectively.

3.6.3. Parameters of Experiments

In this section, we conduct comparative experiments for each module of DiffusionFR. During training, the model parameters of DiffusionFR were continuously adjusted to minimize prediction errors. This was achieved using optimization algorithms and loss functions. After several iterations, the hyperparameters of the DiffusionFR model were determined based on commonly used empirical values. The finalized hyperparameters can be found in Table 4.

Table 4. Optimal hyperparameters.

Input Shape	Lr	Activation Function	Batch Size	Classifier	Optimizer	Epoch
224 × 224	0.002	ReLU	32	Softmax	Adam	100

3.6.4. Schemes of Experiments

(1) Comparison of Backbone Networks

DiffusionFR's backbone network was assessed using the original dataset. The analysis included various backbone networks such as ResNet50, VGG16, MobileNetv3, Tripmix-Net, ResNeXt, DAMNet, ResNet34, ResNet101, EfficientNet [41], neuro-heuristic, bilinear pooling with poisoning detection (BPPD), and CNN(r1, r2).

(2) Comparison of Attention Mechanisms

Comparative experiments were conducted to assess the impact of attention mechanisms on the algorithm. The evaluated attention mechanisms included LAM, CBAM [42], CCA [43], and SE [44].

(3) Comparison of Diffusion Models

To assess the impact of the diffusion model proposed in this paper on the final recognition performance, we conducted a comparative experiment. This experiment involved two deblurring methods: the diffusion module proposed in this paper and the Gaussian denoising module.

(4) Effect of Light Reflection Noise on Recognition Performance

The datasets were labeled according to the light reflection noise added. For instance, D_0E_0 signifies the original dataset without any added noise, while $D_{0.6}E_{100}$ represents the dataset with light reflection noise, where the light diameter is 0.6 cm and the light intensity is 100 Lux, added to D_0E_0. This naming convention is used for other datasets as well.

Nine datasets were created by categorizing the light reflection noise based on different light diameters and intensities. These datasets are named as $D_{0.6}E_{100}$, $D_{0.6}E_{250}$, $D_{0.6}E_{400}$, $D_{0.8}E_{100}$, $D_{0.8}E_{250}$, $D_{0.8}E_{400}$, $D_{1.0}E_{100}$, $D_{1.0}E_{250}$, and $D_{1.0}E_{400}$. Table 5 provides an overview of the data volume for the fish image dataset with added light reflection noise. An example of this dataset is shown in Figure 8.

Table 5. Fish Dataset with Added Light Reflection Noise.

ID	Name of Fish	$D_{0.6}E_{100}$	$D_{0.6}E_{250}$	$D_{0.6}E_{400}$	$D_{0.8}E_{100}$	$D_{0.8}E_{250}$	$D_{0.8}E_{400}$	$D_{1.0}E_{100}$	$D_{1.0}E_{250}$	$D_{1.0}E_{400}$	Total
1	*Dascyllus reticulatus*	115	115	115	115	115	115	115	115	115	1035
2	*Neoniphon sammara*	106	106	106	106	106	106	106	106	106	954
3	*Abudefduf vaigiensis*	105	105	105	105	105	105	105	105	105	945
4	*Canthigaster valentini*	110	110	110	110	110	110	110	110	110	990
5	*Pomacentrus moluccensis*	118	118	118	118	118	118	118	118	118	1062
6	*Zebrasoma scopas*	107	107	107	107	107	107	107	107	107	963
7	*Hemigymnus melapterus*	104	104	104	104	104	104	104	104	104	936
8	*Lutjanus fulvus*	103	103	103	103	103	103	103	103	103	927
9	*Scolopsis bilineata*	108	108	108	108	108	108	108	108	108	972
10	*Scaridae*	114	114	114	114	114	114	114	114	114	1026
11	*Pempheris vanicolensis*	102	102	102	102	102	102	102	102	102	918
12	*Plectroglyphidodon dickii*	107	107	107	107	107	107	107	107	107	963
13	*Zanclus cornutus*	119	119	119	119	119	119	119	119	119	1071
14	*Neoglyphidodon nigroris*	105	105	105	105	105	105	105	105	105	945
15	*Balistapus undulatus*	111	111	111	111	111	111	111	111	111	999
16	*Siganus fuscescens*	114	114	114	114	114	114	114	114	114	1026
17	*Chromis chrysura*	116	116	116	116	116	116	116	116	116	1044
18	*Amphiprion clarkii*	106	106	106	106	106	106	106	106	106	954
19	*Chaetodon lunulatus*	115	115	115	115	115	115	115	115	115	1035
20	*Chaetodon trifascialis*	119	119	119	119	119	119	119	119	119	1071
21	*Myripristis kuntee*	112	112	112	112	112	112	112	112	112	1008
22	*Acanthurus nigrofuscus*	109	109	109	109	109	109	109	109	109	981
23	*Hemigymnus fasciatus*	102	102	102	102	102	102	102	102	102	918
24	*Abactochromis labrosus*	111	111	111	111	111	111	111	111	111	999
25	*Abalistes stellaris*	116	116	116	116	116	116	116	116	116	1044
	Total	2754	2754	2754	2754	2754	2754	2754	2754	2754	24,786

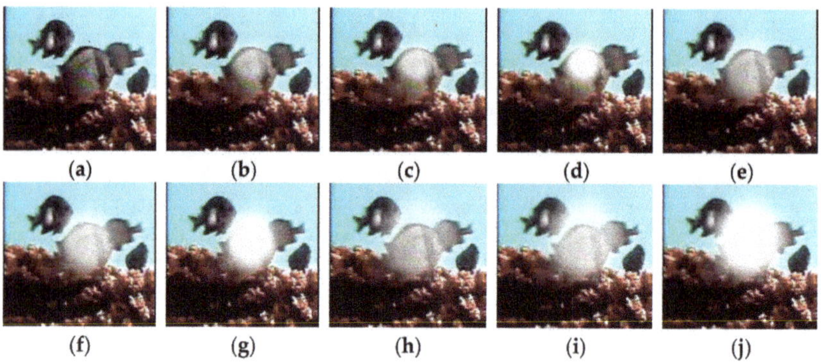

Figure 8. Example of fish pictures with added light reflection noise. (**a**) D_0E_0; (**b**) $D_{0.6}E_{100}$; (**c**) $D_{0.6}E_{250}$; (**d**) $D_{0.6}E_{400}$; (**e**) $D_{0.8}E_{100}$; (**f**) $D_{0.8}E_{250}$; (**g**) $D_{0.8}E_{400}$; (**h**) $D_{1.0}E_{100}$; (**i**) $D_{1.0}E_{250}$; and (**j**) $D_{1.0}E_{400}$.

We conducted a comparative analysis on datasets with light reflection noise to assess the effectiveness of using corrected fish images for species-specific fish recognition.

(5) Effect of Water Ripple Noise on Recognition Performance

To add water ripple noise to the dataset and generate the water ripple effect, the following steps and Equations were used. First, an empty array X of the same size as the original image was created to store the generated water ripple effect. Next, offsets (including offset_x and offset_y) were calculated for each pixel based on the amplitude (A) and frequency (F) by iterating through each pixel in a loop. Then, the pixel values corresponding to these offsets were assigned to each pixel of the empty array X, generating the water ripple effect. Finally, the resulting water ripple effect was overlaid onto the original image, creating the final image with water ripples. Equations involved are shown in (10)–(14).

The offset was calculated using Equations (10) and (11).

$$\mathrm{offset\_x} = \mathrm{A} * \sin(2 * \pi * \mathrm{y_i} * \mathrm{F}) \tag{10}$$

$$\mathrm{offset\_y} = \mathrm{A} * \cos(2 * \pi * \mathrm{x_i} * \mathrm{F}) \tag{11}$$

where $(\mathrm{x_i}, \mathrm{y_i})$ denotes the coordinates of a pixel point in the image, F is the frequency of the water ripple, and A is the amplitude of the water ripple.

The pixel assignment of array X is calculated using Equations (12) and (13).

$$\mathrm{X}[\mathrm{x_i}] = (\mathrm{x_i} + \mathrm{offset\_x})\%\mathrm{width} \tag{12}$$

$$\mathrm{X}[\mathrm{y_i}] = (\mathrm{y_i} + \mathrm{offset\_y})\%\mathrm{height} \tag{13}$$

where width and height are the width and height of the image, respectively.

The final image generation is calculated using Equation (14).

$$\mathrm{img\_with\_ripples} = \mathrm{img} + \mathrm{X} \tag{14}$$

where img denotes the original image, and img_with_ripples denotes the final image with water ripples.

The datasets were labeled according to the water ripple noise added. For instance, F_0A_0 signifies the original dataset without any added noise, while $F_{0.04}A_2$ indicates the dataset with water ripple noise having a frequency of 0.04 and an amplitude of 2 added to F_0A_0. This naming convention is used for other datasets as well.

Water ripple noise can be classified based on the frequency and amplitude of the water ripples. Increasing the frequency and amplitude results in a higher offset and greater oscillation in the generated water waves. In this study, the water ripple noise was categorized into three groups: $F_{0.04}A_2$, $F_{0.06}A_6$, and $F_{0.08}A_{10}$, primarily based on their frequency and amplitude. Table 6 provides an overview of the data volume of the fish image dataset with the addition of water ripple noise, while Figure 9 offers an illustrative example.

Table 6. Fish Dataset with Added Water Ripple Noise.

ID	Name of Fish	$F_{0.04}A_2$	$F_{0.06}A_6$	$F_{0.08}A_{10}$	Total
1	*Dascyllus reticulatus*	115	115	115	345
2	*Neoniphon sammara*	106	106	106	318
3	*Abudefduf vaigiensis*	105	105	105	315
4	*Canthigaster valentini*	110	110	110	330
5	*Pomacentrus moluccensis*	118	118	118	354
6	*Zebrasoma scopas*	107	107	107	321
7	*Hemigymnus melapterus*	104	104	104	312
8	*Lutjanus fulvus*	103	103	103	309
9	*Scolopsis bilineata*	108	108	108	324
10	*Scaridae*	114	114	114	342

Table 6. *Cont.*

ID	Name of Fish	$F_{0.04}A_2$	$F_{0.06}A_6$	$F_{0.08}A_{10}$	Total
11	*Pempheris vanicolensis*	102	102	102	306
12	*Plectroglyphidodon dickii*	107	107	107	321
13	*Zanclus cornutus*	119	119	119	357
14	*Neoglyphidodon nigroris*	105	105	105	315
15	*Balistapus undulatus*	111	111	111	333
16	*Siganus fuscescens*	114	114	114	342
17	*Chromis chrysura*	116	116	116	348
18	*Amphiprion clarkii*	106	106	106	318
19	*Chaetodon lunulatus*	115	115	115	345
20	*Chaetodon trifascialis*	119	119	119	357
21	*Myripristis kuntee*	112	112	112	336
22	*Acanthurus nigrofuscus*	109	109	109	327
23	*Hemigymnus fasciatus*	102	102	102	306
24	*Abactochromis labrosus*	111	111	111	333
25	*Abalistes stellaris*	116	116	116	348
	Total	2754	2754	2754	8262

Figure 9. Example of fish pictures with added water ripple noise. (**a**) F_0A_0; (**b**) $F_{0.04}A_2$; (**c**) $F_{0.06}A_6$; and (**d**) $F_{0.08}A_{10}$.

We conducted a comparative analysis of datasets with water ripple noise to assess the effectiveness of using corrected fish images for species-specific fish recognition.

We conducted Experiments 1 through 5 to assess the impact of different backbone networks, attention mechanisms, diffusion models, as well as light reflection noise and water ripple noise on recognition performance. These experiments were evaluated using three metrics: training accuracy, the Top-1 accuracy test, and the Top-5 accuracy test. The objective was to comprehensively evaluate their recognition performance and analyze the results.

4. Results

In this study, the BlurryFish dataset was used to perform comparative experiments on the key innovations of the proposed methodology.

4.1. Comparison of Backbone Networks

This study compared and analyzed the backbone network of DiffusionFR. For example, DiffusionFR_VGG16 refers to using VGG16 instead of ResNet50 as the backbone network in DiffusionFR. Similar comparisons were made with other backbone networks. The results of these comparisons can be found in Table 7.

Table 7. Accuracies of Different Feature Extraction Networks.

Model	Training (%)	Top-1 Test (%)	Top-5 Test (%)
DiffusionFR	97.55	92.02	95.17
DiffusionFR_VGG16	91.78	86.38	89.48
DiffusionFR_MobileNetv3	93.05	87.55	90.60
DiffusionFR_Tripmix-Net	93.43	88.07	91.21
DiffusionFR_ResNeXt	94.32	88.90	91.98

Table 7. *Cont.*

Model	Training (%)	Top-1 Test (%)	Top-5 Test (%)
DiffusionFR_DAMNet	93.80	88.26	91.38
DiffusionFR_ResNet34	96.22	90.80	93.85
DiffusionFR_ResNet101	94.35	88.92	91.72
DiffusionFR_EfficientNet	96.10	90.48	93.38
DiffusionFR_neuro-heuristic	96.82	91.25	94.27
DiffusionFR_BPPD	97.02	91.49	94.63
DiffusionFR_CNN(r1, r2)	97.33	91.69	94.80

Table 7 displays the performance metrics of DiffusionFR on the original dataset. The training accuracy is 97.55%. The corresponding Top-1 accuracy test score was 92.02%, and the Top-5 accuracy test score was 95.17%. These values indicate that DiffusionFR outperforms other methods in terms of accuracy. These values also demonstrate that DiffusionFR, with ResNet50 as the chosen backbone network, has a higher potential for achieving superior recognition performance.

4.2. Comparison of Attention Mechanisms

To evaluate the impact of the attention mechanism on the algorithm, a comparative experiment was conducted, as shown in Table 8. The experiment compared the performance of DiffusionFR with DiffusionFR without any attention mechanism, referred to as DiffusionFR_noA. Furthermore, classical attention methods were used as substitutes for LAM. For example, DiffusionFR_CBAM incorporated CBAM as the attentional method in DiffusionFR. The results of these comparisons are presented in Table 8.

Table 8. Accuracies of Different Attention Mechanisms for LAMs.

Model	Training (%)	Top-1 Test (%)	Top-5 Test (%)
DiffusionFR	97.55	92.02	95.17
DiffusionFR_noA	95.31	89.98	93.06
DiffusionFR_CBAM	96.50	91.10	94.22
DiffusionFR_CCA	97.03	91.52	94.57
DiffusionFR_SE	96.05	90.59	93.70

Table 8 shows that the training accuracy of DiffusionFR on the original dataset was 97.55%. The corresponding Top-1 accuracy test score was 92.02%, and the Top-5 accuracy test score was 95.17%. It is important to note that all these metrics outperform the performance of other methods. This establishes DiffusionFR as the method with the most effective recognition capability.

4.3. Comparison of Diffusion Models

The final recognition results for the diffusion model proposed in this paper were obtained through experiments, as presented in Table 9. This table includes the performance of DiffusionFR, DiffusionFR_noTSD, and DiffusionFR_Gaussian. DiffusionFR_noTSD refers to the method where the TSD was removed from the proposed method, and DiffusionFR_Gaussian involves using Gaussian denoising [45] instead of the TSD. The results of these methods are compared in Table 9.

Table 9. Accuracies of Different Diffusion Models.

Model	Training (%)	Top-1 Test (%)	Top-5 Test (%)
DiffusionFR	97.55	92.02	95.17
DiffusionFR_noTSD	92.41	89.51	91.96
DiffusionFR_Gaussian	97.20	91.76	93.98

Table 9 presents the performance metrics of DiffusionFR on the original dataset. The training accuracy of DiffusionFR is recorded as 97.55%. The corresponding Top-1 accuracy test and Top-5 accuracy test scores are reported as 92.02% and 95.17%, respectively. It is important to note that all these metrics outperform the performance of other methods. This establishes DiffusionFR as the method with the most effective recognition capability.

4.4. Effect of Light Reflection Noise on Recognition

We performed a comparative analysis using DiffusionFR's backbone network on datasets with light reflection noise to evaluate the usability of corrected fish images for species-specific fish recognition. The results of this analysis are presented in Table 10. TSD's effectiveness in processing fish images with light reflection noise is visually demonstrated in Figure 10. The presence of TSD reduces the noise before deblurring, thereby preserving critical features for accurate recognition. Additionally, TSD performs better in handling light reflection noise compared to water ripple noise.

Table 10. Accuracies for Data with Different Light Reflection Noise Effects.

Model	Indicator	D_0E_0	$D_{0.6}E_{100}$	$D_{0.6}E_{250}$	$D_{0.6}E_{400}$	$D_{0.8}E_{100}$	$D_{0.8}E_{250}$	$D_{0.8}E_{400}$	$D_{1.0}E_{100}$	$D_{1.0}E_{250}$	$D_{1.0}E_{400}$
DiffusionFR	Training (%)	97.55	97.28	97.13	96.91	96.66	96.52	96.28	96.08	95.87	95.74
	Top-1 Test (%)	92.02	91.78	91.58	91.34	91.14	90.95	90.75	90.56	90.37	90.19
	Top-5 Test (%)	95.17	94.90	94.70	94.54	94.31	94.11	93.96	93.70	93.54	93.30
DiffusionFR_VGG16	Training (%)	91.78	91.54	91.33	91.17	90.94	90.72	90.53	90.29	90.16	89.89
	Top-1 Test (%)	86.38	86.13	85.96	85.74	85.49	85.36	85.12	84.94	84.70	84.55
	Top-5 Test (%)	89.48	89.20	89.02	88.82	88.67	88.40	88.20	88.06	87.86	87.64
DiffusionFR_MobileNetv3	Training (%)	93.05	92.82	92.56	92.40	92.18	91.96	91.83	91.63	91.41	91.18
	Top-1 Test (%)	87.55	87.33	87.06	86.90	86.68	86.52	86.30	86.10	85.91	85.74
	Top-5 Test (%)	90.60	90.37	90.15	89.93	89.75	89.56	89.36	89.12	88.93	88.72
DiffusionFR_Tripmix-Net	Training (%)	93.43	93.15	92.97	92.74	92.55	92.41	92.15	91.98	91.76	91.59
	Top-1 Test (%)	88.07	87.81	87.64	87.45	87.25	87.02	86.84	86.58	86.46	86.25
	Top-5 Test (%)	91.21	90.92	90.71	90.52	90.37	90.11	89.92	89.78	89.52	89.35
DiffusionFR_ResNeXt	Training (%)	94.32	94.12	93.89	93.65	93.51	93.27	93.08	92.91	92.63	92.50
	Top-1 Test (%)	88.90	88.69	88.42	88.29	88.07	87.89	87.60	87.49	87.24	87.10
	Top-5 Test (%)	91.98	91.68	91.57	91.38	91.18	90.89	90.76	90.48	90.28	90.13
DiffusionFR_DAMNet	Training (%)	93.80	93.51	93.33	93.18	92.97	92.73	92.60	92.32	92.17	91.90
	Top-1 Test (%)	88.26	88.06	87.86	87.56	87.36	87.26	87.06	86.81	86.62	86.37
	Top-5 Test (%)	91.38	91.14	90.94	90.72	90.51	90.31	90.11	89.91	89.72	89.51
DiffusionFR_ResNet34	Training (%)	96.22	95.98	95.76	95.55	95.36	95.13	94.96	94.76	94.57	94.40
	Top-1 Test (%)	90.80	90.54	90.35	90.17	89.99	89.74	89.59	89.39	89.12	88.95
	Top-5 Test (%)	93.85	93.63	93.44	93.20	93.02	92.81	92.61	92.36	92.24	91.98
DiffusionFR_ResNet101	Training (%)	94.35	94.07	93.86	93.66	93.46	93.33	93.07	92.87	92.66	92.46
	Top-1 Test (%)	88.92	88.65	88.50	88.24	88.04	87.87	87.65	87.48	87.25	87.11
	Top-5 Test (%)	91.72	91.47	91.25	91.11	90.85	90.64	90.49	90.22	90.08	89.89
DiffusionFR_EfficientNet	Training (%)	96.10	95.86	95.61	95.37	95.25	94.94	94.80	94.63	94.43	94.23
	Top-1 Test (%)	90.48	90.43	90.20	89.99	89.87	89.57	89.40	89.26	88.96	88.81
	Top-5 Test (%)	93.38	93.52	93.32	93.03	92.87	92.63	92.48	92.20	92.10	91.87
DiffusionFR_neuro-heuristic	Training (%)	96.82	96.33	96.16	95.93	95.74	95.61	95.35	95.12	94.93	94.75
	Top-1 Test (%)	91.25	90.88	90.58	90.42	90.20	89.99	89.82	89.65	89.39	89.24
	Top-5 Test (%)	94.27	93.98	93.71	93.57	93.31	93.21	93.05	92.77	92.59	92.31
DiffusionFR_BPPD	Training (%)	97.02	96.63	96.48	96.28	96.11	96.01	95.77	95.57	95.40	95.25
	Top-1 Test (%)	91.49	91.40	91.13	90.99	90.80	90.61	90.47	90.32	90.09	89.56
	Top-5 Test (%)	94.63	94.33	94.08	93.97	93.73	93.66	93.52	93.27	93.11	92.86
DiffusionFR_CNN(r1, r2)	Training (%)	97.33	96.89	96.86	96.62	96.41	96.18	96.07	95.76	95.56	95.37
	Top-1 Test (%)	91.69	91.54	91.28	91.01	90.76	90.72	90.40	90.28	90.15	89.83
	Top-5 Test (%)	94.80	94.64	94.33	94.23	93.97	93.88	93.57	93.34	93.28	92.95

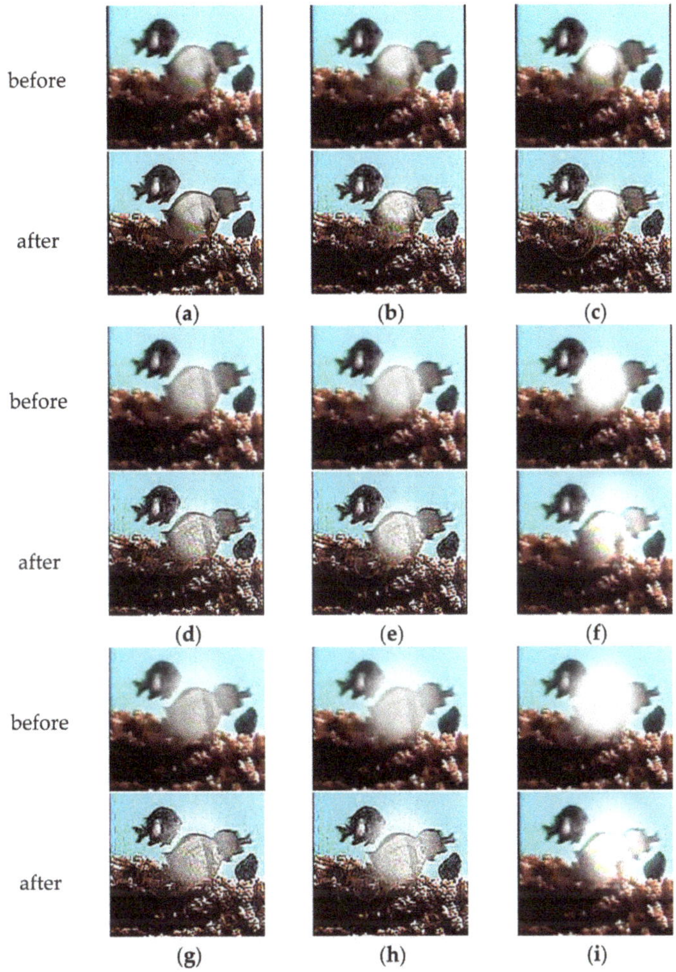

Figure 10. Comparison of images before and after TSD deblurring of light reflection noise. (**a**) $D_{0.6}E_{100}$; (**b**) $D_{0.6}E_{250}$; (**c**) $D_{0.6}E_{400}$; (**d**) $D_{0.8}E_{100}$; (**e**) $D_{0.8}E_{250}$; (**f**) $D_{0.8}E_{400}$; (**g**) $D_{1.0}E_{100}$; (**h**) $D_{1.0}E_{250}$; and (**i**) $D_{1.0}E_{400}$.

In Table 10, the mean value of the training accuracy of DiffusionFR on the nine datasets ($D_{0.6}E_{100}$, $D_{0.6}E_{250}$, $D_{0.6}E_{400}$, $D_{0.8}E_{100}$, $D_{0.8}E_{250}$, $D_{0.8}E_{400}$, $D_{1.0}E_{100}$, $D_{1.0}E_{250}$, and $D_{1.0}E_{400}$) with added light reflection noise was 86.85%. The mean value of the Top-1 accuracy test was 81.87%, and the mean value of the Top-5 accuracy test was 84.71%. These values indicate that DiffusionFR outperforms other methods in terms of accuracy. These values also demonstrate that DiffusionFR, with ResNet50 as the chosen backbone network, has a higher potential for achieving superior recognition performance.

4.5. Effect of Water Ripple Noise on Recognition

We conducted a comparative analysis using DiffusionFR's backbone network on datasets with water ripple noise to evaluate the usability of corrected fish images for species-specific fish recognition. The results of this analysis can be found in Table 11. Figure 11 provides a visual representation of TSD's ability to process fish images containing water ripple noise. TSD effectively reduces the frequency and intensity of water ripple noise in the images before deblurring, mitigating its impact on the critical feature extraction capability of the DiffusionFR model. This ensures that the fish image before deblurring can accurately show ID characters.

Table 11. Accuracies for Data with Different Water Ripple Noise Effects.

Model	Indicator	F_0A_0	$F_{0.04}A_2$	$F_{0.06}A_6$	$F_{0.08}A_{10}$
DiffusionFR	Training (%)	97.55	95.45	95.00	94.56
	Top-1 Test (%)	92.02	89.96	89.55	89.10
	Top-5 Test (%)	95.17	93.12	92.75	92.33
DiffusionFR_VGG16	Training (%)	91.78	89.64	89.19	88.93
	Top-1 Test (%)	86.38	84.35	83.94	83.43
	Top-5 Test (%)	89.48	87.32	86.96	86.58
DiffusionFR_MobileNetv3	Training (%)	93.05	91.05	90.54	90.24
	Top-1 Test (%)	87.55	85.44	85.10	84.58
	Top-5 Test (%)	90.60	88.53	88.11	87.77
DiffusionFR_Tripmix-Net	Training (%)	93.43	91.30	90.86	90.57
	Top-1 Test (%)	88.07	85.88	85.49	85.14
	Top-5 Test (%)	91.21	89.19	88.73	88.39
DiffusionFR_ResNeXt	Training (%)	94.32	92.15	91.91	91.38
	Top-1 Test (%)	88.90	86.75	86.36	86.02
	Top-5 Test (%)	91.98	89.90	89.52	89.02
DiffusionFR_DAMNet	Training (%)	93.80	91.76	91.37	90.91
	Top-1 Test (%)	88.26	86.14	85.76	85.39
	Top-5 Test (%)	91.38	89.20	88.82	88.47
DiffusionFR_ResNet34	Training (%)	96.22	94.10	93.79	93.30
	Top-1 Test (%)	90.80	88.62	88.32	87.95
	Top-5 Test (%)	93.85	91.75	91.33	90.88
DiffusionFR_ResNet101	Training (%)	94.35	92.28	91.79	91.54
	Top-1 Test (%)	88.92	86.77	86.51	86.04
	Top-5 Test (%)	91.72	89.71	89.13	88.77
DiffusionFR_EfficientNet	Training (%)	96.10	94.03	93.77	93.21
	Top-1 Test (%)	90.48	88.61	88.27	87.87
	Top-5 Test (%)	93.38	91.72	91.27	90.84
DiffusionFR_neuro-heuristic	Training (%)	96.82	93.61	93.34	92.76
	Top-1 Test (%)	91.25	88.14	87.78	87.36
	Top-5 Test (%)	94.27	91.19	90.72	90.26
DiffusionFR_BPPD	Training (%)	97.02	94.94	94.43	94.07
	Top-1 Test (%)	91.49	89.46	89.02	88.58
	Top-5 Test (%)	94.63	92.58	92.28	91.81
DiffusionFR_CNN(r1, r2)	Training (%)	97.33	95.23	94.71	94.22
	Top-1 Test (%)	91.69	89.64	89.28	88.85
	Top-5 Test (%)	94.80	92.74	92.40	92.10

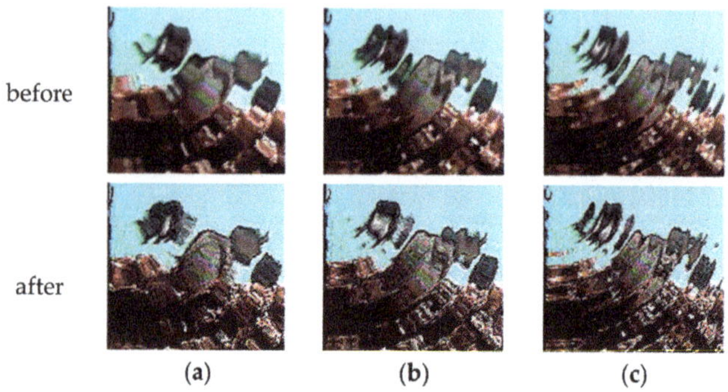

Figure 11. Comparison of images before and after TSD deblurring of water ripple noise. (**a**) $F_{0.04}A_2$; (**b**) $F_{0.06}A_6$; (**c**) $F_{0.08}A_{10}$.

In Table 11, the mean value of the training accuracy of DiffusionFR on the three datasets ($F_{0.04}A_2$, $F_{0.06}A_6$, and $F_{0.08}A_{10}$) with added water ripple noise is 95.00%. The mean value of the Top-1 accuracy test was 89.54%, and the mean value of the Top-5 accuracy test was 92.73%. These values indicate that DiffusionFR outperforms other methods in terms of accuracy. These values also demonstrate that DiffusionFR, with ResNet50 as the chosen backbone network, has a higher potential for achieving superior recognition performance.

5. Discussion

Based on the analysis of Tables 7–11, we have drawn several significant conclusions. Firstly, ResNet50 performs better than other backbone networks when selected as the backbone network for DiffusionFR. Compared to ResNet34 and ResNet101, the deeper network structure of ResNet50 enables a more effective capture of intricate image features and mitigates the risk of gradient vanishing or explosion [46]. Additionally, ResNet50's effective integration of the attention mechanism and the residual network approach contribute to its superior performance in propagating the model gradient.

Furthermore, a comparison between DiffusionFR and DiffusionFR_noA reveals that DiffusionFR outperforms DiffusionFR_noA in terms of training accuracy and accuracy on the test set. This indicates that DiffusionFR is capable of capturing crucial features and achieving more accurate classification and prediction. DiffusionFR also demonstrates superior performance compared to other standard attention methods, further validating the effectiveness of the incorporated LAM.

Moreover, DiffusionFR exhibits remarkable results among the compared methods, achieving superior performance in terms of training accuracy and accuracy on the test set. The proposed TSD approach for fish recognition in blurry scenarios proves to be highly effective. DiffusionFR's end-to-end integrated framework [47] for denoising and recognition surpasses a two-stage scheme by leveraging the interrelationships between these tasks. It enhances accuracy and stability by efficiently handling noise [48] and blur [49] information.

Additionally, the impact of light reflection noise and water ripple noise on recognition performance is evident from the analysis. Increasing light amplitude, light diameter, frequency, and amplitude of water ripples in the datasets leads to a decreasing trend in the training accuracy, Top-1 test accuracy, and Top-5 test accuracy of the same backbone network method. This highlights the significant role of light reflection and water ripples in recognition performance and reinforces the usability of corrected fish images for species-specific recognition even in the presence of these noise scenarios.

In comparing the neuro-heuristic analysis of video and bilinear pooling with poisoning detection (BPPD) to the DiffusionFR method, it becomes clear that DiffusionFR outperforms these approaches. While recent advancements in the neural network field have shown progress, DiffusionFR exhibits superior performance, even when compared to CNN(r1, r2).

6. Conclusions

In this study, we propose a method called DiffusionFR, which combines the diffusion model and attention mechanism to address the challenge of fish image recognition in blurry scenarios. The approach involves deblurring fish scene pictures using a two-stage diffusion network model, TSD, to restore clarity. Furthermore, a learnable attention module, LAM, was incorporated to enhance the accuracy of fish recognition.

DiffusionFR achieves the highest mean values of training accuracy, Top-1 test accuracy, and Top-5 test accuracy, at 94.91% on the original dataset. It also maintains the highest mean values of accuracy at 94.65% on the datasets with added light reflection noise and 92.84% on the datasets with added water ripple noise.

The effectiveness of DiffusionFR is evident from its superior performance compared to other approaches that use different backbone networks, attention mechanisms, and Gaussian denoising. DiffusionFR proves to be more accurate and robust, making it applicable in various underwater applications such as underwater photography, underwater detection, and underwater robotics.

Although this study successfully improves fish image recognition in blurry scenarios, there is still room for improvement due to the complex and uncertain nature of the marine environment. Additionally, the recognition of overlapping and occluded regions in natural fish scenarios needs further exploration. It is essential to construct relevant datasets, refine the network model, and conduct comprehensive studies to contribute effectively to fish conservation and related industries in the future.

Author Contributions: Conceptualization, G.W. and B.S.; data curation, B.S.; formal analysis, G.W. and B.S.; funding acquisition, L.M.; investigation, B.S.; methodology, B.S.; project administration, L.M.; resources, L.M., X.Y., P.W. and L.K.; software, B.S.; supervision, L.M. and L.K.; validation, G.W. and B.S.; visualization, G.W. and B.S.; writing—original draft, B.S.; writing—review and editing, G.W. and B.S. All authors have read and agreed to the published version of the manuscript.

Funding: This research was funded by the Key Research and Development Program of Zhejiang Province (grant number: 2021C02005) and the National Natural Science Foundation of China (grant number: U1809208) and the Zhejiang Philosophy and Social Science Planning Project (grant number: 22NDJC108YB).

Institutional Review Board Statement: Not applicable.

Informed Consent Statement: Not applicable.

Data Availability Statement: The code for our proposed model DiffusionFR and the dataset used in the experiments can be found on GitHub: https://github.com/zafucslab/DiffusionFR (accessed on 30 September 2023).

Conflicts of Interest: The authors declare that they have no known competing financial interests or personal relationships that could have appeared to influence the work reported in this paper.

References

1. Cooke, S.J.; Bergman, J.N.; Twardek, W.M.; Piczak, M.L.; Casselberry, G.A.; Lutek, K.; Dahlmo, L.S.; Birnie-Gauvin, K.; Griffin, L.P.; Brownscombe, J.W.; et al. The movement ecology of fishes. *J. Fish Biol.* **2022**, *101*, 756–779. [CrossRef]
2. Rees, S.E.; Sheehan, E.V.; Stewart, B.D.; Clark, R.; Appleby, T.; Attrill, M.J.; Jones, P.J.; Johnson, D.; Bradshaw, N.; Pittman, S.; et al. Emerging themes to support ambitious UK marine biodiversity conservation. *Mar. Policy* **2020**, *117*, 103864. [CrossRef]
3. Chen, Y.; He, G.; Yin, R.; Zheng, K.; Wang, G. Comparative Study of Marine Ranching Recognition in Multi-Temporal High-Resolution Remote Sensing Images Based on DeepLab-v3+ and U-Net. *Remote Sens.* **2022**, *14*, 5654. [CrossRef]
4. Follana-Berná, G.; Palmer, M.; Campos-Candela, A.; Arechavala-Lopez, P.; Diaz-Gil, C.; Alós, J.; Catalan, I.; Balle, S.; Coll, J.; Morey, G.; et al. Estimating the density of resident coastal fish using underwater cameras: Accounting for individual detectability. *Mar. Ecol. Prog. Ser.* **2019**, *615*, 177–188. [CrossRef]
5. Ben Tamou, A.; Benzinou, A.; Nasreddine, K. Targeted Data Augmentation and Hierarchical Classification with Deep Learning for Fish Species Identification in Underwater Images. *J. Imaging* **2022**, *8*, 214. [CrossRef] [PubMed]
6. Zhao, Y.; Shen, Q.; Wang, Q.; Yang, F.; Wang, S.; Li, J.; Zhang, F.; Yao, Y. Recognition of water colour anomaly by using hue angle and Sentinel 2 image. *Remote Sens.* **2020**, *12*, 716. [CrossRef]
7. Salman, A.; Jalal, A.; Shafait, F.; Mian, A.; Shortis, M.; Seager, J.; Harvey, E. Fish species classification in unconstrained underwater environments based on deep learning. *Limnol. Oceanogr. Methods* **2016**, *14*, 570–585. [CrossRef]
8. Marini, S.; Fanelli, E.; Sbragaglia, V.; Azzurro, E.; Fernandez, J.D.R.; Aguzzi, J. Tracking fish abundance by underwater image recognition. *Sci. Rep.* **2018**, *8*, 13748. [CrossRef] [PubMed]
9. Castillo, G.C.; Sandford, M.E.; Hung, T.; Tigan, G.; Lindberg, J.C.; Yang, W.; Van Nieuwenhuyse, E.E. Using natural marks to identify individual cultured adult delta smelt. *N. Am. J. Fish. Manag.* **2018**, *38*, 698–705. [CrossRef]
10. Hong, S.; Lee, G.; Jang, W.; Kim, S. Improving sample quality of diffusion models using self-attention guidance. In Proceedings of the IEEE/CVF International Conference on Computer Vision, Paris, France, 2–6 October 2023; pp. 7462–7471.
11. Zhang, Y.; Huang, N.; Tang, F.; Huang, H.; Ma, C.; Dong, W.; Xu, C. Inversion-based style transfer with diffusion models. In Proceedings of the IEEE/CVF Conference on Computer Vision and Pattern Recognition, Vancouver, BC, Canada, 17–21 June 2023; pp. 10146–10156.
12. Ruan, L.; Ma, Y.; Yang, H.; He, H.; Liu, B.; Fu, J.; Yuan, N.J.; Jin, Q.; Guo, B. Mm-diffusion: Learning multi-modal diffusion models for joint audio and video generation. In Proceedings of the IEEE/CVF Conference on Computer Vision and Pattern Recognition, Vancouver, BC, Canada, 17–21 June 2023; pp. 10219–10228.
13. Villon, S.; Mouillot, D.; Chaumont, M.; Darling, E.S.; Subsol, G.; Claverie, T.; Villéger, S. A deep learning method for accurate and fast identification of coral reef fishes in underwater images. *Ecol. Inform.* **2018**, *48*, 238–244. [CrossRef]
14. Mannocci, L.; Villon, S.; Chaumont, M.; Guellati, N.; Mouquet, N.; Iovan, C.; Vigliola, L.; Mouillot, D. Leveraging social media and deep learning to detect rare megafauna in video surveys. *Conserv. Biol.* **2022**, *36*, e13798. [CrossRef] [PubMed]
15. Li, L.; Shi, F.; Wang, C. Fish image recognition method based on multi-layer feature fusion convolutional network. *Ecol. Inform.* **2022**, *72*, 101873. [CrossRef]

16. Qu, P.; Li, T.; Zhou, L.; Jin, S.; Liang, Z.; Zhao, W.; Zhang, W. DAMNet: Dual attention mechanism deep neural network for underwater biological image classification. *IEEE Access* **2022**, *11*, 6000–6009. [CrossRef]
17. Alaba, S.Y.; Nabi, M.M.; Shah, C.; Prior, J.; Campbell, M.D.; Wallace, F.; Ball, J.E.; Moorhead, R. Class-aware fish species recognition using deep learning for an imbalanced dataset. *Sensors* **2022**, *22*, 8268. [CrossRef] [PubMed]
18. Liang, J.M.; Mishra, S.; Cheng, Y.L. Applying Image Recognition and Tracking Methods for Fish Physiology Detection Based on a Visual Sensor. *Sensors* **2022**, *22*, 5545. [CrossRef]
19. Yoshimura, M.; Otsuka, J.; Irie, A.; Ohashi, T. Rawgment: Noise-accounted raw augmentation enables recognition in a wide variety of environments. In Proceedings of the IEEE/CVF Conference on Computer Vision and Pattern Recognition, Vancouver, BC, Canada, 17–21 June 2023; pp. 14007–14017.
20. Liang, Y.; Liang, W. ResWCAE: Biometric Pattern Image Denoising Using Residual Wavelet-Conditioned Autoencoder. *arXiv* **2023**, arXiv:2307.12255.
21. Peng, L.; Zhu, C.; Bian, L. U-shape transformer for underwater image enhancement. *IEEE Trans. Image Process.* **2023**, 3066–3079. [CrossRef] [PubMed]
22. Hong, G. Image fusion, image registration and radiometric normalization for high resolution image processing. In *Technical Report*; University of New Brunswick: Fredericton, NB, Canada, 2023.
23. Połap, D. Neuro-heuristic analysis of surveillance video in a centralized IoT system. *ISA Trans.* **2023**, *140*, 402–411. [CrossRef] [PubMed]
24. Połap, D.; Jaszcz, A.; Wawrzyniak, N.; Zaniewicz, G. Bilinear pooling with poisoning detection module for automatic side scan sonar data analysis. *IEEE Access* **2023**, *11*, 72477–72484. [CrossRef]
25. Raavi, S.; Chandu, P.B.; SudalaiMuthu, T. Automated Recognition of Underwater Objects using Deep Learning. In Proceedings of the 2023 7th International Conference on Trends in Electronics and Informatics (ICOEI), Tirunelveli, India, 11–13 April 2023; pp. 1055–1059.
26. Dhariwal, P.; Nichol, A. Diffusion models beat gans on image synthesis. *Adv. Neural Inf. Process. Syst.* **2021**, *34*, 8780–8794.
27. Ho, J.; Jain, A.; Abbeel, P. Denoising diffusion probabilistic models. *Adv. Neural Inf. Process. Syst.* **2020**, *33*, 6840–6851.
28. Sohl-Dickstein, J.; Weiss, E.; Maheswaranathan, N.; Ganguli, S. Deep unsupervised learning using nonequilibrium thermodynamics. In Proceedings of the International Conference on Machine Learning, Lille, France, 6–11 July 2015; PMLR: New York, NY, USA, 2015; pp. 2256–2265.
29. Ronneberger, O.; Fischer, P.; Brox, T. U-net: Convolutional networks for biomedical image segmentation. In Proceedings of the Medical Image Computing and Computer-Assisted Intervention–MICCAI 2015: 18th International Conference, Munich, Germany, 5–9 October 2015; Proceedings, Part III 18. Springer International Publishing: Hershey, Switzerland, 2015; pp. 234–241.
30. Wang, R.; An, S.; Liu, W.; Li, L. Invertible Residual Blocks in Deep Learning Networks. *IEEE Trans. Neural Netw. Learn. Syst.* **2023**, 1–7. [CrossRef]
31. Lin, J.; Liu, D.; Yang, H.; Li, H.; Wu, F. Convolutional neural network-based block up-sampling for HEVC. *IEEE Trans. Circuits Syst. Video Technol.* **2018**, *29*, 3701–3715. [CrossRef]
32. Mao, G.; Liao, G.; Zhu, H.; Sun, B. Multibranch attention mechanism based on channel and spatial attention fusion. *Mathematics* **2022**, *10*, 4150. [CrossRef]
33. Wang, L.; Li, M. The quantitative application of channel importance in movement intention decoding. *Biocybern. Biomed. Eng.* **2022**, *42*, 630–645. [CrossRef]
34. Cui, X.; Zou, C.; Wang, Z. Remote sensing image recognition based on dual-channel deep learning network. *Multimed. Tools Appl.* **2021**, *80*, 27683–27699. [CrossRef]
35. Yuan, Y.; Guo, H.; Bai, H.; Qin, W. Adaptive weighted multiscale feature fusion for small drone object detection. *J. Appl. Remote Sens.* **2022**, *16*, 034517. [CrossRef]
36. Koonce, B.; Koonce, B. ResNet 34. In *Convolutional Neural Networks with Swift for Tensorflow: Image Recognition and Dataset Categorization*; Apress: Berkeley, CA, USA, 2021; pp. 51–61.
37. He, K.; Zhang, X.; Ren, S.; Sun, J. Deep residual learning for image recognition. In Proceedings of the IEEE Conference on Computer Vision and Pattern Recognition, Las Vegas, NV, USA, 27–30 June 2016; pp. 770–778.
38. Zhang, Q. A novel ResNet101 model based on dense dilated convolution for image classification. *SN Appl. Sci.* **2022**, *4*, 1–13. [CrossRef]
39. Xu, Y.; Goodacre, R. On splitting training and validation set: A comparative study of cross-validation, bootstrap and systematic sampling for estimating the generalization performance of supervised learning. *J. Anal. Test.* **2018**, *2*, 249–262. [CrossRef] [PubMed]
40. Guo, L.; Huang, P.; Huang, D.; Li, Z.; She, C.; Guo, Q.; Zhang, Q.; Li, J.; Ma, Q.; Li, J. A classification method to classify bone marrow cells with class imbalance problem. *Biomed. Signal Process. Control.* **2022**, *72*, 103296. [CrossRef]
41. Tan, M.; Le, Q. Efficientnet: Rethinking model scaling for convolutional neural networks. In Proceedings of the International Conference on Machine Learning, Long Beach, CA, USA, 10–15 June 2019; PMLR: New York, NY, USA, 2019; pp. 6105–6114.
42. Woo, S.; Park, J.; Lee, J.Y.; Kweon, I.S. Cbam: Convolutional block attention module. In Proceedings of the European Conference on Computer Vision (ECCV), Munich, Germany, 8–14 September 2018; pp. 3–19.
43. Huang, Z.; Wang, X.; Huang, L.; Huang, C.; Wei, Y.; Liu, W. Ccnet: Criss-cross attention for semantic segmentation. In Proceedings of the IEEE/CVF International Conference on Computer Vision, Seoul, Republic of Korea, 27 October–2 November 2019; pp. 603–612.

44. Hu, J.; Shen, L.; Sun, G. Squeeze-and-excitation networks. In Proceedings of the IEEE Conference on Computer Vision and Pattern Recognition, Salt Lake City, UT, USA, 18–23 June 2018; pp. 7132–7141.
45. Zhang, K.; Zuo, W.; Chen, Y.; Meng, D.; Zhang, L. Beyond a gaussian denoiser: Residual learning of deep cnn for image denoising. *IEEE Trans. Image Process.* **2017**, *26*, 3142–3155. [CrossRef] [PubMed]
46. Ravikumar, A.; Sriraman, H. Mitigating Vanishing Gradient in SGD Optimization in Neural Networks. In *International Conference on Information, Communication and Computing Technology;* Springer Nature Singapore: Singapore, 2023; pp. 1–11.
47. Ye, Z.; Tan, X.; Dai, M.; Lin, Y.; Chen, X.; Nie, P.; Ruan, Y.; Kong, D. Estimation of rice seedling growth traits with an end-to-end multi-objective deep learning framework. *Front. Plant Sci.* **2023**, *14*, 1165552. [CrossRef] [PubMed]
48. Elad, M.; Kawar, B.; Vaksman, G. Image denoising: The deep learning revolution and beyond—A survey paper. *SIAM J. Imaging Sci.* **2023**, *16*, 1594–1654. [CrossRef]
49. Zhang, X.; Cui, J.; Jia, Y.; Zhang, P.; Song, F.; Cao, X.; Zhang, J.; Zhang, L.; Zhang, G. Image restoration for blurry optical images caused by photon diffusion with deep learning. *J. Opt. Soc. Am. A* **2023**, *40*, 96–107. [CrossRef] [PubMed]

animals

MDPI

Conference Report

Welfare and Enrichment of Managed Nocturnal Species, Supported by Technology

Fiona French [1,*], Paige Bwye [2], Laura Carrigan [3], Jon Charles Coe [4,†], Robert Kelly [5], Tiff Leek [6], Emily C. Lynch [7], Eric Mahan [7] and Cathy Mingee [7]

1 School of Computing and Digital Media, London Metropolitan University, 166-220 Holloway Road, London N7 8DB, UK
2 Bristol Zoological Society, Hollywood Lane, Bristol BS10 7TW, UK; pbwye@bzsociety.org.uk
3 Zoological Society of London, London N1 4RY, UK; laura.carrigan@zsl.org
4 Independent Researcher, Cambridge, VIC 3777, Australia
5 Centre for Research in Animal Behaviour, University of Exeter, Rennes Drive, Exeter EX4 4RN, UK; rk528@exeter.ac.uk
6 Faculty of Science, Technology, Engineering and Maths, The Open University, Milton Keynes MK7 6AA, UK; tiff.leek@open.ac.uk
7 North Carolina Zoo, 4401 Zoo Parkway, Asheboro, NC 27205, USA; emily.lynch@nczoo.org (E.C.L.); eric.mahan@nczoo.org (E.M.); cathy.mingee@nczoo.org (C.M.)
* Correspondence: f.french@londonmet.ac.uk
† Jon Coe Design was retired.

Simple Summary: The behaviours and needs of nocturnal animals can be overlooked by humans, potentially because of our poor night vision and diurnal waking hours. Despite certain challenges in studying many nocturnal animals, appropriate provisions for their welfare should be supported in both wild and managed environments. To investigate this issue and explore ways to offer technology-enhanced welfare, husbandry practices and enrichment opportunities for nocturnal species, we conducted a multidisciplinary workshop (Moon Jam). During the event, species experts provided animal welfare briefs that related to specific challenges for nocturnal animals in different contexts. Teams of participants addressed these challenges in collaborative design sessions, producing a collection of hand-crafted models to share their ideas. An important aspect of the workshop was to be inclusive of all the stakeholders involved, including zoo management teams, animal stewards and zoo visitors, as well as the individual species. In this paper, we present our reflections on managed nocturnal animal welfare, framing these within current practices and Moon Jam workshop outputs. We contribute a set of guidelines for those involved with caring for zoo-housed nocturnal species, emphasizing the provision of technology-enhanced husbandry and enrichment opportunities.

Abstract: This paper addresses the potential for technology to support husbandry and enrichment opportunities that enhance the welfare of zoo and sanctuary-housed nocturnal and crepuscular species. This topic was investigated through the medium of a multidisciplinary workshop (Moon Jam) that brought together species experts, zoo designers, Animal-Computer Interaction researchers and post-graduate students in collaborative discussions and design sessions. We explain the context through an examination of existing research and current practices, and report on specific challenges raised and addressed during the Moon Jam, highlighting and discussing key themes that emerged. Finally, we offer a set of guidelines to support the integration of technology into the design of animal husbandry and enrichment that support wellbeing, to advance the best practices in keeping and managing nocturnal and crepuscular animals.

Keywords: nocturnal; environmental enrichment; animal-computer interaction; collaborative design; potto; armadillo; aye-aye; bushbaby; coral; vampire bat

1. Introduction

Human interest in other species has motivated a significant amount of research into their cognitive, behavioural and physical characteristics and wellbeing. Scientific endeavour has built a robust knowledge base describing many non-human biological traits. Some non-human animals (hereinafter 'animals') have been harder for scientists to investigate in their natural settings than others, because of their environment, lifestyle and associated sensory perceptions. For example, dark environments, i.e., deep sea, underground and at night [1], can be difficult to navigate and access, which may be an added challenge for researchers to obtain relevant information. People rely on human-specific sensory modalities to understand and interact with the world, which can result in information on animals that are imperceptible to human sensory capabilities potentially being overlooked without support by technology [2]. Moreover, researchers may require technological solutions to facilitate a complete conceptual picture of a dark or restricted location, whereas the inhabitants will have evolved senses that enable them to thrive in such an environment, such as adaptive echolocation in greater mouse-eared bats (*Myotis myotis*) for hunting efficiency [3]

In this paper, we focus on the lives of nocturnal species in managed environments, considering what we know of their usual behaviours in their natural habitats, and reflecting on what kinds of structural habitats and enrichment can encourage the expression of these behaviours in a zoo or sanctuary setting.

Our method for investigating this topic involved holding a dedicated Moon Jam workshop, where participants were provided briefs from nocturnal animal welfare experts to address husbandry and enrichment challenges and were invited to respond with novel design solutions [4]. Participants included animal experts, computer science and interaction design researchers and postgraduates studying engineering and design disciplines. The Moon Jam was part of a series of multidisciplinary, collaborative design workshops (zoo jams) that provide opportunities to share skills and knowledge while discussing different themes around animal welfare in a logical, creative and open-ended way. By bringing together participants with diverse skills and shared interests in other species, the zoo jams aim to expand designers' fields of reference and lever technology in support of welfare and enrichment goals [5]. Ultimately, the goal is to take concepts forward into action plans and evaluate them with the intended users (animal and human stakeholders).

This paper provides contextual background and offers a discussion around key topics that are perceived as being challenging for animal husbandry, with examples showing how technology can support husbandry and enrichment solutions. We share briefs from the Moon Jam to illustrate specific themes and provide topical context.

2. Background

Perhaps surprisingly, around 69% of described mammals are nocturnal, 20% diurnal, 2.5% crepuscular and 8.5% cathemeral [6]. Due to artificial light, humans are considered facultatively cathemeral, despite activity being concentrated during the daytime [7]. Nocturnal species evolved to rely on senses in addition to sight. For instance, some nocturnal mammals have more species-specific chemoreceptor genes and more complicated olfactory organs in comparison to diurnal mammals [8], which are responsible for smell and taste. Enhanced sensitivities may also relate to vibrations, electro-magnetic fields or air pressure. Raising awareness of nocturnal species and their specific needs is important so that humans learn how to co-exist with wildlife in their natural environments, as well as how to provide appropriate settings that support welfare in managed settings. This need is arguably increasing, where human population growth may put further pressure on resources and affect natural land use in different ways [9], for example, by reducing available habitat, introducing new dangers and causing pollution. Pollution includes not only plastic waste and chemical spills but also vibro-acoustic pollution, air pollution and light pollution. To reduce negative impacts on animals' welfare as much as possible, we need to understand more about their lifestyles and behaviours, their modes of perception and communication and their assorted capabilities.

The difficulty of finding or watching animals in the dark can create an additional challenge in gaining knowledge on nocturnal species and their environmental and management needs. Fortunately, recent advances in remote sensing technology have provided further information from both wild populations and their natural habitats and managed animals and their artificial habitats [10]. These tools and findings can be used to optimise natural behavioural opportunities supporting improved animal welfare.

'Enrichment' can loosely be defined as a practice that provides additional stimuli to an animal with the aim of increasing its physical and/or mental wellbeing, and can be categorised into cognitive, food, physical, sensory and social forms [11]. The use of enrichment for animals in human care is motivated by the importance of enhancing welfare through encouraging species-specific behaviours, the need to reduce undesirable behaviours and the duty to provide good healthcare [12]. For example, a suitable intervention might offer foraging opportunities that practise skills used in the wild. This could occupy a significant portion of the animal's time budget and mental attention, while simultaneously supporting physical objectives, i.e., body condition or weight management. Offering control and choice (some autonomy) to animals in restricted environments has been shown to reduce stress levels [13–15] and has other potential benefits, such as enabling the development of competence, preparing species for reintroduction to the wild and giving researchers opportunities to investigate the animals' preferences [16]. This could lead to a better understanding of animals' cognitive, perceptive and physical abilities, which in turn help us to consider their perspectives.

To further investigate this theme of nocturnal enrichment in managed environments with visitors, we organised a multidisciplinary workshop (Moon Jam) to explore technology-enhanced enrichment strategies for nocturnal species. We brought together a diverse group of participants, including species experts, zoo designers, Animal–Computer Interaction researchers and postgraduate students, to discuss some of the challenges faced by both animals and humans.

3. Method

The workshop type used for investigating enrichment and husbandry strategies for nocturnal animals was a zoo jam, which is characterised as a multidisciplinary, collaborative design event where participants network, sharing skills and ideas. Key features of a zoo jam are (i) to have one theme with multiple associated challenges, (ii) species-specific briefs provided by animal experts, (iii) expert feedback on concepts, (iv) time-constrained activities with clear goals, (v) co-crafting and presentation of rough prototypes and (vi) open dissemination of outputs and issues raised [5].

The Moon Jam workshop [4] was held over one day in December 2023, as part of the 10th Animal–Computer Interaction Conference [17], hosted at North Carolina State University, USA. The main aims were (i) to explore husbandry and enrichment opportunities for nocturnal species through group discussions and collaborative design sessions, (ii) to address a series of animal welfare briefs provided by experts, by producing focused technology-enhanced designs for husbandry and enrichment opportunities, (iii) to incorporate animal-centred design principles, in order to extend the reach of human design and (iv) to maintain a multispecies, multi-stakeholder perspective throughout. Data collection included participant observations, feedback from species experts and analysis of the proposed designs. The methodology ensured a comprehensive evaluation of the effectiveness of these enrichment strategies in promoting natural behaviours and improving animal welfare.

The interdisciplinary nature of the Moon Jam was crucial for bringing together both expertise in animal welfare and stewardship, in particular of nocturnal species, and expertise in using technology as a means for developing complex systems to support behavioural management and to offer enhanced control of the environment, applicable to both humans and animals.

Altogether, there were 18 co-located participants, comprising 6 Animal–Computer Interaction (ACI) researchers, 2 species experts, 10 postgrad students and 3 remote animal experts taking part on that day, while 12 participants (postgrad students) took part in a half-day Mini-Moon-Jam at London Metropolitan University, working with a small subset of the briefs. Some of the briefs were general, concerning challenges that face all managed nocturnal species, such as ways to provide appropriate lighting and access to fresh air. Others were specifically related to the following species: aye-ayes (*Daubentonia madagascariensis*), pottos (*Perodicticus potto*), Mohol bushbabies (*Galago moholi*), Southern three-banded armadillos (*Tolypeutes matacus*), common vampire bats (*Desmodus rotundus*) and living coral (mixed species). We explain these briefs, the associated workshop responses and the subsequent analyses in the discussion section that follows. This is presented as seven key themes: (4.1) Lighting, (4.2) Natural Experiences, (4.3) Space and Socialisation, (4.4) Foraging, (4.5) Specificity, (4.6) Multisensory Modalities and (4.7) Stakeholders.

4. Results and Discussion

In this section, we discuss key topics relevant to nocturnal animal husbandry in managed environments. We consider environmental conditions such as lighting and ventilation, the importance of experiencing natural events, animal behavioural needs that extend over 24 h, difficulties associated with enabling normal feeding behaviours and the potential impact of visitors in a managed environment. In addition, we consider all the stakeholders involved in a complex organisation such as a zoo, considering the educational, entertainment, research and financial sufficiency objectives that are critical aspects of a zoo's mandate.

Each theme is introduced through one of the briefs we received for the Moon Jam, and then discussed through the lens of participants' concepts and reflections, with expert responses.

4.1. Lighting

The Moon Jam brief for mixed species of living coral, supplied by Greensboro Science Centre (GSC) in Greensboro, NC 27455 (USA), exemplifies the willingness of animal stewards to offer the best husbandry possible, as well as showing how complex it can be to provide an appropriate environment for a species that inhabits a relatively inaccessible location—in this case, the ocean floor.

4.1.1. Brief: Living Coral

Supplied by Lindsay Zarecky and Jessica Hoffman at GSC

Most organisms utilise a variety of external cues to fulfil their biological needs. Reef organisms, like fish and especially corals, utilise both the solar and lunar cycle. Solar light is key for food production while the lunar cycle is used to determine the best time to reproduce [18].

In addition to light levels, corals are also exposed to several environmental changes including daily tidal cycles, varying levels of light (solar and lunar) intensity, seasonal temperature swings, food concentrations and periodic intense storm surges. How these environmental changes impact coral health is not well understood but could play a role in their overall health and wellbeing.

The living corals exhibit at GSC shows visitors a growing Great Barrier Reef coral reef ecosystem that includes 30+ different species of corals, coral-friendly fish and other invertebrates (Figure 1). The live coral exhibit is part of the "Communities Connected" gallery, and the tank volume is 1018 gallons (3.85 cubic metres), with a filtration system, circulation pump, heating and lighting. The substrate is crushed Aragonite gravel.

Figure 1. Living coral exhibit at Greensboro Science Centre. Image courtesy of GSC.

Coral health and wellbeing are important to GSC for a multitude of reasons. It is incredibly challenging to offer corals poignant enrichment. A variety of natural live and frozen foods as well as manufactured foods are offered, but it has so far not been possible to capitalise on bigger experiential enrichment that is created for other animals. In addition, replicating the corals' natural environment to a better extent could increase the welfare and wellbeing of the corals being kept. Ultimately, the aim is to enable the corals to reproduce in captivity, thus increasing the knowledge aquarists and scientists have of a very fragile ecosystem.

Although coral care at GSC focuses a great deal on meeting their solar needs, there are not yet methods to effectively meet other environmental conditions. This brief requires teams to develop scenarios that would better mimic the environmental conditions corals experience in the wild, such as lunar light cycles and tidal shifts, without compromising husbandry needs and considering the visitor experience. Solutions should reference the environmental conditions of the Great Barrier Reef since this was the original habitat of these types of coral, but ideally, there would be some level of flexibility to adapt to Florida corals for the future.

4.1.2. Responses: Living Coral

Participants found this a complex challenge, but there were several suggestions that deployed technology to address the lighting requirements of the brief. In the natural environment, the transition from day to night and vice-versa happens gradually over a prolonged period of dusk or dawn. Nocturnal animals may not necessarily live in a 24 h cycle of darkness, and creating natural sunlight patterns for them when they should be asleep will help them sense what time of year it is, as well as when it is time to be awake and asleep. In the absence of regular light exposure, such as for species living underground, the temperature can module surface emergence behaviours [19], facilitating an interrelationship with the circadian rhythm and photoperiod.

In managed environments, dimmer switches could be applied to the lighting systems of managed species, although husbandry and enrichment solutions light management should consider and value zoo staff time. A technology-mediated solution could automatically imitate natural lighting for zoo-housed species by using bulbs that can fade across different Kelvin temperatures and lux levels in order to match the qualities of the sun over the course of a day. Additionally, the photoperiod and directionality of light should also be considered to match natural conditions in the species' home range. Feedback from zoo staff and species experts was positive on the theme of technology enabling gradual dawn–dusk lighting systems and associated changes in wavelengths. This might be a useful intervention to offer a more realistic experience to animals, which in turn from a visitor's perspective may provide a more engaging experience allowing the observation of nocturnal animals across differing light periods.

In support of such an initiative, Moonshine is an example LED control system [19] that enables users to mimic habitat-specific conditions or certain types of light pollution. Moonshine can colour-shift LED lights to recreate natural moonlight cycles and predict illuminance in different locations during the year. Its intended use is for field ecologists and researchers investigating the effects of light pollution. However, at present, there are limitations on its wavelengths at both ends of the visible spectrum, including in the near-UV and far red, in comparison to moonlight [20], which may not be suited to nocturnal animals utilising UV vision.

In managed environments, the effect of artificial light and what might be considered optimum requirements for nocturnal species is beginning to be investigated [21]. The colour wavelengths produced through light are important because they can influence melatonin production, a hormone concerning the circadian rhythm of species by acting as a signal to synchronise biochemical, physiological and behavioural processes, in humans [22] and non-human species [21]. Simultaneously, light pollution in wild environments creates the need to reduce disturbance to wildlife, such as through the use of amber lights, emitting no blue wavelengths, as opposed to traditional white lights which emit blue wavelengths, which may be the better choice for minimalising disruption in some nocturnal insects [23]. Since the temporal changes around light intensity, direction and wavelength are relatively predictable, many animals have evolved to use this information to regulate their behaviour. This can include seasonal adaptations, such as moulting, the timing of reproductive activities, as well as modifying food intake, activity and immune function [24].

Moreover, it is important to note that the quality of light (wavelength, intensity, duration) may all make a difference to animals in managed settings, and that appropriate light settings differ across species. More research is needed to understand and inform optimum light conditions at a species level, including their elasticity to cope outside of optimal ranges. Fortunately, through the growing application of technology in zoo research and management of animals, there are better opportunities to investigate the lighting conditions of nocturnal species in zoos. The use of artificial UV-B can be used to encourage basking behaviours of species to support vitamin D production. In fact, exposure to appropriate UV-B is now common practice for certain animals, primarily reptiles [25], and within callitrichid captive management [26]. Comparatively, less attention has been given to providing artificial UV-B to many nocturnal species that may naturally receive UV-B whilst they sleep in the wild. However, considering that the sun emits different wavelengths all working together, it is important to think about visible and infrared light along with the UV light. Moreover, the use of more natural full-spectrum light in managed environments could be utilised further to promote plant growth in nocturnal exhibits, providing more natural stimuli within indoor-only environments.

In relation to the requirement to mimic environmental conditions for mixed species of living coral, suggestions from Moon Jam participants included adjusting the water temperature dynamically using a system that tracked and automatically responded to open weather data. In parallel with this, good husbandry would try and match the current flow in the tank with live data relating to the Great Barrier Reef or Florida tides, available via the Australian Institute of Marine Science reefs portal [27] and the National Oceanic and Atmospheric Administration website [28]. The tank pump system could supply variations in nutrients with different currents, so the corals could anticipate food arrival. This would create a situation-related positive affective state in the coral, in this case, associated with having an opportunity to feed. The inclusion of such opportunities within husbandry practices is linked to the enhancement of animal welfare [29].

This introduces the second key theme that emerged, relating to natural experiences for nocturnal and crepuscular animals that are housed in indoor environments and is explored in the following section.

4.2. *Natural Experiences*

The Fresh Air brief relates to land animals housed indoors that in the wild might receive information from moving air such as chemical signals from their conspecifics, scents relating to predators and prey, indications of food and water availability, as well as imminent changes in weather conditions.

4.2.1. Brief: Air Flow

Supplied by Jon Coe

Due to the typical need of day–night reversal systems to create an enclosed artificially lit environment, this might affect experiences of changing environmental conditions that are associated with outdoor air and scent sensations. Teams should think of ways to address this, considering natural experience opportunities for managed nocturnal species.

4.2.2. Responses: Air Flow

This brief generated several possible concepts that deployed technology to enable a simulation of environmental conditions, again, mapping these to real-world conditions in the animals' native habitats.

In general, all climate control could be managed using technology inside the enclosure, dynamically influencing factors such as humidity and temperature throughout the day and night, including providing natural gradients in microclimates. For example, artificial snow could be used to simulate winter and might be a novel and exciting substrate that encourages playful behaviour in some species. Artificial rain, mist and fog could also be provided. Conversely, basking lamps or heat sources without light such as embedded heating cables could be animal-triggered to create changing microclimates within enclosures.

Enriching olfaction was a more challenging problem, as it required the introduction of new smells in the atmosphere—chemical signals that are often imperceptible to humans. One suggestion was to install ventilation systems that drew in naturally scented air from outside directly into the enclosure, although the use of ventilation systems is arguably more of a welfare requirement than an enriching opportunity. Another suggestion involved creating an external enclosure using breathable blackout fibres for the roof and parts of the wall, allowing air to circulate but no light to penetrate. It would also be possible, albeit expensive, to deploy photochromic ventilated glass walls, which zoo staff could control remotely, using an app. During the evening time, when the artificial light was switched on, these same features would stop light pollution from the enclosure affecting the external environment.

Since novelty is known to provoke interest in many animals, a simple low-tech idea to stimulate olfaction was the introduction of new logs or other furnishings to an enclosure. This would be likely to encourage fresh scent marking by some species. Another solution for fresh air provision that relied less on technology was to build a tunnel allowing crepuscular animals to go outside their enclosure at dawn and enter an external compound that was shared between different species at different times during the day or night. The crepuscular animals would be able to smell the scents left behind by other animals as the dew evaporated. This would also be a time before visitors arrived, so it would be quieter, and they might be less fearful. The assumption was that as the sun rose, they would naturally return to their dark setting indoors. This leads to the next important theme, around the provision of space.

4.3. Space and Socialisation

This Moon Jam brief relates to the exhibiting of mixed species, providing an introduction to the advantages and challenges associated with this aspect of husbandry, and with managing space in general. Addressing the use of space within enclosures could change social behaviours within and between species, affording positive experiences.

4.3.1. Brief: Pottos and Bushbabies

Supplied by Laura Carrigan, London Zoo

The animals in London Zoo's Night Zone are kept on reverse lighting so their nighttime (when they are awake) is during human daytime, and then their daylight (when they go to bed) is when people have left for the day. All the animals have microchips in their shoulders.

Mixed species exhibits enable the maximisation of space by linking exhibits and providing natural enrichment for the animals, as well as offering a better experience for visitors as there is more activity to view. In addition to housing three Moholi bushbabies and two pottos all together, the largest exhibit in the Night Zone also includes two Malagasy giant rats (*Hypogeomys antimena*) (Figure 2).

Figure 2. Male and female pottos, juvenile Moholi bushbaby and Malagasy giant jumping rat in London Zoo Nocturnal House. Images courtesy of Laura Carrigan, London Zoo.

The enclosure is approximately 3 m (H) × 11 m (L) × 5 m (W), which is about 165 cubic metres or 5827 cubic feet, with a coir/bark chip substrate and branches/liana structures throughout. Some of the structures are fixed, and others move when the animal uses them. This mixed species exhibit enables the animals to have more space compared to being housed separately, and the bushbabies have developed some form of relationship with the female potto. However, there is one issue, which is that it is difficult to control their access to food, so the older male bushbaby has gained weight. To try and counteract this, it is necessary to hand feed the pottos their favourite food (insects), because if it is scattered or placed in enrichment, the bushbabies are too quick and manage to steal everything.

Bushbabies can jump over two metres in one leap and can cover distances in seconds, which makes them much more agile and speedy than the other animals they live with. As arboreal animals, they will generally spend most of their time off the ground and traverse by jumping and running along branches. They are fed pellets, insects and gum, but they help themselves to the pottos' vegetables as well.

By contrast, the male and female potto move slowly across branches, although if threatened or angry, can move quickly. They cannot jump from branch to branch and will not go down to the ground for anything. The giant rat will occasionally steal live food that is scattered around the enclosure, but not enough to be an issue. Sometimes it is also necessary to hand-feed her treats like nuts and avocado since if the bushbabies like it, they will take that too.

The challenge for teams is to overcome the husbandry challenge of food provision in a mixed species exhibit by designing a feeding device that is accessible to the pottos but inaccessible to the bushbabies.

4.3.2. Responses: Pottos and Bushbabies

Two possible solutions were proposed: low tech and high tech. The low-tech version took account of the fact that pottos have longer arms than bushbabies; therefore, food could be in a space with tubular access that could only be reached by the pottos (Figure 3A). The simplicity of this design makes it very appealing. There are few parts to break, and it would be cheap and easy to place multiple versions around the enclosure.

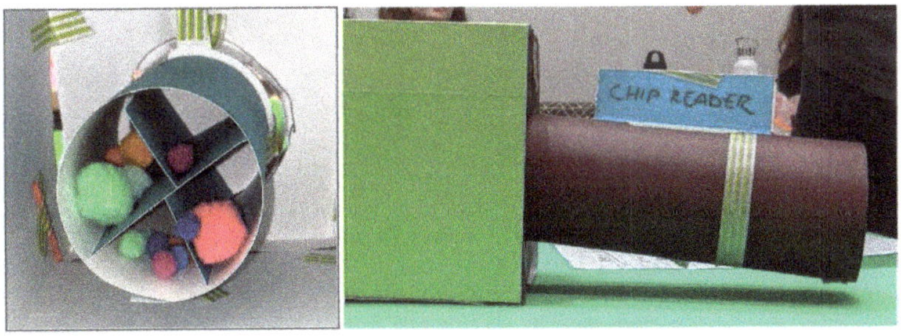

(A) (B)

Figure 3. Cardboard mock-ups of tunnel system with rotating food dispenser (**A**) and higher-tech tunnel with microchip program access (**B**). Images from Moon Jam.

On the other hand, since the low-tech solution was purely mechanical, it would not be able to discriminate between individuals, so could not be used for remotely monitoring access, and it would be species specific. Also, it would not dispense food over time, meaning that care staff would always need to be involved in provisioning. The higher tech solution required power and electronics but had more functionality and would be flexible for different mixed species enclosures. It could be set with a timer to dispense different food items and could discriminate between micro-chipped individuals (Figure 3B), enabling a clear record of which animal accessed which kind of food at what time. This solution also used the long tunnel concept to restrict bushbaby access.

Alternatively, bushbabies could be lured to their feeding portal and held inside while pottos had time to forage. However, this concept would be more complex to build and potentially less practical for other exhibits. Other proposals included setting aside separate smart-gated spaces in the enclosure for food delivery, with the gate operated by reading animals' microchips [30,31]. A problem with this idea was that the bushbabies could jump nimbly through the gate when opened by a potto. Nonetheless, in general, the concept of animal-controlled 'smart gates' that facilitated the animals' ability to freely access areas within an enclosure, using a microchip-controlled system, was regarded positively by zoo designers and staff.

Enclosure area standards are an important benchmark, and it was noted that existing guidelines represent minimum standards. The clear trend is to enlarge these areas in zoos and related facilities, especially as animals can become more physically fit and intellectually motivated. Broad advice from experts is to aim to exceed today's standards and best practices. At the Moon Jam, the recommendation was to use space creatively and with regard to natural behavioural patterns, aiming to maximise and optimise what was available to use by the animal. One example was to have maximum rather than minimum branching cover for climbing/arboreal species, thus providing a wider surface area for spreading out food or food devices to promote exploration and foraging. In general, the more complex the environment, the more opportunity there is to include branching routes and structures that support species-specific mating, scent marking and social and toilet-related behaviours.

While it is possible to develop enriching features within existing enclosures, there are many opportunities that can be offered in the design of new enclosures as well. A single space or area might not represent the best design, although combining exhibits can offer new challenges, as demonstrated by the mixed species enclosure housing pottos and bushbabies. Having the choice, control and ability to move from place to place is enriching, especially if each place offers complimentary resources. Long connections between different accessible enclosures (trailways) can also provide interest and stimulate exploration. The Center for Great Apes in Florida (USA) constructed one of the first of these for their primates, and the concept has been taken up by other zoos. An ambitious example is Philadelphia Zoo's Big Cat Falls Trailway which allows the felids to walk over visitors' heads as they travel to different but connected spaces [32]. It gives the animals a rich perspective on the environment around them while protecting other species from predation. Such features can also be popular with visitors. While these examples were developed for diurnal species, enclosed indoor trails or flyways (either for single or mixed species) could be developed for animals housed in large nocturnal houses.

A potential complication associated with rotating or alternating species through enclosures is the risk of bacterial or viral transmission from one to another. Albeit, these risks could be reduced through regular animal health faecal screening methods. In addition, staff should wear sterile boots and gloves to clear enclosures and handle animals. In cases where infections are confirmed, zoos may increase their biosecurity practices, which extends to limiting the use of mixed species exhibits. This issue potentially exists within any open enclosure, because if the carrier was a wild mouse or bird, droppings could be present in any accessible browse.

Despite some challenges, particularly in enabling allocated food provision for species, carefully monitored mixed species exhibits can provide food and sensory-based enrichment without the occurrence of intra or interspecies conflict [33]. The following briefs also incorporate food provision, with a focus on natural foraging behaviours.

4.4. Foraging

Nearly all the Moon Jam briefs are related in some part to feeding—access to food, behavioural repertoires associated with foraging and hunting and managing change, such as age-related conditions.

Food-based enrichment can support natural feeding opportunities and increase species-specific behaviours in acquiring food resources. For instance, burying food or placing food in Kong pet store toys could generally increase the frequency of digging and rooting behaviours. Specific examples in nocturnal species include the use of artificial termite mounds increasing sit-and-wait predation behaviours in bushbabies [34]. Promoting foraging behaviour is arguably important for species to take control over their feeding experiences, which is likely to be intrinsically connected to their wellbeing and survival. Certainly, in the wild, predator species can incur a risk of mortality by starvation [35,36], which may be increased without the skills to acquire food or an opportunity to express them. If species in managed environments are to be considered for wild release programmes, which may be a growing priority for zoo organisations and rehabilitation centres in the future, behaviours directly associated with a species' ability to secure food in the wild should be strongly encouraged in managed care. As a case in point, abnormal dentition, gouging behaviour and frequency were reported in a translocated Javan slow loris that was later found deceased with a deformed jaw [37], indicative that these deviations in feeding behaviours may have contributed to its mortality.

The following brief from North Carolina Zoo illustrates some of the challenges associated with enabling natural feeding behaviours for nocturnal species kept in managed environments.

4.4.1. Brief: Common Vampire Bats

Supplied by North Carolina Zoo

North Carolina Zoo is home to a colony of 66 common vampire bats, which live in a nocturnal cave area. The habitat is on a reverse light cycle, creating nighttime lighting conditions for the bats during daytime hours, which encourages wakeful periods for the bats during operating hours.

Common vampire bats are considered sanguinivores, or animals that consume only blood. They are the only species of bat that only consumes mammalian blood, and in a natural habitat, feed from a wide range of sleeping mammals. Using echolocation, along with specialised scent detection, vampire bats find the warmest spots on the mammals, where blood runs closest to the skin. The bats typically land near their prey and then walk and climb to the best feeding location. A small hole (approximately 0.5 mm) is punctured with their sharp teeth and the anticoagulant contained in their saliva ensures blood flow. This process can take around 20 min but rarely wakes the prey animal. Vampire bats can consume about 2 tablespoons (35.5 mL) of blood at a time and must eat at least every other day.

At the North Carolina Zoo, vampire bats are fed harvested cow blood (Figure 4). The blood is treated to reduce coagulation and stored in the fridge to keep it fresh. Although this is a common management practice, it restricts how blood can be presented to bats. First, the blood cannot be warmed before presentation, as this would reduce its quality and duration of freshness. Second, the blood is usually placed in open dishes. These factors both reduce the natural hunting and feeding behaviours of vampire bats.

Figure 4. Vampire bat feeding on cow's blood. Image courtesy of the North Carolina Zoo.

The challenge for teams was to design a system for feeding that supports the expression of the bats' natural foraging behaviours.

4.4.2. Responses: Common Vampire Bats

The main concept from participants involved creating an artificial system for presenting fresh, warm blood that circulated between a set of blood bags (Figure 5). The bags would be made from a biocompatible polymer so the bats could pierce them safely—similar to the natural casing used on sausages. More research would be required to test whether the bats could smell the blood through this artificial skin and to find an appropriate method for maintaining suitable pressure within the system.

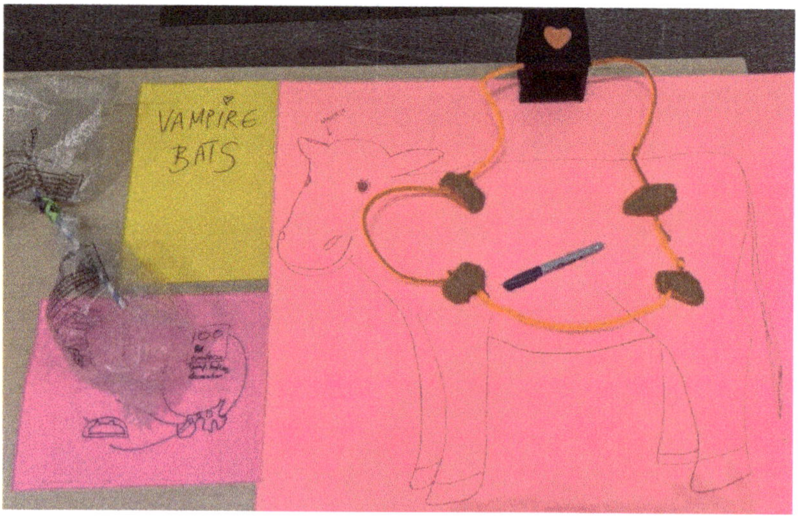

Figure 5. Physical mock-up of compression system comprising blood bags holding cow's blood. Image from Moon Jam.

Animal care staff found this concept to be very creative and reflective of vampire bat behaviour in the wild.

Food-based enrichment can form part of the overall healthcare management of animals such as by increasing activity budgets, though these effects can differ between species [38]. For nocturnal reptiles, examples of enrichment include stimulating the olfactory senses and associated behaviours of tongue touching, lip-licking and sniffing in leopard geckos [39]. Mimicking wild food presentation may also be important to improve food consumption, such as manual shaking of mice to encourage a strike response in snakes. Hanging up meat could also encourage more active and extended feeding behaviour durations, which may be important in species prone to obesity and muscle wastage, or for species that otherwise spend a large amount of their time in resting behaviours [40].

Visual, olfactory and auditory senses may also be stimulated through the enrichment design (materials, size, shape, complexity). The enrichment could simply be a novel food source and/or changes in its presentation, such as the use of live fish, which has been demonstrated to reduce pacing in tigers [41]. Moreover, by increasing the diversity of invertebrate food sources, different feeding behaviours can be promoted, such as digging or object manipulation to obtain burrowing mealworms, compared with the need for running, climbing and catching to capture locusts or crickets.

Another aim for food-based enrichment may be to help reduce undesirable or stereotypy behaviours, such as excessive pacing in felid species [42] or stereotypy swimming patterns in Vietnamese pond turtles [43]. More research is needed to investigate any physiological responses to enrichment, including food-based enrichment. Nonetheless, preliminary evidence has identified reduced faecal corticosterone following combined enrichment use (manipulable, sensory, and feed) in Asiatic lions [44], suggesting that in this case, the enrichment had lowered stress hormones within a captive environment.

Notably, enrichment can be manipulated to increase the level of difficulty, depending on whether the desired goal is to extend the duration of activity, to suit individual considerations (age and health status) or to address individual response variability to enrichment [45].

This highlights another important theme, which is the provision of specificity—solutions for individual members of a species. As with humans, there is rarely a 'one-size-fits-all' answer that works for everyone.

4.5. Specificity

The next Moon Jam brief featuring food provision illustrates the potential need to tailor husbandry practices and enrichment for a specific individual depending on their circumstances. The challenge of attending to the welfare of a particular animal involves learning about individual as well as species characteristics, to provide a personalised experience for that animal. Designing enrichment for one individual can give rise to insights pertaining to all members of that species, as well as to other animals experiencing similar life conditions.

4.5.1. Brief: Aye-Aye

Supplied by Paige Bwye

Aye-ayes are a nocturnal lemur from Madagascar possessing unique adaptations including their large ears, their thinner and elongated middle digits on each hand, and their overbuilt masticatory apparatus [46], to assist specialised tree-gouging behaviour. As percussive foragers, aye-ayes use their specialised middle digit (Figure 6A) to tap on bark while thought to be listening to vibrations within their auditory range (Figure 6B), indicating invertebrates inside. Once identified, they gnaw into the bark to access their prey with their continuously growing incisors (Figure 6C). Using their middle digits, they then excavate the prey deep within the hollow crevices. When they are not eating insects, they also forage on the ground for seeds and fallen fruits.

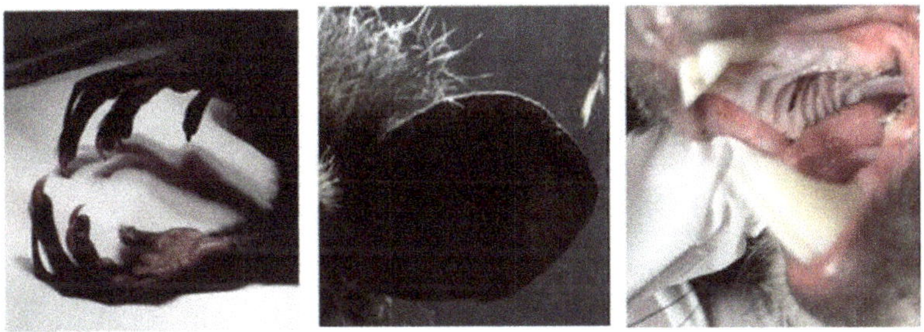

(A) (B) (C)

Figure 6. Aye-aye features—hands (**A**), ears (**B**) and teeth (**C**). Image courtesy of Paige Bwye.

For the purpose of the brief, the aye-aye enclosure dimensions to consider are 6 × 4 × 6 m (approx. 150 cubic metres or 1766 cubic feet), featuring horizontal and vertical branching, elevated nest boxes and rotten logs at ground level. Aye-ayes are generally destructive in captive environments because they can chew through strong surfaces, rendering many enrichment materials unsuitable. Moreover, they are a socially dispersed species and typically housed solitary in zoos.

The aye-aye used for this brief is beginning to show age-related changes, including cataracts in both eyes (Figure 7). Although his mobility is good, he cannot see individual insects in his environment. For this reason, it would benefit him to make his food provision easier to access to ensure consumption, but overtime this could lead to his teeth overgrowing from reduced gnawing opportunities. Teams were asked to apply what they had learnt about aye-ayes and this particular individual's background information into consideration to produce a technology-based solution that promotes species-specific foraging and incorporates dental health (serving as enrichment, but also as a required husbandry intervention).

Figure 7. Aye-aye with cataracts. Image courtesy of Paige Bwye.

4.5.2. Responses: Aye-Aye

The simplest, non-technical concepts produced by teams were chew toys made from hard hollowed wood, such as bamboo, filled with termites and sawdust, then sealed at the ends, so the aye-aye would need to use his teeth to access the food. The lengths of wood would be secured in a log holder. An alternative, more proactive suggestion was to use a scented coconut shell to hold food, since the shell would be abrasive and wear down his teeth.

Teams spent a significant amount of time designing a reusable feeder that could track bite activity. This concept used flat plates of hardwood inside a tough edible tunnel that the aye-aye had to bite, so the tunnel was disposable, but the centre part could be refilled (Figure 8). Insects were delivered on the plates and a sensor placed in the central section could measure bite activity. In a situation where the aye-aye could not hear the insects, a vibromotor would be used to simulate the sound of moving food; alternatively, it would be possible to amplify the sounds of insects, so they were easier to find. Expert responses identified that the use of any sensor materials would have to be inaccessible to the aye-aye directly to prevent the consumption of non-edible material.

For species that require regular mastication good husbandry practices should seek to provide food-based enrichment which enables gnawing, which may contribute to preventative and reactive dental care. For instance, the provision of hardwood sticks resolved malocclusions by improving molar occlusal wear in pine voles [47]. While the teams' enrichment concepts worked towards meeting the dental health goal that was set in the brief, the motivation to forage and feed might not have been fully stimulated if the visually impaired aye-aye had been presented with a concealed pipe containing insects. However, as his hearing is unimpaired, acoustic enrichment could complement food-based enrichment in this scenario using technological 'lures' in the form of insect sounds in different locations around his enclosure to encourage a foraging response.

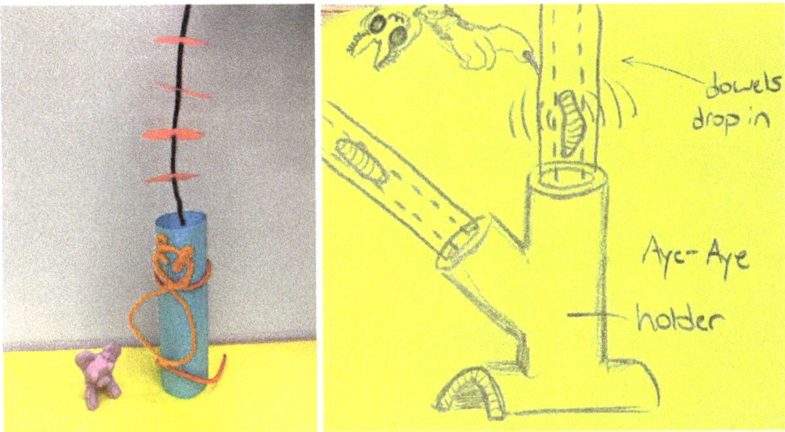

Figure 8. Two versions of a food holder—removable plate system and hollow dowel. Images from Moon Jam.

This leads us to the next theme, which relates further to foraging solutions taking into consideration the sensory modalities used by another nocturnal species.

4.6. Multisensory Modalities

As mentioned in the Background section, nocturnal species do not typically rely on vision, but also use acoustic, olfactory and tactile signals, and may have the ability to perceive their world using other sensory modalities such as sensitivity to electromagnetic fields, humidity and air or water pressure changes. The Moon Jam brief that introduces this topic focuses on the foraging behaviour of an armadillo.

4.6.1. Brief: Southern Three-Banded Armadillo

Supplied by Robert Kelly

The southern three-banded armadillo is a nocturnal species native to Central South America, inhabiting savanna and dry forest. They can use their distinctively long and powerful claws (Figure 9A) to dig through tree bark and termite mounds to forage for insects. Like their anteater relatives, these armadillos possess long and sticky tongues (Figure 9B), and shovel-like snouts to extract termites from small crevices and to root around in the forest floor for other insect prey. Due to their nocturnal traits, this species compensates for poor eyesight by being equipped with well-developed auditory and olfactory senses.

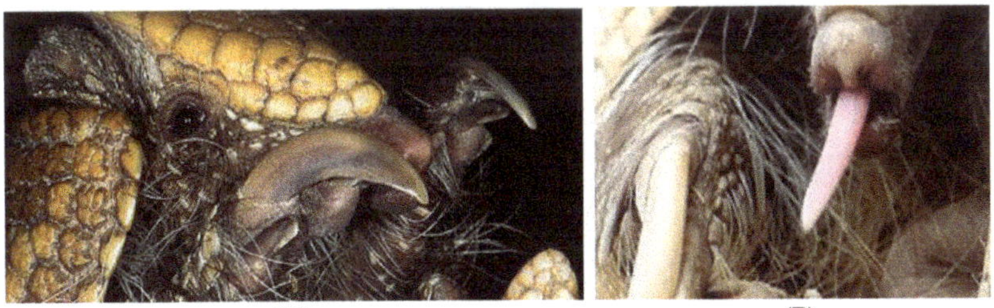

(A) (B)

Figure 9. Armadillo claws (**A**) and tongue (**B**). Image courtesy of Robert Kelly.

Despite being a commonly housed zoo animal, armadillos are comparatively understudied. They are primarily solitary but may often be housed in breeding pairs in the zoo.

Heated nest-boxes are normally provided as sleeping quarters and refuges. Armadillos are typically housed in reversed day–night systems to promote activity levels, although differences amongst species exist. Variation in ambient temperature influences levels of activity in armadillos. For example, lower air temperature in the southern three-banded armadillo is associated with decreased activity, whereas the inverse is observed in the six-banded armadillo (*Euphractus sexcinctus*), which becomes more active with decreasing temperatures [48].

As a terrestrial species, substrate provision is an important consideration. Typically bark chips are used which can promote natural digging and rooting behaviours. However, this means that smaller enrichment devices can sometimes be displaced or accidentally dug beneath the substrate. Armadillo claws are very powerful, and they can be destructive animals—they have the potential to dig through small, discrete gaps and are strong enough to excavate concrete, so teams should bear this in mind when considering construction materials and device design. Being accommodated in darkened environments, and due to their poor eyesight, armadillos may struggle to locate enrichment without some form of olfactory cue.

An adult male southern three-banded armadillo is currently housed solitary at Amazon World Zoo Park on the Isle of Wight (Figure 10). He occupies an enclosure of 34 cubic m (1200 cubic feet) with a coarse bark chip substrate. Logs are spaced randomly around the enclosure with a heated nest box at either end. Due to his old age, his eyesight is very poor, and he is particularly sensitive to loud noises and vibrations. Can teams devise an idea that promotes foraging activity, taking his age and requirements into account?

Figure 10. Geriatric southern three-banded armadillo. Photo courtesy of Robert Kelly.

4.6.2. Responses: Southern Three-Banded Armadillo

Since armadillos spend the majority of their time underground, this concept involved a modular raised tunnel system (Figure 11). The sections were joined with junctions, so staff could alter the routes periodically to make it more dynamic. There would be puzzle food boxes located at junctions, easy to fill from above. In theory, the armadillo would be able to smell which ones had treats, using his tongue to reach the food inside the containers. Sensors could monitor his behaviour and light up LEDs on the top of the tunnel as he passed so that visitors knew where he was at any given moment. Another possibility would be to build the tunnel from clear acrylic so people could watch in low light. This assumes that if the individual cannot see very well, the tactile and physical character of the tunnel would reassure him.

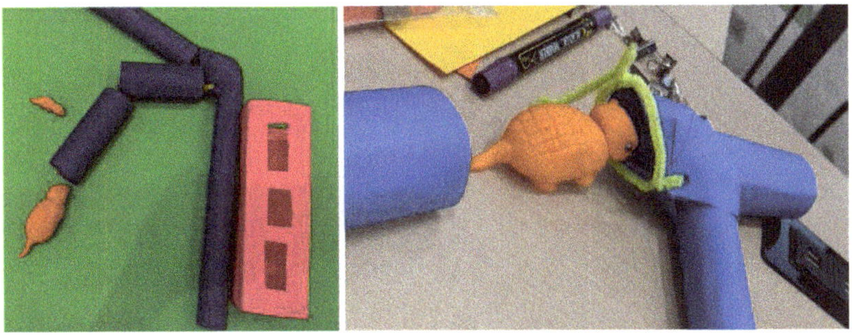

Figure 11. Model of the armadillo investigating a tunnel system with an adjacent play tunnel. Images from Moon Jam.

A solution to lengthening claws due to inactivity could also be to coat the architectural flooring surface with an epoxy-bonded abrasive grit such as 'stonhard' flooring beneath the bark mulch. To promote activity levels in zoo-housed armadillos and encourage animals to engage in species-specific behaviours, husbandry and enrichment programmes have been devised, but not consistently put into practice for a number of reasons, including complex organisational policies, excessive caution and difficulties judging effectiveness [49]. A comparison of the nine-banded armadillo (*Dasypus novemcinctus*), the Llanos long-nosed armadillo (*D. sabanicola*) and the southern naked-tailed armadillo (*Cabassous unicinctus*) suggests that food-based enrichment may be ineffective at changing armadillo activity periods, but is able to reduce abnormal behaviour and increase foraging behaviour [50], although further research using larger sample sizes is required to confirm this. Similarly, food-based enrichment was considered ineffective at promoting activity in the six-banded armadillo (*E.sexcinctus*), the large hairy armadillo (*Chaetophractus villosus*), and the southern three-banded armadillo (*T. matacus*) [51]. Armadillos compensate for poor eyesight with keen olfactory senses and can respond and discriminate between different scent-based cues [52]. Thus, scent-based elements could be considered within the enrichment design.

The Moon Jam brief for the armadillo generated a lot of ideas around potential solutions that could simultaneously engage zoo visitors. One idea involved a perforated tunnel wall that ran adjacent to a section of his tunnel—aimed at encouraging children to enter the darkened armadillo world, where they could both smell each other, for mutual olfactory stimulation. However, criticism of this was that 'parallel play' often results in loud, disruptive behaviour by participating children, who become lost in their own interactions, and, therefore, miss the educational value of the experience.

Other physical games included keeping completely still in a specific location for a short duration, to trigger a display of live infrared camera footage and various acoustic games that involved intense listening. There were also ideas for potential mobile or touchscreen games, such as armadillo maze puzzles and 'Spot the Armadillo', which involved trying to guess his location correctly as he moved along the tunnel system (when the LEDs dimmed). The computer games were all apps that could be developed without introducing anything novel to the armadillo enclosure, as a way of maintaining visitor engagement.

The hearing capability of armadillos is thought to be sensitive, and, currently, we have a limited understanding of their responses to auditory stimuli in captive environments. Therefore, acoustic enrichment would be a useful area of research to develop. Human sound pollution presents a different challenge within managed environments. Special design and construction materials are required to lessen external auditory disruption that might be caused by loud zoo visitors, ventilation and pumping equipment, after-hours concerts or occasional nearby construction projects [53–55]. Zoo-housed armadillos can be susceptible to stressors including increased handling for education purposes [56] and visitor presence.

We argue that more attention should turn to understudied nocturnal species, such as the armadillo, to ensure that husbandry and management practices to promote welfare can be established. This involves the cooperation and collaboration of zoo staff, managers, stewards, designers and researchers, and ideally would raise awareness of these animals to the wider public through visitor engagement. The following theme speaks to the importance of taking all stakeholders into account (humans and animals) when undertaking design projects.

4.7. Stakeholders

Developing artificial habitat design solutions requires an in-depth knowledge of the context, which in the case of zoos involves understanding the requirements and perspectives of staff (managers and stewards), visitors, designers, researchers and non-human animals. Moon Jam participants primarily focused on the animal briefs they were given, acknowledging after the workshop that they had considered nocturnal husbandry and enrichment goals and more-than-human aesthetics, notably sensory modalities, cognitive and physical characteristics, and social and environmental preferences of the species in question. In addition, they were tasked with producing a focused, feasible design that considered context (physical, cultural, geographical, social environment, time constraints and ease of use for human carers), logistics (skills and resources required, financial and time implications), zoo mission statements (education, conservation and entertainment objectives for visitors) and research potential (including design evaluation, iteration and testing with users, and publication).

The authors concluded that the length of the Moon Jam event precluded such a detailed analysis of the context for each brief, but that these important considerations would be part of future plans when moving forward from conceptual design to prototyping.

In relation to zoo management, the viability of enrichment should be assessed in terms of inputs (e.g., cost to build and maintain, including staff time) and outputs (e.g., the measurable effect on the species or individual, subsequent visitor engagement, opportunities for publicity around welfare, research dissemination and contribution to wildlife initiatives). The challenges, opportunities and needs of researchers can be supported through access to technology such as infrared video, tracking and recording devices [10], as well as developing their ideas for experimental design. Animal stewards are crucial to the success of any project since they are the fundamental link between humans and other species; they implement initiatives, collect data, interpret affective states and offer feedback on systems, as well as undertaking their usual caring responsibilities. They should therefore be included as contributors, along with animal carers, designers and researchers to any enrichment design discussion.

As mentioned earlier, visitor engagement is critical for a number of reasons: (i) it facilitates access to animals so that people better understand other species that share our world; (ii) it increases the footfall required to maintain the zoo as a viable financially independent organisation; (iii) it supports the funding of research and conservation projects and (iv) it can indirectly contribute to advertising since people will share their experiences on social media.

The Moon Jam brief Stop the Flashes aims to identify alternative schemes for supporting the visibility of nocturnal species in day–night reversal to visitors, which excludes the use of visitors' personal flashlights.

4.7.1. Brief: Stop the Flashes

Balancing the lighting needs of visitors and nocturnal animals under day–night reversal can create a challenge in managed environments which could amplify the use of visitors using phone flashes/flashlights. In some incidents, visitors may turn on their own lights to be able to move around more easily in dim light areas and we must acknowledge that not all visitors' eyesight will be the same as one another. In other cases, visitors could be tempted to use their phone lights to improve the visibility of species and proceed to take photos with flash, failing to understand the negative implications of such bright lights. It is important to balance the lighting needs of the visitors for safety and experience satisfaction whilst maintaining species' requirements for appropriate lighting and light cycles, which are critical for their biological rhythms, contributing to health and wellbeing.

4.7.2. Responses: Stop the Flashes

Participants drew on a combination of technological intervention and gamification techniques to dissuade visitors from using their phones. Two main concepts were suggested, the first of which involved installing special photochromic glass viewing windows with controllable properties. A bright light would trigger the window to go dark, simultaneously stopping everyone from taking photos. The use of peer pressure to stop people from being antisocial was deemed to be more effective than notices, so this concept worked as social engineering. This is an example of gamification techniques being deployed to manipulate human behaviour and was supported by zoo design colleagues.

The second idea was to develop an app that visitors could install on their phones. It would have access to the phone settings (such as location-based tracking) so it automatically restricted flash usage when they were inside the nocturnal house, but it had exciting benefits too. For example, visitors could access remote low-light cameras to observe animal activity from inside the enclosure on their phones or pads.

Visitor experience could also be enhanced if humans were able to transition more gradually from daylight to the interior of a nocturnal exhibit, by passing through a darkening corridor that enabled their eyes to adjust normally. One possible solution to extending the visitor eye adjustment time would be to organize the nocturnal species encountered, such that the route began with the most light-tolerant species and progressed to the species needing the darkest environments. Another way to accommodate human eye adjustment to darkened nocturnal areas in new constructions is to locate nocturnal exhibits connected to indoor mid-light level facilities such as reptile houses or museum areas. Visitors gradually adjust to mid-light levels while viewing reptiles, then adapt to lower light levels viewing nocturnal species. Returning via the mid-light reptile displays allows visitors to readjust their vision before exiting to outdoor sunlight levels. Alternatively, staff, interactive activities, or videos could be strategically placed at the beginning of exhibits to increase visitor time in adjustment zones. Ultimately, when nocturnal exhibits are designed to incorporate animal welfare, theming and visitor experience, attraction ratings are likely to reflect this. An example is the 85.2% 'very good' visitor rating of Singapore's night safari, an attraction dedicated to nocturnal species, exceeding ratings over ten other nature-based attractions in Singapore [57].

It is also important to consider how to handle visitor expectations, so staff need to find ways to be clever in how to offer engagement with nocturnal species whether that is through allowing visitors to view species across artificial sunrise and sunset times when these animals are more visible as previously discussed, by providing animal talks on these species from a trained member of staff who is able to locate the animal, or using video screens to show the behaviours of these animals that may or may not be live footage (such as cameras placed in view of enrichment).

5. Guidelines for Nocturnal Enrichment

This section offers a brief overview of the technology-enabled ideas we have explored that can support the wellbeing of nocturnal species, and some guidelines for developers.

5.1. Overview

The examples in the Moon Jam briefs (Figure 12) and associated responses in the Discussion section have illustrated how nocturnal species rely extensively on tactile, olfactory and auditory senses, are highly sensitive to variations in wavelength, intensity and direction of light, and can perceive fluctuations in heat, humidity, pressure and direction within the medium they inhabit (air or water) and also through their substrate and other environmental features. There are also other senses available to non-humans but apparently undetectable by humans, such as perception of electromagnetic fields, and senses not previously mentioned, such as taste and proprioception.

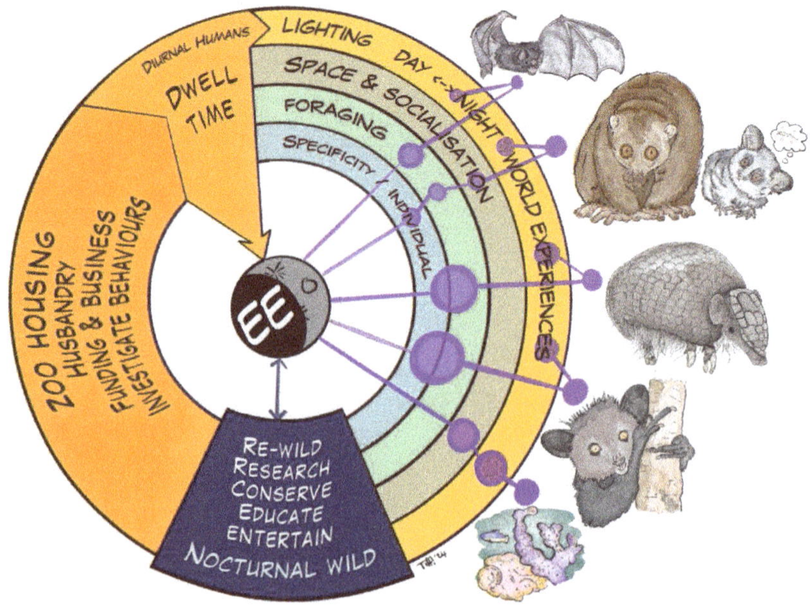

Figure 12. Husbandry and enrichment design framework for nocturnal species. The purple circular points correspond to the focal points of each species' brief (the larger the size, the more targeted the brief was to the consideration). Image designed by Tiff Leek.

In relation to lighting, technology can enable auto-transitions across day and night, simulating real-world conditions. Sunlight and moonlight can be emulated, with wavelength, angle and duration corresponding to species-specific global locations. Photochromic glass can be controlled to achieve different effects, from masking light to masking the presence of visitors on the other side of a barrier. In these scenarios, the technology supports both stewards and animals, by automating procedures to save human time and by using sophisticated techniques to recreate natural experiences. Other environmental changes, such as ocean tides and weather variability, can also be recreated using technology, to support realistic behavioural responses from the animals involved.

Tech can also be used to sense and track animals' use of space, through the use of radiofrequency identification (RFID) with tagged individuals, while gated systems can limit or permit access to certain locations. Sensors could, for example, reveal the whereabouts of animals underground to visitors, using lights above a tunnel, or through streaming infrared camera footage. Many species have round-the-clock behavioural needs, so cameras could be used to monitor activity budgets over a 24 h period.

In relation to foraging, devices can release food randomly, provide specific nutrition for particular individuals or be programmed to be triggered by both animals and humans. Tech would also enable the development of a system that mimicked live animals, to feed warm blood to bats, and the generation of artificial acoustic signals to lure animals to different resources and encourage activity. Many animals have individual needs, and technology can also be used to monitor their health, such as capturing data about bite strength or body temperature. The Moon Jam briefs did not include nocturnal avian and reptilian species, however, the ways in which technology can support the welfare of nocturnal species can apply to these animal groups.

5.2. Multispecies Interaction Design

It is important to note that animals co-evolved with environmental conditions directly affecting their welfare, by influencing their evolutionary strategies for survival and reproduction, notably feeding, social interactions and exploratory behaviour. Moreover, all living organisms experience and make sense of their world through their sensory perceptions, and many have sufficient autonomy to enact choices based on what they perceive. The opportunity to make meaningful choices within complex and manipulable artificial habitats offers managed animals control over their life experiences, which is central to animal welfare [58] and is, therefore, an important aspect of an enrichment plan. While some reactions may be 'hard-wired' (driven by evolved instincts), others may be based on individual preferences (e.g., favourite treats) or neurological reinforcement that has been established through previous experience (e.g., positive or negative reactions from conspecifics). In the wild, animals have the ability to choose between a wide range of different resources and experiences including microclimates, light levels and many more. They should have access to similar choices while in managed care environments.

The ability to perceive phenomena, discriminate and choose between different sensorial experiences and make these decisions based on previous and immediate personal experience is indicative of aesthetic sensibility [59,60]. This suggests that environments, resources and interactions with others can be more or less pleasurable for animals, depending on their perceptions and preferences. We therefore argue that designers should carefully consider species-specific characteristics in relation to the aesthetics of system design when developing new features or experiences for enclosures. This includes features such as the smell, taste, texture, malleability, colour, shape, sound, position and interaction associated with any device.

Games, toys and control systems with species-specific interfaces offer opportunities for cognitive and sensory stimulation to animals within managed environments, as well as autonomy and the chance to gain competence [61]. This applies to both humans and non-humans, providing zoo staff with interactive features they can control, visitors with engaging apps that educate in entertaining ways, researchers with ways to investigate animal preferences and capabilities and the animals with different means to learn new skills, make relevant decisions and work for rewards (contra-freeloading) within their enclosures.

Aside from feasibility and cost, key considerations to be made around the use of technology are related to (1) Ethics, (2) Messaging, (3) Usability, (4) Teamwork, (5) Futureproofing, (6) Biocentric design, (7) Evaluation and (8) Contextual relevance and broader applications.

1. Ethical issues are complex and a discussion of the many perspectives on animal welfare and management is beyond the scope of this paper. We point to the field of Animal–Computer Interaction, where there are many examples of literature that defines, and projects that exemplify animal-centred design principles. Examples include descriptions of design methodologies and frameworks that enable animals to be involved in the design process as contributors [62] and discussions around values beyond welfare, ergonomics or usability, such as privacy and consent [63]. Ultimately, it is the design team's responsibility to find ways to communicate the team's intentions with client animals and to interpret the animal's resulting responses.
2. Messaging relates to direct visitor–animal experiences as well as to signage and apps for zoo visitors, which can give a powerful signal about the attitudes and priorities of the establishment. For example, do the display techniques demonstrate human dominance over animals and the environment, or represent humans and animals as equally entitled residents of Earth? Could animals, apps and games be mistaken for human children's entertainment? Does the signage empower or trivialise human endeavours to support animals' lives or desires?
3. Usable systems are a fundamental requirement, whether being used by humans or animals. For humans, technology needs to be easy to learn and use. For non-humans, it is crucial for designers to gain a deep understanding of the species' natural history

and an individual animal's personal history. With such knowledge it should be possible to lever the animal's usual behaviour—the affordance of the system should include mechanisms, presentation, and aesthetic qualities. Being able to work with prototypes is essential, so as to iteratively test designs and modify them based on the animal's actions and reactions.

4. Simplicity in design is desirable but can be hard to achieve, and it may seem easier to rely on known technological solutions. Often, however, collaboratively working on challenges within a multidisciplinary team can generate simple, non-technical solutions that are cheap and easy to implement, such as the 'potto sleeves' that restrict bushbaby access to treats. Teamwork offers participants ownership of the design since everyone participates in its creation; this, in turn, is motivating for stakeholders and facilitates future deployment.
5. Futureproofing involves ensuring that technological solutions do not quickly become obsolete. It is appropriate to design flexible and adaptable systems that can easily be maintained or adjusted, provide support for human users and be willing to make changes as new knowledge becomes available or enclosures are updated. This concept also applies to advocacy around welfare and enrichment, meaning that today's 'best practice' may be considered a very low threshold in the future.
6. It is important for humans to accept that we do not know everything about other species. The state of knowledge we have today will inevitably be superseded in a few years, and corresponding welfare standards and practices for managing animals will also change. As it stands, biocentric design is the optimum approach for creating zoo enclosures—in other words, trying to recreate the environment in which the animal evolved, with as many of the relevant experiential features as possible. This involves the human design team trying to 'see the world with new eyes', which may involve the use of technology to expand our limited perceptions. Immersion in the umwelt of another species is an exciting prospect that can give researchers and designers insights that transcend their original context and facilitate the development of new knowledge and opportunities to better understand our ecology.
7. Feedback from Moon Jam participants suggested that there is a lack of available information on suitable methods for evaluating enrichment designs. Planning a research study requires viable and specific research questions. For example, we cannot ask: 'Is it successful?' about a new enrichment device without first defining our measure of 'success'. Traditional scientific papers are heavily biased towards collecting and interpreting quantitative data, but there is a growing appreciation of qualitative research, particularly in the early stages of a project, for identifying and refining problems, and later on, to collect stakeholders' perspectives, for example. Data does not need to represent a large population to be valid—investigating a small sample of a species, such as those individuals housed in one zoo, can lead to a greater understanding of that species and their needs. Mellen and MacPhee offered a framework for environment enrichment in 2001 involving Setting goals, Planning, Implementing, Documenting, Evaluating, and Readjusting (referred to as 'SPIDER'). Since then, evaluation techniques have been highlighted in the species-specific context of cheetahs [64], lemurs [65] and lizards [66], and for training animals [67]. Alligood and Leighty [68] discuss different trends and the UK organisation National Centre for the Replacement, Refinement and Reduction of Animals in Research (NC3Rs) offers a valuable husbandry guide for researchers [69].
8. Finally, everything we learn about a small sample of a species housed in a zoo will inform research and conservation projects with its wild counterparts and have relevance to similar animals in other contexts. For example, in domestic environments, many rodents are frequently kept as companion animals. Mice, chinchillas, rats and hamsters are all nocturnal, very sensitive to light and noise, and usually active at night and around dawn and dusk. They need safe places to hide since they are prey animals, and they are highly sociable in the wild (except for Syrian hamsters). The RSPCA

offers guidance on how best to look after rodents [70], but there is minimal legislation to ensure that pet owners treat their animals responsibly and provide appropriate welfare. Zoos offer an ideal opportunity to share information about nocturnal species with the wider public. Moreover, zoos' commitment to conservation, research, visitor education and the wellbeing of the species they house has the potential to make a global impact in a wide range of contexts for both wild and managed animals.

6. Conclusions

The lives of nocturnal and crepuscular animals in nature and in managed care have long been a mystery to scientific observers, caregivers and nature lovers. Their activities have been obscured from our human senses in darkness. Today, new and upgraded technologies such as infrared and motion-activated sensors, recorders, and cameras, coordinated with RFID identification, tracking and imaging systems, and Animal-Computer Interaction programs are improving our understanding of how nocturnal species live in their wild and managed environments and what they need to thrive. This information can be transferred into action to support good animal welfare and public display possibilities in captive environments. Technology-assisted husbandry and enrichment can be used as an interdisciplinary approach to benefit the wellbeing of nocturnal species whilst simultaneously educating human stakeholders on species' behaviour and ecology. Managed environments of nocturnal species may particularly benefit from artificial lighting and cameras as resources that permit and monitor the natural circadian cycle of nocturnal species, both when human carers are around and when they are not. Alongside technology, the continuing acquisition of new animal knowledge from wild research, evidence-based animal management, more detailed animal welfare assessments, and animal-centred habitat design frameworks, all suggest exciting opportunities to meet the needs of nocturnal species.

We explored some of these issues and techniques using the Moon Jam workshop process that brought together diverse participants to discuss and reflect on real challenges for nocturnal species housed in zoos. Workshop outcomes showed that a greater understanding of both species and individual animal senses, needs, preferences and motivations quickly led to testable concepts for improving nocturnal environments and offering enriching opportunities. Some were simple, such as potto arm-length sleeve feeders. Others could use existing commercial technology such as RFID smart pet gates and feeders. Still, others would require advanced technical design and testing. However, with the integration of Wi-Fi timers and applications synchronising weather data to outlet controls, these intricate systems may become as straightforward as downloading the appropriate app to manage all functionalities conveniently in the palm of one's hand. This user-friendly approach is certainly foreseeable soon.

The use of newer technology to enhance visitor experiences was not discussed as much in the workshop, but wearable infrared and starlight visors presently in use by the military and hunters could be adapted for use in zoos, sanctuaries and aquariums. Live infrared projections could reveal hidden animal activities, and now some smartphone cameras are adapted for extreme low-light conditions. Improved nocturnal habitats and management programs are likely to increase animals' natural activity, with the potential for improving both the health and wellbeing of the animals and the engagement of visitors.

In addition, rapid advancements in artificial intelligence techniques enable automated surveillance of animals. This decreases the need for human involvement, which can be stressful for other species, and also has the potential to offer positive welfare benefits. Machine Learning is being widely used to enable AI systems to recognise patterns in collections of data, therefore automating processes that previously required a human to spend a significant amount of time undertaking observations and analysing video recordings. Examples include (i) enhanced diagnoses of health conditions, (ii) the ability to monitor and interpret social behaviours and group dynamics and (iii) longitudinal passive data collection and analysis, to investigate seasonal variation. Moreover, facial or body recognition of individuals can enable the automation of bespoke feeding arrangements and access to areas of an enclosure, removing the requirement to tag animals.

While managed nocturnal animals were our focus, essential knowledge we used was gained in field studies. Continued threats to wild nocturnal species globally, namely habitat loss and climate change, signify the growing importance of improving the lives of managed animals such as through developing animal-centred technology to safeguard these species. Moreover, by overcoming the current limitations of managed environments, there may be more scope to support the foundations of zoo, sanctuary and aquarium-based breed-and-release programs for rewilding endangered nocturnal species to suitable protected areas in the future.

Author Contributions: Conceptualization, F.F., T.L., P.B., R.K. and J.C.C.; methodology, F.F., T.L., P.B., R.K. and J.C.C.; formal analysis, F.F., T.L., P.B., R.K. and J.C.C.; investigation, F.F., T.L., P.B., R.K. and J.C.C.; resources, F.F., T.L., P.B., R.K., J.C.C., L.C., E.C.L., E.M. and C.M.; writing—original draft preparation, F.F., T.L., P.B., R.K. and J.C.C.; writing—review and editing, F.F., P.B., R.K., J.C.C., T.L., L.C., E.M. and C.M.; visualization, F.F. and T.L.; project administration, F.F. All authors have read and agreed to the published version of the manuscript.

Funding: This research received no external funding.

Informed Consent Statement: Not applicable.

Data Availability Statement: The original contributions presented in the study are included in the article, further inquiries can be directed to the corresponding author/s.

Acknowledgments: We would like to thank the Moon Jam and Mini-Moon-Jam participants, without whose contributions this paper would not have been possible: ACI researchers Ceara Byrne (MIT), Ilyena Hirsky-Douglas (University of Glasgow), Clara Mancini (Open University), and Dave Roberts (NCSU); amazing postgraduate students Jiaqi Wang (University of Glasgow), Allison Able, Janet Brock, Grant Forbes, Nitish Gupta, Colt Nichols, Alex Raposo, Yusuf Satici, George Wang, and Yifan Wu (all NCSU), and Amina Anwar, Tehseen Fatima, Anoshia Fayaz, Matu Ferdinandus, Zahra Hajali, Ali Hassan, Nachiketa Jangid, Carla Joseph, Gurpreet Kaur, Yelena Makarov, Evropi Syriopoulou, and Sarp Uzulmez (London Metropolitan University). The briefs and corresponding photographs were provided by Paige Bwye, Robert Kelly, Jon Coe, Laura Carrigan (London Zoo), Cathy Mingee (GSC), Lindsey Zarecky and Jessica Hoffman (GSC), while Eric Mahan and Emily Lynch (GSC) contributed expert feedback and advice on outputs during the main Moon Jam event.

Conflicts of Interest: The authors declare no conflicts of interest.

References

1. Owens, A.C.; Lewis, S.M. The impact of artificial light at night on nocturnal insects: A review and synthesis. *Ecol. Evol.* **2018**, *8*, 11337–11358. [CrossRef]
2. Geerah, D.R.; O'Hagan, R.P.; Wirdateti, W.; Nekaris, K.A.I. The use of ultrasonic communication to maintain social cohesion in the Javan slow loris *(Nycticebus javanicus)*. *Folia Primatol.* **2019**, *90*, 392–403. [CrossRef] [PubMed]
3. Stidsholt, L.; Greif, S.; Goerlitz, H.R.; Beedholm, K.; Macaulay, J.; Johnson, M.; Madsen, P.T. Hunting bats adjust their echolocation to receive weak prey echoes for clutter reduction. *Sci. Adv.* **2021**, *7*, eabf1367. [CrossRef]
4. Moon Jam: Enhancing Welfare for Nocturnal Species. Available online: https://www.zoojam.org/moon (accessed on 1 May 2024).
5. ZooJam: Technology Supporting Enrichment. Available online: https://www.zoojam.org/ (accessed on 1 May 2024).
6. Bennie, J.J.; Duffy, J.P.; Inger, R.; Gaston, K.J. Biogeography of time partitioning in mammals. *Proc. Natl. Acad. Sci. USA* **2014**, *111*, 13727–13732. [CrossRef] [PubMed]
7. Cox, D.T.; Gaston, K.J. Cathemerality: A key temporal niche. *Biol. Rev.* **2024**, *99*, 329–347. [CrossRef] [PubMed]
8. Wang, G.; Zhu, Z.; Shi, P.; Zhang, Y. Comparative genomic analysis reveals more functional nasal chemoreceptors in nocturnal mammals than in diurnal mammals. *Chin. Sci. Bull.* **2010**, *55*, 3901–3910. [CrossRef]
9. Meyer, W.B.; Turner, B.L. Human population growth and global land-use/cover change. *Annu. Rev. Ecol. Syst.* **1992**, *23*, 39–61. [CrossRef]
10. Whitham, J.C.; Miller, L.J. Using technology to monitor and improve zoo animal welfare. *Anim. Welf.* **2016**, *25*, 395–409. [CrossRef]
11. Voorhees, C. Animal creating stimulating enrichment: Functional enrichment and for captive animals. Observing and assessing their use. In *Exploring Animal Behavior in Laboratory and Field*; Zimbler-DeLorenzo, H., Margulis, S.W., Eds.; Academic Press: Amsterdam, The Netherlands, 2021; pp. 337–339.
12. Podturkin, A.A. In search of the optimal enrichment program for zoo-housed animals. *Zoo Biol.* **2021**, *40*, 527–540. [CrossRef]

13. Cannon, T.H.; Heistermann, M.; Hankison, S.J.; Hockings, K.J.; McLennan, M.R. Tailored Enrichment Strategies and Stereotypic Behavior in Captive Individually Housed Macaques (*Macaca* spp.). *J. Appl. Anim. Welf. Sci.* **2016**, *19*, 171–182. [CrossRef]
14. de Azevedo, C.S.; Lima, M.F.F.; Cipreste, C.F.; Young, R.J.; Rodrigues, M. Short Report: Belo Horizonte Zoo: Greater Rhea: Environmental Enrichment & Behavioural Changes. *Int. Zoo Yearb.* **2013**, *47*, 163–170. [CrossRef]
15. Hamilton, J.; Fuller, G.; Allard, S. Evaluation of the Impact of Behavioral Opportunities on Four Zoo-Housed Aardvarks (*Orycteropus afer*). *Animals* **2020**, *10*, 1433. [CrossRef]
16. Learmonth, M.J.; Sherwen, S.; Hemsworth, P.H. Assessing preferences of two zoo-housed Aldabran giant tortoises (*Aldabrachelys gigantea*) for three stimuli using a novel preference test. *Zoo Biol.* **2021**, *40*, 98–106. [CrossRef] [PubMed]
17. Animal-Computer Interaction Conference. Available online: https://www.aciconf.org/aci2023 (accessed on 2 May 2024).
18. Lin, C.; Takahashi, A.; Mulla, A.; Nozawa, Y. Moonrise timing is key for synchronized spawning in coral. *Biol. Sci.* **2021**, *118*, e2101985118. [CrossRef]
19. Flôres, D.E.; Jannetti, M.G.; Improta, G.C.; Tachinardi, P.; Valentinuzzi, V.S.; Oda, G.A. Telling the seasons underground: The circadian clock and ambient temperature shape light exposure and photoperiodism in a subterranean rodent. *Front. Physiol.* **2021**, *12*, 738471. [CrossRef] [PubMed]
20. Poon, L.; Jenks, I.T.; Crampton WG, R. MoonShine: A software-hardware system for simulating moonlight ground illuminance and re-creating artificial moonlight cycles in a laboratory environment. *Methods Ecol. Evol.* **2024**, *15*, 701–715. [CrossRef]
21. Fuller, G.; Raghanti, M.A.; Dennis, P.M.; Kuhar, C.W.; Willis, M.A.; Schook, M.W.; Lukas, K.E. A comparison of nocturnal primate behavior in exhibits illuminated with red and blue light. *Appl. Anim. Behav. Sci.* **2016**, *184*, 126–134. [CrossRef]
22. Shechter, A.; Kim, E.W.; St-Onge, M.; Westwood, A. Blocking nocturnal blue light for insomnia: A randomized controlled trial. *J. Psychiatr. Res.* **2018**, *96*, 196–202. [CrossRef] [PubMed]
23. Deichmann, J.L.; Gatty, C.A.; Navarro, J.M.A.; Alonso, A.; Linares-Palomino, R.; Longcore, T. Reducing the blue spectrum of artificial light at night minimises insect attraction in tropical lowland forest. *Insect Conserv. Divers.* **2021**, *14*, 247–259. [CrossRef]
24. Liu, J.A.; Meléndez-Fernández, O.H.; Bumgarner, J.R.; Nelson, R.J. Effects of light pollution on photoperiod-driven seasonality. *Horm. Behav.* **2022**, *141*, 105150. [CrossRef]
25. Baines, F.M.; Chattell, J.; Dale, J.; Garrick, D.; Gill, I.; Goetz, M.; Skelton, T.; Swatman, M. How much UVB does my reptile need? The UV-Tool, a guide to the selection of UV lighting for reptiles and amphibians in captivity. *J. Zoo Aquar. Res.* **2016**, *4*, 42–63.
26. Lopez, J.; Wormell, D.; Rodríguez, A. Preliminary evaluation of the efficacy and safety of a UVB lamp used to prevent metabolic bone disease in pied tamarins *Saguinus bicolor* at Jersey Zoo. *Dodo (Trinity)* **2001**, *37*, 41–49.
27. EReefs Portal. Available online: https://ereefs.aims.gov.au/ereefs-aims/gbr4/temp-wind-salt-current (accessed on 2 May 2024).
28. NOAA. Available online: https://tidesandcurrents.noaa.gov/products.html (accessed on 2 May 2024).
29. Mellor, D.J. Enhancing animal welfare by creating opportunities for positive affective engagement. *N. Z. Vet. J.* **2015**, *63*, 3–8. [CrossRef]
30. Hoy, J.M.; Murray, P.J.; Tribe, A. The potential for microchip-automated technology to improve enrichment practices. *Zoo Biol.* **2010**, *29*, 586–599. [CrossRef]
31. Coe, J.C.; Hoy, J. Choice, control, and computers: Empowering wildlife in human care. *Multimodal Technol. Interact.* **2020**, *4*, 92. [CrossRef]
32. Kirchgessner, M.L.; Sewall, B.J. The impact of environmental, social, and animal factors on visitor stay times at big cat exhibits. *Visit. Stud.* **2015**, *18*, 150–167. [CrossRef]
33. White, B.C.; Houser, L.A.; Taylor, S.; Elliott, J.L.L. Activity-Based Exhibition of Five Mammalian Species: Evaluation of Behavioral Changes. *Zoo Biol.* **2003**, *22*, 269–285. [CrossRef]
34. Clark, F.E.; Melfi, V.A. Environmental enrichment for a mixed-species nocturnal mammal exhibit. *Zoo Biol.* **2012**, *31*, 397–413. [CrossRef]
35. Andrén, H.; Linnell, J.D.; Liberg, O.; Andersen, R.; Danell, A.; Karlsson, J.; Odden, J.; Moa, P.F.; Ahlqvist, P.; Kvam, T.; et al. Survival rates and causes of mortality in Eurasian lynx (*Lynx lynx*) in multi-use landscapes. *Biol. Conserv.* **2006**, *131*, 23–32. [CrossRef]
36. Pereira, J.A.; Fracassi, N.G.; Rago, V.; Ferreyra, H.; Marull, C.A.; McAloose, D.; Uhart, M.M. Causes of mortality in a Geoffroy's cat population—A long-term survey using diverse recording methods. *Eur. J. Wildl. Res.* **2010**, *56*, 939–942. [CrossRef]
37. Campera, M.; Brown, E.; Imron, M.A.; Nekaris, K.A.I. Unmonitored releases of small animals? The importance of considering natural dispersal, health, and human habituation when releasing a territorial mammal threatened by wildlife trade. *Biol. Conserv.* **2020**, *242*, 108404. [CrossRef]
38. Puehringer-Sturmayr, V.; Fiby, M.; Bachmann, S.; Filz, S.; Grassmann, I.; Hoi, T.; Janiczek, C.; Frigerio, D. Effects of food-based enrichment on enclosure use and behavioral patterns in captive mammalian predators: A case study from an Austrian wildlife park. *PeerJ* **2023**, *11*, e16091. [CrossRef]
39. Bashaw, M.J.; Gibson, M.D.; Schowe, D.M.; Kucher, A.S. Does enrichment improve reptile welfare? Leopard geckos (*Eublepharis macularius*) respond to five types of environmental enrichment. *Appl. Anim. Behav. Sci.* **2016**, *184*, 150–160. [CrossRef]
40. Seyrling, I.; Dierkes, P.W.; Burger, A.L. Diurnal and Nocturnal Behaviour of Cheetahs (*Acinonyx jubatus*) and Lions (*Panthera Leo*) in Zoos. *Animals* **2022**, *12*, 2367. [CrossRef]
41. Bashaw, M.J.; Bloomsmith, M.A.; Marr, M.J.; Maple, T.L. To hunt or not to hunt? A feeding enrichment experiment with captive large felids. *Zoo Biol. Publ. Affil. Am. Zoo Aquar. Assoc.* **2003**, *22*, 189–198. [CrossRef]

42. Ruskell, A.D.; Meiers, S.T.; Jenkins, S.E.; Santymire, R.M. Effect of bungee-carcass enrichment on behavior and fecal glucocorticoid metabolites in two species of zoo-housed felids. *Zoo Biol.* **2015**, *34*, 170–177. [CrossRef]
43. Johansson, E. *The Impact of Food Enrichment on the Behaviour of Turtles in Captivity*; Swedish University of Agricultural Sciences: Uppsala, Sweden, 2017; pp. 1–39.
44. Goswami, S.; Patel, S.K.; Kadivar, R.; Tyagi, P.C.; Malik, P.K.; Mondol, S. Effects of a combined enrichment intervention on the behavioural and physiological welfare of captive Asiatic lions (*Panthera leo persica*). *Appl. Anim. Behav. Sci.* **2021**, *236*, 105222. [CrossRef]
45. Boccacino, D.; Maia, C.M.; Santos, E.F.; Santori, R.T. Effects of environmental enrichments on the behaviors of four captive jaguars: Individuality matters. *Oecol. Aust.* **2018**, *22*, 63–73. [CrossRef]
46. Dickinson, E.; Kolli, S.; Schwenk, A.; Davis, C.E.; Hartstone-Rose, A. DiceCT analysis of the extreme gouging adaptations within the masticatory apparatus of the aye-aye (*Daubentonia madagascariensis*). *Anat. Rec.* **2020**, *303*, 282–294. [CrossRef]
47. Harvey, S.B.; Alworth, L.C.; Blas-Machado, U. Molar malocclusions in pine voles (*Microtus pinetorum*). *J. Am. Assoc. Lab. Anim. Sci.* **2009**, *48*, 412–415.
48. Attias, N.; Oliveira-Santosa, L.G.R.; Fagan, W.F.; Mourão, G. Effects of air temperature on habitat selection and activity patterns of two tropical imperfect homeotherms. *Anim. Behav.* **2018**, *140*, 129–140. [CrossRef]
49. Tuite, E.K.; Moss, S.A.; Phillips, C.J.; Ward, S.J. Why Are Enrichment Practices in Zoos Difficult to Implement Effectively? *Animals* **2022**, *12*, 554. [CrossRef] [PubMed]
50. Duarte, A.C.; Trujillo, F.; Superina, M. Behavioral responses of three armadillo species (Mammalia: Xenarthra) to an environmental enrichment program in Villavicencio, Colombia. *Zoo Biol.* **2016**, *35*, 304–312. [CrossRef] [PubMed]
51. Kelly, R.; Rose, P.E. Assessing the impact of environmental enrichment on behavior in understudied armadillo species: A case study. *Zoo Biol.* **2024**, *43*, 100–109. [CrossRef] [PubMed]
52. Vincent, J.L.; Vonk, J. Aroma-dillo or Area-dillo? An examination of armadillos' sensory modality bias. *Behav. Process.* **2022**, *202*, 104751. [CrossRef] [PubMed]
53. Orban, D.A.; Soltis, J.; Perkins, L.; Mellen, J.D. Sound at the zoo: Using animal monitoring, sound measurement, and noise reduction in zoo animal management. *Zoo Biol.* **2017**, *36*, 231–236. [CrossRef] [PubMed]
54. Cronin, K.A.; Bethell, E.J.; Jacobson, S.L.; Egelkamp, C.; Hopper, L.M.; Ross, S.R. Evaluating mood changes in response to anthropogenic noise with a response-slowing task in three species of zoo-housed primates. *Anim. Behav. Cogn.* **2018**, *5*, 209–221. [CrossRef]
55. Soltis, J.; Orban, D.; Adams, K.; Mellen, J.; Perkins, L. Sensory environments for zoo animals: Measuring anthropogenic sounds, light, and ground vibration in a multi-species exhibit. In Proceedings of the Jacksonville Zoo Animal Welfare Symposium, Jacksonville, FL, USA, 22–27 January 2018.
56. Baird, B.A.; Kuhar, C.W.; Lukas, K.E.; Amendolagine, L.A.; Fuller, G.A.; Nemet, J.; Willis, M.A.; Schook, M.W. Program animal welfare: Using behavioral and physiological measures to assess the well-being of animals used for education programs in zoos. *Appl. Anim. Behav. Sci.* **2016**, *176*, 150–162. [CrossRef]
57. Henderson, J.C.; Koh, A.; Soh, S.; Sallim, M.Y. Urban environments and nature-based attractions: Green tourism in Singapore. *Tour. Recreat. Res.* **2001**, *26*, 71–78. [CrossRef]
58. Kagan, R.; Veasey, J. Challenges of zoo animal welfare. In *Wild Mammals in Captivity: Principles and Techniques for Zoo Management*; Kleiman, D.G., Thompson, K.V., Kirk Baer, C., Eds.; University of Chicago Press: Chicago, IL, USA, 2010; pp. 11–21.
59. French, F. Expanding aesthetics. *Front. Vet. Sci.* **2022**, *9*, 855087. [CrossRef]
60. Webber, S.; Cobb, M.L.; Coe, J.C. Welfare through Competence: A Framework for Animal-Centric Technology Design. *Front. Vet. Sci.* **2022**, *9*, 885973. [CrossRef] [PubMed]
61. Kelling, N.J.; Gaalema, D.E.; Kelling, A.S. A modified operational sequence methodology for zoo exhibit design and renovation: Conceptualising animals, staff, and visitors as interdependent coworkers. *Zoo Biol.* **2014**, *33*, 336–348. [CrossRef]
62. van der Linden, D. Animal-centered design needs dignity: A critical essay on ACI's core concept. In *ACI '22, Proceedings of the Ninth International Conference on Animal-Computer Interaction, Newcastle-upon-Tyne, UK, 5–8 December 2022*; ACM: New York, NY, USA, 2022; p. 2. ISBN 9781450398305.
63. Ruge, L.; Mancini, C. An Ethics Toolkit to Support Animal Centered Research and Design. *Front. Vet. Sci.* **2022**, *9*, 891493. [CrossRef] [PubMed]
64. Quirke, T.; O'Riordan, R.M. Evaluation and interpretation of the effects of environmental enrichment utilizing varying degrees of sampling effort. *Zoo Biol.* **2013**, *32*, 262–268. [CrossRef] [PubMed]
65. Fernandez, E.J.; Timberlake, W. Selecting and testing environmental enrichment in lemurs. *Front. Psychol.* **2019**, *10*, 482182. [CrossRef]
66. Waterman, J.O.; McNally, R.; Harrold, D.; Cook, M.; Garcia, G.; Fidgett, A.L.; Holmes, L. Evaluating environmental enrichment methods in three zoo-housed Varanidae lizard species. *J. Zool. Bot. Gard.* **2021**, *2*, 716–727. [CrossRef]
67. Fernandez, E.J. Training as enrichment: A critical review. *Anim. Welf.* **2022**, *31*, 1–12. [CrossRef]
68. Alligood, C.; Leighty, K. Putting the "E" in SPIDER: Evolving trends in the evaluation of environmental enrichment efficacy in zoological settings. *Anim. Behav. Cogn.* **2015**, *2*, 200–217. [CrossRef]

69. NC3Rs. Available online: https://nc3rs.org.uk/3rs-resources/evaluating-environmental-enrichment (accessed on 2 May 2024).
70. RSPCA Rodents. Available online: https://www.rspca.org.uk/adviceandwelfare/pets/rodents (accessed on 2 May 2024).

animals

MDPI

Communication

Behavioral Coding of Captive African Elephants (*Loxodonta africana*): Utilizing DeepLabCut and Create ML for Nocturnal Activity Tracking

Silje Marquardsen Lund [1,*,†], Jonas Nielsen [1,†], Frej Gammelgård [1,*,†], Maria Gytkjær Nielsen [1,†], Trine Hammer Jensen [1,2] and Cino Pertoldi [1,2]

1 Department of Chemistry and Bioscience, Aalborg University, Frederik Bajers Vej 7H, 9220 Aalborg, Denmark; jnielc21@student.aau.dk (J.N.); maniek21@student.aau.dk (M.G.N.); cp@bio.aau.dk (C.P.)
2 Aalborg Zoo, Mølleparkvej 63, 9000 Aalborg, Denmark
* Correspondence: siljemarquardsenlund@gmail.com (S.M.L.); fgamme21@student.aau.dk (F.G.)
† These authors contributed equally to this work.

Simple Summary: This paper presents a way to automate computer vision processes applied to behavior recognition on closed-circuit television (CCTV) footage of two captive African elephants. Object detection software using both Create ML and DeepLabCut was used to control the accuracy of using such models, and those models were subsequently used to analyze seven days' worth of nighttime footage to assess the general behavioral patterns of the elephants, showcasing the possibility of using automated tools for behavioral analysis.

Abstract: This study investigates the possibility of using machine learning models created in DeepLabCut and Create ML to automate aspects of behavioral coding and aid in behavioral analysis. Two models with different capabilities and complexities were constructed and compared to a manually observed control period. The accuracy of the models was assessed by comparison with manually scoring, before being applied to seven nights of footage of the nocturnal behavior of two African elephants (*Loxodonta africana*). The resulting data were used to draw conclusions regarding behavioral differences between the two elephants and between individually observed nights, thus proving that such models can aid researchers in behavioral analysis. The models were capable of tracking simple behaviors with high accuracy, but had certain limitations regarding detection of complex behaviors, such as the stereotyped behavior sway, and displayed confusion when deciding between visually similar behaviors. Further expansion of such models may be desired to create a more capable aid with the possibility of automating behavioral coding.

Keywords: machine learning; nocturnal behavior; computer vision; captive elephants

1. Introduction

1.1. *Objectives of Wildlife Conservation*

The World Association of Zoos and Aquaria (WAZA) aims to conserve endangered species through breeding programs and exchange of captive animals. This association requires certain standards of its members and emphasizes the importance of welfare among captive animals [1]. Based on consistent and current research, good physical and psychological welfare among captive animals must be maintained and therefore associations such as this play an important role in conservation [1–4].

Because of the importance of animal welfare, it is essential to ensure that captive animals are consistently studied in relation to their behavioral reactions to different aspects of their captive lives, such as enclosure design, enrichment, and much more [5–8]. An example of an animal that may require such studies to properly conserve the species is the

African elephant (*Loxodonta* sp.), which has faced great declines in population size all over Africa [5,9].

1.2. Captive Elephant Behavior and Welfare

To accommodate the welfare needs of captive elephants, normal behaviors must first be monitored and understood [6,10,11]. Behaviors such as foraging, locomotion, social behavior, etc., likely influence the welfare of elephants and help understand undesired behaviors that may indicate stress [12–14]. It is important to also address the nocturnal behavior of elephants since night behavior can differ from behaviors observed during the day. An example of one such behavior is recumbent sleep, where the elephants lay down and sleep, which exclusively occurs during the night [14,15]. On average, captive African elephants lie down to sleep around two hours per night and tend to lie down more if their bedding is comfortable [14–17]. Besides rest, feeding and atypical behaviors are the behaviors most commonly observed in captive elephants, yet the activity level is lower at night compared to during the day [14,18].

Atypical behaviors observed in animals kept in captivity are usually those that deviate from the norm for their species and are not commonly observed in their natural habitat. These behaviors are frequently regarded as signs of compromised welfare [19–22]. A type of stereotypic behavior is characterized by the consistent and inappropriate repetition of specific movements or body postures. These actions seemingly lack any purpose or function and appear to be coping mechanisms to reduce stress, but the exact causes remain unclear [21–25]. Different forms of stereotypic behavior have been observed in elephants, including whole-body movements [10,13,22,26,27]. Among whole-body movements, 'swaying' is most common and is defined as a rhythmic side-to-side movement of the body, typically observed while standing [12,18,22]. Consistent observations of elephants may be helpful in handling these behaviors when they arise.

1.3. Machine Learning as a Tool for Behavioral Analysis

One widely used method for observation of animal behavior is videography [14,28], although this method can be highly time-consuming [29–32]. Manual scoring is limited by human capabilities, such as the observer not recognizing behavioral patterns or failing to spot new patterns. At the same time, it is difficult to standardize the scoring of behaviors by human observers due to subjectivity [33]. Inconsistency between different observers can therefore not be avoided completely [31]. Furthermore, it is challenging to track multiple animals and behaviors at the same time, despite the use of video material [29,31]. Some of these logistical problems with behavioral analysis may be aided with the use of machine learning [32].

Machine learning used in video and image analysis (computer vision) has been explored in recent years in application to a variety of purposes [34,35]. The use of object detection as a machine learning tool to find and recognize a given object has been investigated previously and used to recognize animals and their behaviors [32,36–38]. Tools such as this may prove useful as a way of automating behavioral analysis in the near future, which may reduce the workhours required of the researcher [30,32]. However, these uses and methods are still in their infancy which necessitates further investigation of different machine learning models, methods, and implementations [36,37,39].

DeepLabCut is a Machine Learning software that specializes in pose estimation in video material, by using points for tracking specific body parts [40,41]. This is utilized in behavior tracking by marking body parts of interest on a relatively small dataset of images showing a diverse range of behaviors by the subject of interest [32]. Constructing a DeepLabCut model capable of accurately tracking body parts of interest may allow for automatization of behavior coding in behavioral studies [42].

Create ML is a built-in application for iOS products with the ability to train custom machine learning models such as object detection models with no code. Create ML models can be trained to detect and recognize objects of interest, such as an elephant executing a specific behavior, by annotating a relatively small and diverse dataset. This annotation can be used in various ways, such as using the image annotation tool RectLabel for marking boxes, polygons, or skeletons on the subject, and categorizing the behavior. This may allow for construction of a simple model that can recognize simple behaviors with relatively user-friendly software [43].

1.4. Aim of This Paper

This paper aims to use DeepLabCut and Create ML to construct models capable of tracking selected body parts and classifying elephant behaviors, aiming to streamline and automate behavioral analysis processes, thus ultimately alleviating the workload of researchers and zookeepers, and standardizing behavioral coding. This study simultaneously examines nocturnal activity of two captive African elephants with the following hypotheses:

- This study expects that the machine learning models can predict selected behaviors on the same level as manual scoring;
- This study expects that behavioral differences between the elephants and behavioral differences between days can be demonstrated using selected computer vision models.

2. Materials and Methods

2.1. Subjects and Enclosure

The behavior of two captive female African elephants, exhibited in Aalborg Zoo, Denmark, was examined. Both elephants were born wild in South Africa around 1982 and relocated to Aalborg Zoo in 1985. In this study, the elephants are referred to as Subject A and B.

The elephant enclosure consisted of an indoor area and an outdoor area. The elephants did not have access to the outdoor enclosure at night during the examined period. The indoor enclosure consisted of concrete floors and walls with metal wires towards the visitor area (see Appendix A). The two elephants were able to have physical contact during the night through metal bars between enclosures E1 and E2. Subject A had access to enclosure E1, and a corridor attached to the enclosure (56 square meters). Subject B had access to enclosures E2 and E3 as well as a corridor measuring a total of 116 square meters.

The elephants' diet consisted of branches, seed grass hanging from nets in the inside and outside enclosures, and concentrate pellets that would periodically be released into the enclosure from a timer-automated mechanism, as a type of enrichment. Fresh fruits and vegetables were spread around their outdoor enclosure daily, which allowed for foraging behaviors. Foraging boxes, accessible with use of their trunks at the back wall of their enclosure, were also opened periodically throughout each night, controlled by a timer.

2.2. Data Collection

Prior to data collection, an ethogram with selected behaviors was made. The behaviors were determined based on similar behavioral studies of the same elephants in previous publications, namely Bertelsen et al. (2020) and Andersen et al. (2020), and were modified for the purpose of this study [14,44]. This ethogram was used to conduct a manual control, using researchers experienced in behavioral coding, where the researchers were assigned a subject each. The behaviors were coded continually on a second-by-second basis. These scorings were used to compare manual coding with the models; see Table 1.

Table 1. Ethogram used for behavioral coding of the two subjects, used for both manual and automatic coding.

Behavior	Description
Standing	The elephant is standing or walking. This behavior is the default if no other selected behavior is taking place.
Lying down	The elephant is lying down on the floor of the enclosure.
Drinking	The elephant is drinking from a water bowl.
Foraging	The elephant is using the foraging boxes, accessed using trunks in the holes at the back of the enclosure.
Hay-net	The elephant is using its trunk to reach the hay-net at the top of the enclosure.
Swaying	The elephant is swaying from side to side for at least 5 s.
Out of view	The elephant is out of view of the camera. This may also include falsely unlabeled frames by the machine learning models.

The data were collected from the 12th to 18th of March 2024 to use for model creation, and from the 16th to the 22nd of April 2024 for analysis purposes, using three cameras (ABUS, 25 FPS). The three cameras were placed at the visitor viewing side facing towards each enclosure (see Appendix A). The two elephants were observed during the night for seven hours from 22:00 to 05:00 (DST).

2.3. Data Analysis

The analyzed days in this experiment provided data which were compared with statistical tests using Excel (Version 2404 Build 16.0.17531.20120) and RStudio (R version 4.1.2 (1 November 2021)).

For body part tracking, DeepLabCut (version 2.3.9.) was used [32,42]. Specifically, 244 frames taken from 70 30-min videos were labelled (95% were used for training), and no preprocessing was performed. A ResNet-50-based neural network was used with the parameters set to 400,000 training iterations. Validation was carried out with a single shuffle, and the test error found was 17.44 pixels, train 2.4 pixels (image size for creating the model was 1920 by 1080). A p-cutoff of 0.5 was used to condition the x- and y-coordinates for future analysis. DeepLabCut does not provide annotations of behavior, and it is required of the user to manually define and interpret the coordinates of the results. In this study, the behaviors were classified in Excel using parameters set by comparing the coordinates of relevant body parts with the manually recorded data and videos from the control period (April 16). Each parameter consisted of distinct coordinate limits and requirements to fit the ethogram, resulting in frame-by-frame behavioral coding. The locations of the selected behaviors can be seen in Appendix B. The algorithm for classifying behaviors using DeepLabCut can be seen below in Algorithm 1.

Algorithm 1. DeepLabCut operation

Input: Video containing the subject to be tracked
1. **Load dataset.** Load the video into DeepLabCut;
2. **Define keypoints.** Specify keypoints of interest (body parts like head, tail, limbs);
3. **Annotate frames.** Annotate a subset of frames manually by marking the keypoints;
4. **Model training.** Use annotated frames to train the pose estimation model;
5. **Pose estimation.** Apply the trained model to new video material;
6. **Refine model (optional).** Correct predictions and retrain the model;
7. **Process coordinates.** Extract CSV file and filter the coordinates for desired body parts;
8. **Classify behaviors.** Set limits for each coordinate corresponding to a desired behavior and filter frames that fulfill the criteria.

Output: Subset of data points that can be classified as a specific behavior.

Difficulties arose with defining some of the parameters, such as 'Drinking', which proved undefinable for both subjects as the label on the trunk tip was unstable. Furthermore, no distinct parameter was definable for the behavior 'Hay-net' for Subject B as the foraging box and the hay net were located at approximately the same place in the video frame.

For object detection, Create ML (version 5.0 (121.1)) was used [43]. Specifically, 370 image frames were extracted from the model creation period and annotated using bounding boxes in RectLabel Pro (version 2023.11.19). Each frame was annotated with a selected behavior as seen in Table 1. 'Swaying' was not labelled for the object detection part. The behavior 'Standing' was labelled 232 times, 'Foraging' was labelled 42 times, 'Lying down' was labelled 39 times, 'Drinking' was labelled 39 times, and 'Hay-net' was labelled 18 times. The dataset was split into two sets, one containing images for training and one containing images for validation. The model was trained with 6000 iterations, and the training set had an accuracy of 95% whilst the validation set had an accuracy of 75%. Create ML automatically classifies the behavior using bounding boxes resulting in a frame-by-frame behavioral coding output. The algorithm for classifying behaviors using Create ML is seen below in Algorithm 2.

Algorithm 2. Create ML operation

Input: Video containing the subject to be tracked.
1. Load dataset. Load the video into RectLabel and extract images;
2. Define bounding boxes. Specify categories of each bounding box of interest (behaviors, such as lying down or standing);
3. Annotate frames. Annotate a subset of frames manually by drawing bounding boxes;
4. Model training. Use annotated frames to train the model;
5. Pose estimation. Apply the trained model to new video material;
6. Refine model (optional). Correct predictions and retrain the model;
7. Process coordinates. Extract CSV file containing frames annotated with behaviors.
Output: Dataset containing the predicted behaviors at all analyzed frames.

To appraise the accuracy of the models designed with Create ML and DeepLabCut, a control was analyzed manually from the video footage from the 16th of April and used to compare the models' results. To test the accuracy of the models compared to the control, a confusion matrix was conducted. In this case, a multiple class confusion matrix was produced and analyzed [45]. This will give insight into where the model performs well and has a high accuracy, as well as where mistakes occur, such as when the model mislabels a behavior. The columns of such a matrix represent the manually observed behavior of the individual while the rows represent the predicted behavior from the models. Time budgets and cumulative graphs were also used to further appraise the models [46]. Furthermore, Kendall's Coefficient of Concordance was used to measure the agreement between the models and the control [47].

The behavioral analysis consisted of time budgets, cumulative graphs, and Kendall's Coefficient of Concordance between days. Time budgets were made for each elephant each day and for the whole study period. This was carried out using the sums, transformed into percentages, of observed time spent on each behavior, where the out of view percentage made up the time where no behavior was observed or classified. Time budgets for each day were used to investigate daily differences in behavior. The time budgets for the whole study period were used to see how much time was spent on each behavior in total and to compare the different models with the control. Kendall's Coefficient of Concordances were used on the data from the time budgets to analyze the similarity between the observed behavior between different days.

Cumulative graphs were made for each behavior each day. The graphs were made with both the manually recorded data and the model data for the control day, while the rest of the study period only had cumulative graphs made for the Create ML model.

A Spearman rank correlation test was used to investigate correlations between the subject's 'Lying Down' behavior from night to night. Correlations were also used to test the similarities between the subjects and if they exhibited similar behavioral patterns throughout the night.

The possibility to observe the stereotypic behavior 'Swaying' was also examined. This was accomplished by calculating the Euclidian distance between a given point of the trunk root, labelled by the DeepLabCut model, and the succeeding point [48]. The distances between the points were calculated and plotted as a cumulative graph together with the actual cumulative time spent on the sway behavior, so that sway could be observed as steep increases in the cumulative sum.

3. Results

3.1. Comparability of Manual and Automatic Behavioural Observations

3.1.1. A General Overview

First, the capabilities of the two machine learning models, compared to a manually conducted analysis using an ethogram, have been displayed using time budgets, illustrated below in Figure 1.

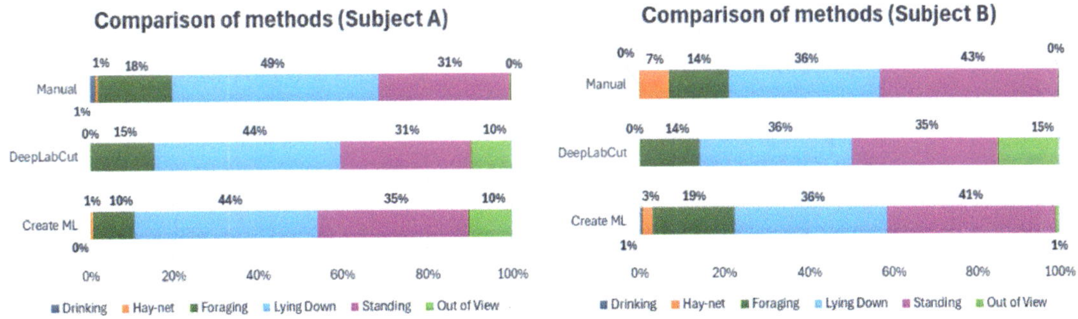

Figure 1. Time budgets for Subjects A and B of the 7 h control period, comparing manual observations with a DeepLabCut and a Create ML model. 'Drinking' is excluded for the DeepLabCut model.

The time budgets for both subjects, using both models, showed similar percentages to the manual observations regarding standing and lying down with relatively small differences. For Subject A, both models had a 10% out of view percentage. Since the manual observations showed an out of view percentage close to zero, this indicates that these were caused by uncertainty by the models, causing them to not label the subject. Both models for Subject A also displayed a lower percentage for 'Foraging' than the manual observations. For Subject B there were notable differences in 'Foraging' and 'Hay-net', which is likely to be caused by the hay-net being close to the foraging boxes, thus showing overlapping coordinates when observed by the DeepLabCut model. The Create ML model also had a lower 'Hay-net' percentage, but contrarily a higher 'Foraging' percentage than the manual observations. The out of view percentage for the DeepLabCut model was noticeably higher than the Create ML model, considering that the manual observations showed an out of view percentage close to zero. This was likely also caused by lack of labelling by the model. 'Drinking' was left out of the DeepLabCut model due to limitations in defining the parameters of this behavior, caused by a lack of consistent appropriate labels needed to properly categorize the behavior.

To further investigate the similarities between the machine learning models and the manually conducted observations, cumulative graphs for each subject, showing each behavior tracked with each method, have been constructed and displayed below in Figure 2.

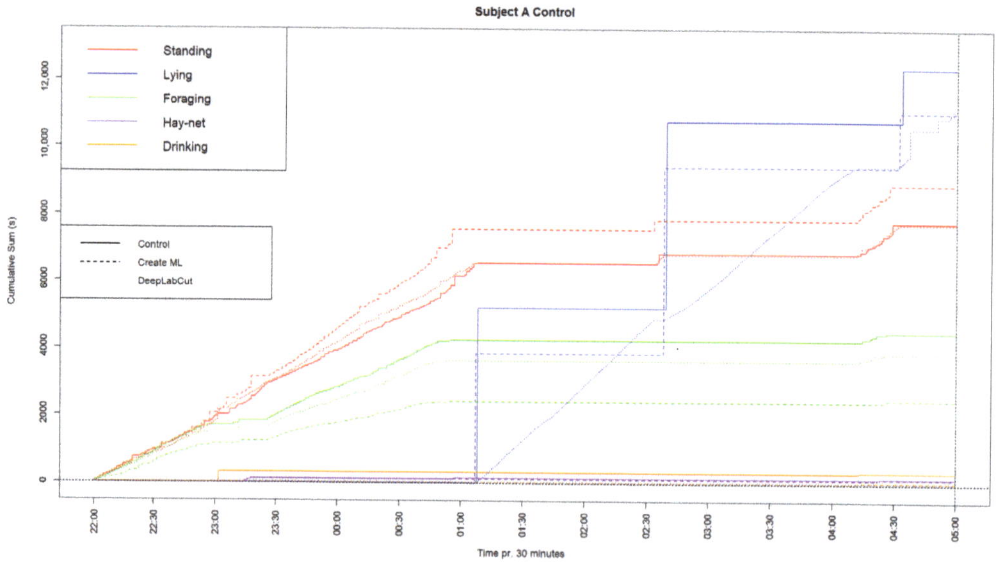

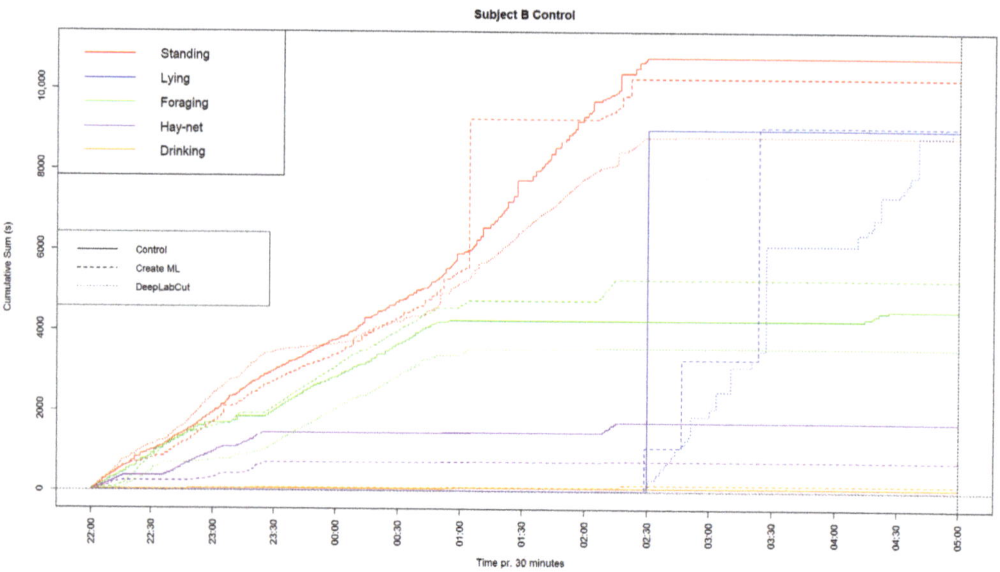

Figure 2. Graph showing the cumulative sums of each behavior for Subjects A and B during the 7 h control period, observed manually, using Create ML and DeepLabCut. Behaviors are distinguished by color and method.

The cumulative graph for Subject A showed a lower sum but similar shape for 'Lying down', which indicates that a period of observations of this behavior went unlabeled for both models. 'Standing' had a similar shape for all methods and a very similar sum for the DeepLabCut model, whereas the Create ML model had a higher sum. 'Foraging' had similar shapes for all methods; however, the sums were generally lower for the models. 'Hay-net' and 'Drinking' were difficult to distinguish clearly, due to low values.

The cumulative graph for Subject B had very similar values for 'Lying down' for all methods. All methods showed similar shapes for 'Standing', although the sums were lower for both models, most noticeably for the DeepLabCut model. 'Foraging' also had similar shapes but higher and lower sums for the Create ML model and DeepLabCut model, respectively. 'Hay-net' also showed a somewhat similar shape to the manual observations for the Create ML model, but this model had a lower sum. The DeepLabCut model showed a 'Hay-net' sum close to zero, due to difficulty in defining these parameters in the enclosure.

Similarly to the time budgets, 'Drinking' was left out of the cumulative graphs for the DeepLabCut model due to limitations in defining the parameters. Furthermore, the shapes in 'Lying down' for both subjects seem largely different; however, this is caused by the chosen type of cumulative graph.

3.1.2. Investigating the Reliability of Two Machine Learning Models

To investigate the reliability of the two models, Kendall's Coefficient of Concordance was utilized. The concordance (W-value) between the models and the control was found to be 0.85 with a p-value of 0.026 for Subject A, and 0.90 with a p-value of 0.019 for Subject B. This indicates a high concordance between the models that is not stochastic for both subjects. This high concordance means that the models and the manual scoring mostly agree on the observed behavior. However, this concordance is not perfect, so a slight disagreement is present.

To test the accuracy of the models, a confusion matrix for the control period was made and the models were compared to the manually observed values and normalized (Appendix C). As seen in Tables A1–A4, both models predicted highly similar values for the behavior 'Lying Down' for both subjects compared to the manually observed values. For the DeepLabCut model, the predicted values for 'Standing' and 'Foraging' were highly similar to the observed values for Subject A; but for Subject B, the model was more inaccurate. The opposite applies for Create ML, where the model was generally most accurate for Subject B. For the behavior 'Hay-net', the DeepLabCut model struggled to predict the correct behavior for Subject A, and for Subject B the classification parameters were not defined, and therefore no value was predicted for this behavior. The Create ML model for the behavior 'Hay-net' predicted highly similar values for Subject A, but for Subject B the behavior was often misclassified as 'Foraging'. For the behavior 'Drinking', the Create ML model often predicted the behavior as 'Standing' for both subjects and for the DeepLabCut model the classification parameters were not defined and therefore no values were predicted correctly. Finally, the predicted values by both models for out of view for Subject B were highly similar to the observed value but it is notable that the total manually observed value for this behavior is 17 s out of 7 h of observation time and is therefore arguably negligible.

3.2. *Using Machine Learning Models for Behavioural Analysis*

3.2.1. Assessing Behavioral Differences

To analyze behavioral differences between the two subjects, two time budgets for the total sums of each behavior for seven nights were constructed for both machine learning models, as is seen below in Figure 3.

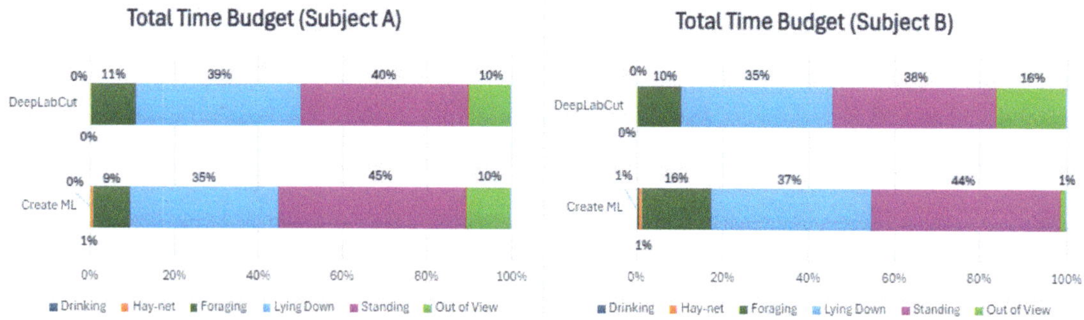

Figure 3. Time budgets for the total time spent on each behavior for Subjects A and B during all seven observed nights, for both ML models. The different colors show different behaviors.

The time budgets display some differences between the models, especially noticeable in the out of view percentages for Subject B. 'Standing' and 'Lying down' were similar for both models for both subjects. 'Foraging' was similar for Subject A in both models but was slightly higher for the Create ML model for Subject B, possibly due to the lower out of view percentage. The DeepLabCut model did not measure the 'Drinking' behavior for either subject. Comparing the total time budgets for the period between the two subjects only showed slight differences.

To further investigate the behaviors of both subjects during the observed period, the sums of behaviors for all individual days have been shown as time budgets in Appendix D. The time budgets for Subject A showed some variation in the behaviors, especially in 'Foraging'. 'Standing' and 'Lying down' also varied somewhat from night to night. Subject B also showed variation in 'Foraging', but generally less so than Subject A. 'Standing' and 'Lying down' varied somewhat for Subject B. There was a noticeable difference in out of view percentages for Subject B, depending on the model, with a consistently much lower percentage for Create ML. The behavioral differences were examined further using cumulative graphs (Appendix E).

Kendall's Coefficient of Concordance was used to examine if the amount of time spent on each behavior was the same each day. The analysis was conducted on the results of both the DeepLabCut and the Create ML model. Subject A showed a concordance of 0.935 with a p-value of 4.29×10^{-6} for the DeepLabCut model and 0.865 with a p-value of 1.31×10^{-5} for the Create ML model. Subject B showed a concordance of 0.951 with a p-value of 3.3×10^{-6} for the DeepLabCut model and 0.869 with a p-value of 1.21×10^{-5} for the Create ML model. All the concordance values were high with a significant p-value indicating that the high concordance is not stochastic. This high concordance indicates an agreement in the observed time a subject spends on different behaviors from day to day. This concordance is not perfect meaning some variations are still present in the subjects' nocturnal behavior.

Spearman's rank correlation for the behavior 'Lying down' was investigated for both models, and the analysis was split between days and individuals (Appendix F).

The analysis between days resulted in mainly positive correlations. Subject A had correlations between 0.965 and −0.053 for DeepLabCut and 0.970 and −0.135 for Create ML. Negative correlations were observed between the 20th and 22nd of April for DeepLabCut and the 19th and 20th of April for Create ML. Subject B had correlations between 0.999 and 0.019 for DeepLabCut and 1.000 and 0.311 for Create ML. No negative correlations were found for Subject B.

The analysis between the two subjects also resulted in mainly positive correlations (Appendix G). The results of the DeepLabCut model had correlations between 0.975 and −0.361. A single negative correlation was found between the 20th of April for Subject A and the 19th of April for Subject B. The results of the Create ML model had correlations between 0.977 and 0.052. No negative correlations were found for the Create ML model.

3.2.2. Investigating Further Applications of Automatic Behavioral Coding

Certain behaviors of a more complex character may potentially be assessed using machine learning methods for behavioral coding. One such behavior is the stereotyped behavior 'Swaying', which is largely relevant for elephants [14]. This behavior is difficult to categorize simply, as has been carried out for the previously mentioned behaviors, since swaying can happen anywhere in the frame and is primarily observable through a side-to-side motion of the elephant's trunk and head. To visualize this behavior using data from the DeepLabCut model, the cumulative distance moved by the point labelled at the trunk root of Subject B was plotted, along with the actual sway noted manually in the control period as a cumulative graph in Figure 4.

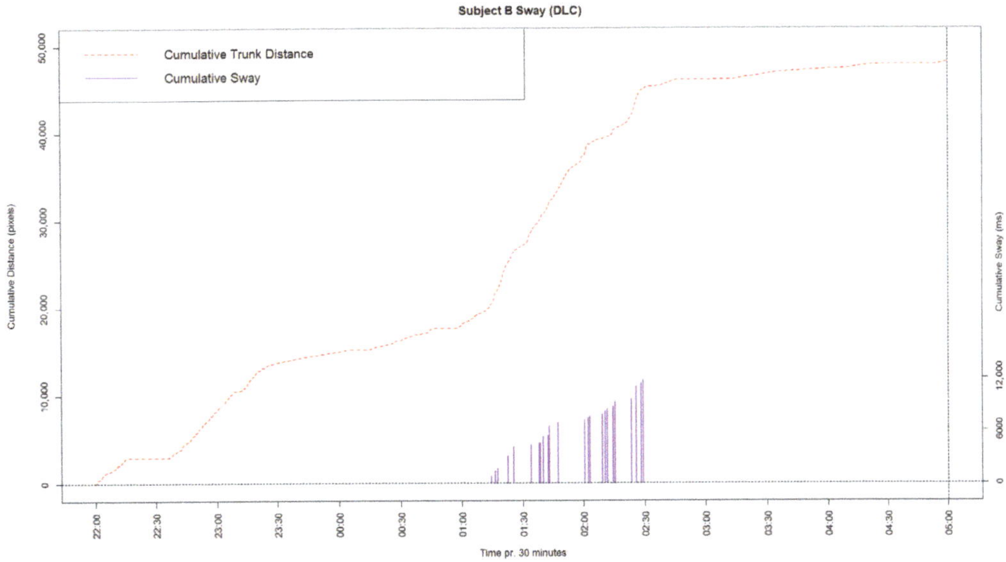

Figure 4. Cumulative graph of trunk root movement (red) and manually observed sway (purple) in the control period. The left y-axis shows cumulative pixel movement; the right y-axis shows the manually coded sway behavior in milliseconds.

As seen in the cumulative graph, a steep increase in the trunk distance appears around 01:15, which approximately matches with the manually observed sway behavior occurring at this time. This is because a steep increase in distance moved by the trunk root will occur as a result of the sway behavior.

4. Discussion

4.1. Performance and Limitations of the Two Machine Learning Models

Before using the two constructed machine learning models, it must first be investigated how accurate they are compared to manually recorded observations. The concordance test between the models showed a high agreement between the models and manual observations (W = 0.848), although there is seemingly room for improvement. The confusion matrixes for each model compared to the manual observations also displayed high accuracy in detecting some behaviors, such as 'Standing' and 'Lying down', although with certain challenges, such as confusing the 'Hay-net' behavior with 'Foraging' for Subject B. This suggests that there may be a need for improving the parameters of classifying each behavior or training the models with better or more material to account for different environments, footage qualities, and a broader range of behaviors. Mathis et al. (2018) discussed the capabilities and limitations of DeepLabCut for various behaviors, noting similar challenges in behavior recognition and out of view instances [32].

It must, however, be noted that the accuracy of these models is based on comparison with manual observations, which itself has inherent problems. Manual observations are not entirely accurate, since they may lack precision in noting the exact time a behavior takes place, and there may be differences in how a sequence of behaviors is coded by different researchers, which usually necessitates inter-rater reliability tests [41,49,50]. These issues should not be present in machine learning models since a behavior is coded at the exact frame and can be standardized across studies. The capabilities of models constructed in both Create ML and DeepLabCut thus emphasize the potential of machine learning models to complement and enhance traditional manual observations in behavioral studies. Further optimization of such computer vision models may also include image processing such as exploring different color spaces and image augmentations [51,52].

4.2. Nocturnal Behavioral Differences of the Two Subjects

The two machine learning models were used to automatically observe the two subjects of the study, with the aim of assessing whether the nocturnal behaviors varied between each subject and individually across each observed night. This analysis was carried out using time budgets, cumulative graphs, and correlations.

Firstly, the total time budgets for each subject across all nights displayed only slight differences between the subjects, most notably in foraging behavior, which might be higher for Subject B. This slight difference is supported by the high concordance values between the days, meaning the observed behavior only differs slightly from day to day. It is, however, inconclusive whether Subject B generally carries out more foraging behavior, due to the differing out of view percentages caused by the lack of confidence by the models. With a closer look at the behavioral patterns using the cumulative graphs (Appendix E), it does, however, appear that the two subjects have some differences. Once again, it appears that 'Foraging' is generally higher for Subject B, along with 'Drinking'. From the 'Lying down' cumulative graph it also appears that sleeping patterns may be somewhat different, since Subject A appears to wake up and walk around more commonly throughout the night, whereas Subject B appears to be lying down for longer periods at a time. The correlations calculated regarding the sleeping patterns compared between the two subjects also showed some correlation, indicating that the subjects go to sleep at similar times, although this varies each night. This shows that the two elephants differ from each other in the nocturnal behavioral patterns; however, there do not appear to be large differences, and overall, the subjects typically carry out somewhat similar behavioral patterns throughout the night. This is in accordance with Bertelsen et al. (2020) which studied the same subjects and found some personality differences displayed through behavior, but similarly the differences were relatively small [14]. A study by Rees (2009) also found behavioral differences on an individual basis in captive Asian elephants (*Elephas maximus*) [10]. This is similar to Tobler (1992), Holdgate et al. (2016), and Schiffmann et al. (2023) who examined recumbent sleep behavior in zoo-housed Asian and African elephants, and also found differences on an individual basis [15,17,53].

It was investigated whether the individual subjects differed in their behavioral patterns from night to night, using time budgets for each night, along with cumulative graphs and correlations of their sleeping behavior. From the time budgets, both subjects appear to vary from night to night in all behaviors. 'Foraging' ranges from very low percentages (1–2%) to high percentages (18–22%), which is also apparent from the cumulative graph. This is in accordance with the study by Finch et al. (2021) who found varying feeding behavior in their nocturnal activity budgets for zoo-housed Asian elephants [54]. 'Standing' and 'Lying down' for both subjects also differed across nights with a range of approximately 20% difference for both behaviors. Further investigation of sleeping patterns using Spearman rank correlations showed that most days had very highly correlated values, indicating a general circadian rhythm; this is in accordance with a study by Casares et al. (2016) that investigated cortisol levels to establish the circadian rhythm of African elephants [55]. However, some days showed much weaker correlation, suggesting differences in night-to-night sleeping patterns caused by the elephants going to sleep at different times. This confirms the hypothesis that computer vision models are capable of demonstrating that the nocturnal behavioral patterns differ from night to night for both individuals, although there is some uncertainty of exactly how much the behaviors differ, due to the out of view percentages. This result is, however, in accordance with the studies by Rees (2009) and Holdgate et al. (2016) who found considerable day-to-day variation in activity budgets for a group of captive Asian elephants and for African and Asian elephants, respectively [10,15].

This study showed that the subjects were lying down approximately 35–39% of the observed time, or just over 2.5 h until 5:00, at which point they would still be lying down, as is seen in the cumulative graphs. This is in accordance with studies such as Holdgate et al. (2016) and Schiffmann et al. (2023) who found that elephants in captivity generally tend to lie down for a similar amount of time during the night; although they also note that the elephants may not be sleeping throughout all of this time [15,17].

4.3. Other Applications of Machine Learning Models for Behavioral Coding

As was displayed in the results regarding the sway behavior, it may be difficult to address complex behaviors, even though it may be possible through different techniques. One such technique was displayed by plotting the distance moved by a point on the trunk root of Subject B. The steep incline on the graph largely matches with the manually observed sway during the night, which indicates that using such a measurement might be useful for observing 'Swaying'. Currently, the presentation of this behavior is, however, primarily visual since it may provide further challenges to precisely define the parameters capable of properly discerning when the gathered data should be categorized as sway behavior. Other challenges related to addressing complex behaviors, such as stereotypic and obsessive self-grooming in primates, face similar challenges since this behavior is also classified by a consistent repetition, which a computer vision model would have difficulties identifying on a frame-by-frame level. However, a study by Yin et al. (2024) is somewhat successful in showing distinct motion trajectories exhibited by a variety of different animals including tigers (*Panthera tigris*), bears (*Ursidae*), and wolves (*Canis lupus*), and classifying this as stereotypic behavior by assessing repetitive patterns [56]. Tackling issues that may arise using computer vision for behavior recognition is somewhat a case-by-case problem, where camera settings, image processing, and other issues should be considered, in order to fit the models appropriately to the research. However, developing such parameters and applying them to similar machine learning model data in the future would prove useful to quickly and accurately find stereotyped behavior to gain insight into the welfare of individual animals.

5. Conclusions

It is apparent from this study that using machine learning models from DeepLabCut and Create ML provides a capable tool for aiding or even replacing certain aspects of behavioral studies. The models could detect simple behaviors with high accuracy, although limitations were met when assessing repetitive behaviors such as 'Swaying'. Similar complex behaviors, such as certain stereotypic behaviors, in other animals may prove equally challenging and the detection of it may require further work to be adequate.

Applying the models to seven nights of footage of nocturnal behavior provided general insight into behavioral patterns and differences between the two studied subjects, as well as differences between individually observed days. This showed that the constructed computer vision models can effectively aid in behavioral analyses, and further expansion and adjustments may be desired. This could potentially be achieved through exploring image augmentation, classification of complex behavioral patterns, or implementing such models to be readily available as a tool for zoological gardens.

Author Contributions: Conceptualization, S.M.L., J.N., F.G., M.G.N., C.P. and T.H.J.; methodology, S.M.L., J.N., F.G. and M.G.N.; validation, S.M.L., J.N., F.G. and M.G.N.; formal analysis, S.M.L., J.N., F.G. and M.G.N.; investigation, S.M.L., J.N., F.G. and M.G.N.; data curation, S.M.L., J.N., F.G. and M.G.N.; writing—original draft preparation, S.M.L., J.N., F.G. and M.G.N.; writing—review and editing, S.M.L., J.N., F.G., M.G.N., C.P. and T.H.J.; visualization, S.M.L., J.N., F.G. and M.G.N.; supervision, C.P. and T.H.J. All authors have read and agreed to the published version of the manuscript.

Funding: Funding for this study was provided by the Aalborg Zoo Conservation Foundation (AZCF; grant number 07-2024).

Institutional Review Board Statement: The Ethical Review Board was not consulted for the purposes of this study, as this study did not interfere with the daily routines of the studied subjects, and solely involved passive observation through video footage.

Informed Consent Statement: We obtained approval from Aalborg Zoo, and the study guarantees all work was carried out within good animal welfare and ethical circumstances. There was no change in daily routines for the animals of concern.

Data Availability Statement: The data presented in this study are available on request from the corresponding author.

Acknowledgments: We would like to thank the employees at Aalborg Zoo for facilitating this study, especially Anders Rasmussen, Paw Fonager Gosmer, and Marianne (My) Eskelund Reetz. Special thanks to Kasper Kystol Andersen for assistance with the technical aspects. Lastly, we would like to thank Simeon Lucas Dahl for technical support.

Conflicts of Interest: The authors declare no conflicts of interest.

Appendix A. Floor Plan of Elephant Enclosure

Figure A1. Illustration of the elephant enclosure in Aalborg Zoo. The indoor enclosures are colored yellow and have the enclosure names written with the dimensions. The dashed lines indicate the outdoor enclosure and the part of the wall between enclosures E1 and E2 where the subjects can have physical contact at night. The locations and positions of the cameras are indicated with the camera icons. The illustration is a modification from Bertelsen et al., 2020 [14].

Appendix B. Selected Behaviors Displayed in Subject Enclosures

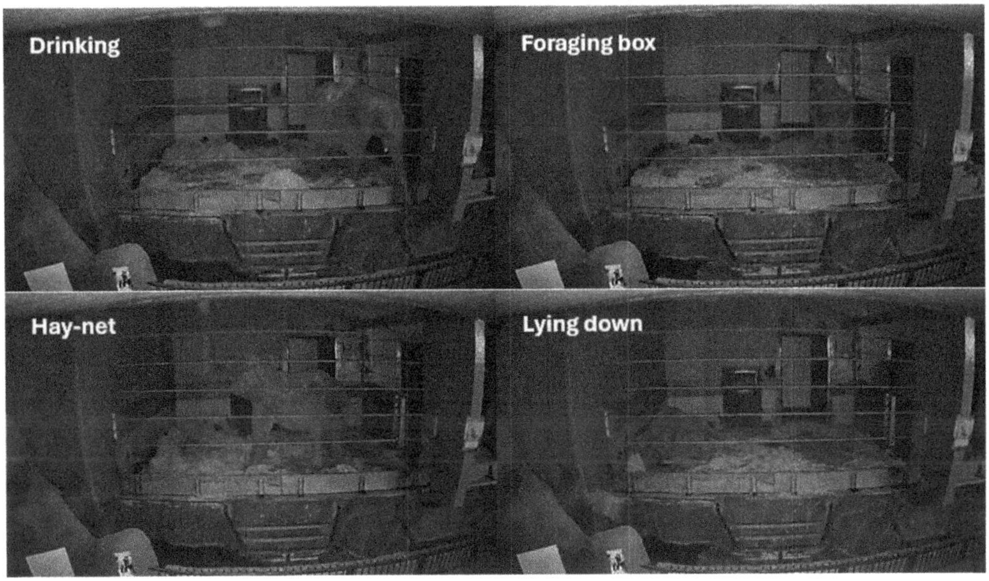

Figure A2. Selected behaviors displayed in the subject enclosure E1 as observed through video monitoring. The images showcase four distinct behaviors: drinking, foraging at the box, using the hay-net, and lying down.

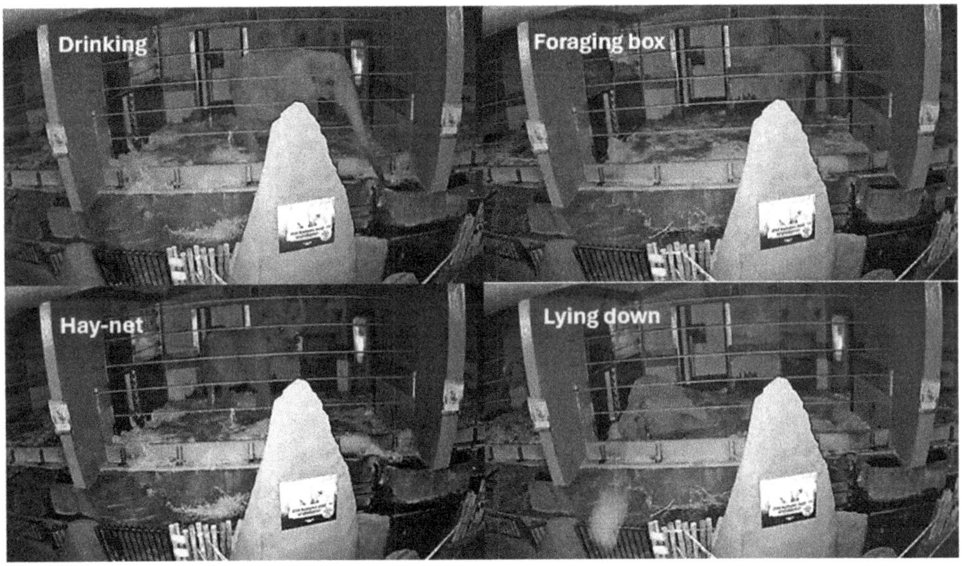

Figure A3. Selected behaviors displayed in the subject enclosure E2 as observed through video monitoring. The images showcase four distinct behaviors: drinking, foraging at the box, using the hay-net, and lying down.

Appendix C. Confusion Matrix Comparison of Models

Table A1. Confusion matrix produced from the manually observed values compared to the predicted values from the DeepLabCut model for Subject A. Darker blue indicates better prediction.

Subject A		**Manually Observed**					
	Behavior:	**Standing**	**Lying down**	**Foraging**	**Drinking**	**Hay-net**	**Out of view**
Predicted by DeepLabCut model	**Standing**	0.86820	0.00016	0.13258	1.00000	0.39926	
	Lying down	0.00036	0.89418				
	Foraging	0.00701		0.84651			
	Drinking						
	Hay-net	0.00005				0.16883	
	Out of view	0.12437	0.10567	0.02091		0.43190	

Table A2. Confusion matrix produced from the manually observed values compared to the predicted values from the DeepLabCut model for Subject B. Darker blue indicates better prediction.

Subject B		**Manually Observed**					
	Behavior:	**Standing**	**Lying down**	**Foraging**	**Drinking**	**Hay-net**	**Out of view**
Predicted by DeepLabCut model	**Standing**	0.64320	0.00170	0.10185	0.76200	0.58339	0.18203
	Lying down	0.02517	0.99657		0.06867	0.00271	
	Foraging	0.02384		0.66607		0.14721	
	Drinking						
	Hay-net	0.00016					
	Out of view	0.30763	0.00173	0.23209	0.16933	0.26669	0.81797

Table A3. Confusion matrix produced from the manually observed values compared to the predicted values from the Create ML model for Subject A. Darker blue indicates better prediction.

Subject A		**Manually Observed**					
	Behavior:	**Standing**	**Lying down**	**Foraging**	**Drinking**	**Hay-net**	**Out of view**
Predicted by Create ML model	**Standing**	0.94253		0.29061	0.84503	0.07534	
	Lying down	0.00051	0.88842				
	Foraging	0.00718		0.53114			
	Drinking	0.00038			0.07602		
	Hay-net	0.00051				0.91096	
	Out of view	0.04887	0.11158	0.17825	0.07895	0.01370	

Table A4. Confusion matrix produced from the manually observed values compared to the predicted values from the Create ML model for Subject B. Darker blue indicates better prediction.

Subject B		**Manually Observed**					
	Behavior:	**Standing**	**Lying down**	**Foraging**	**Drinking**	**Hay-net**	**Out of view**
Predicted by Create ML model	**Standing**	0.91524	0.00121	0.04310	0.36667	0.06907	0.29412
	Lying down		0.99868				
	Foraging	0.05509		0.94939		0.50390	
	Drinking	0.00890	0.00011		0.61667		
	Hay-net	0.00130			0.01667	0.38559	
	Out of view	0.01948		0.00751		0.04144	0.70588

Appendix D. Time Budgets for Each Night

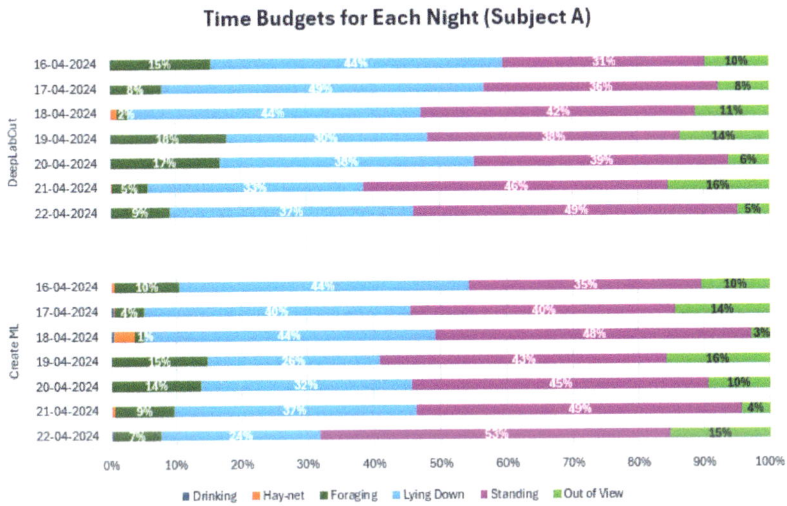

Figure A4. Time budgets for both models across each observed night for Subject A. Each color represents a behavior. Percentages were left out for 'Drinking' and 'Hay-net' due to low values.

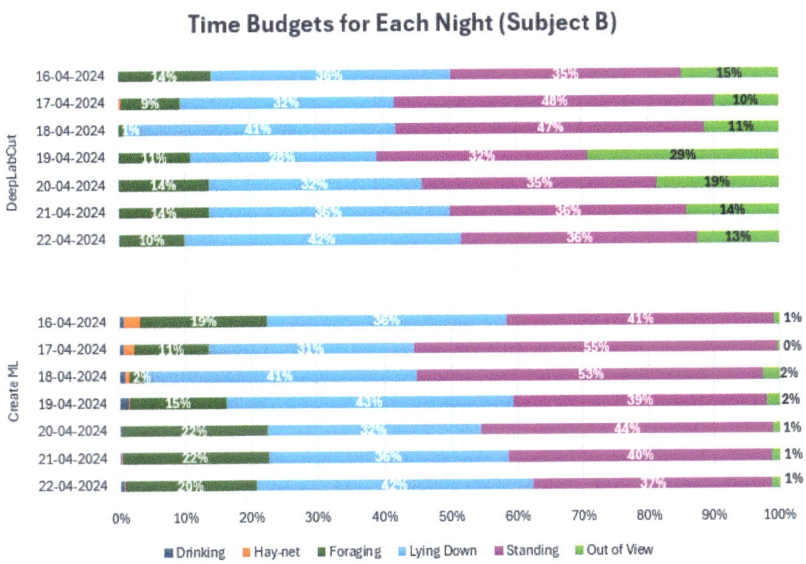

Figure A5. Time budgets for both models across each observed night for Subject B. Each color represents a behavior. Percentages were left out for 'Drinking' and 'Hay-net' due to low values.

Appendix E. Cumulative Graphs for Every Behavior Each Night

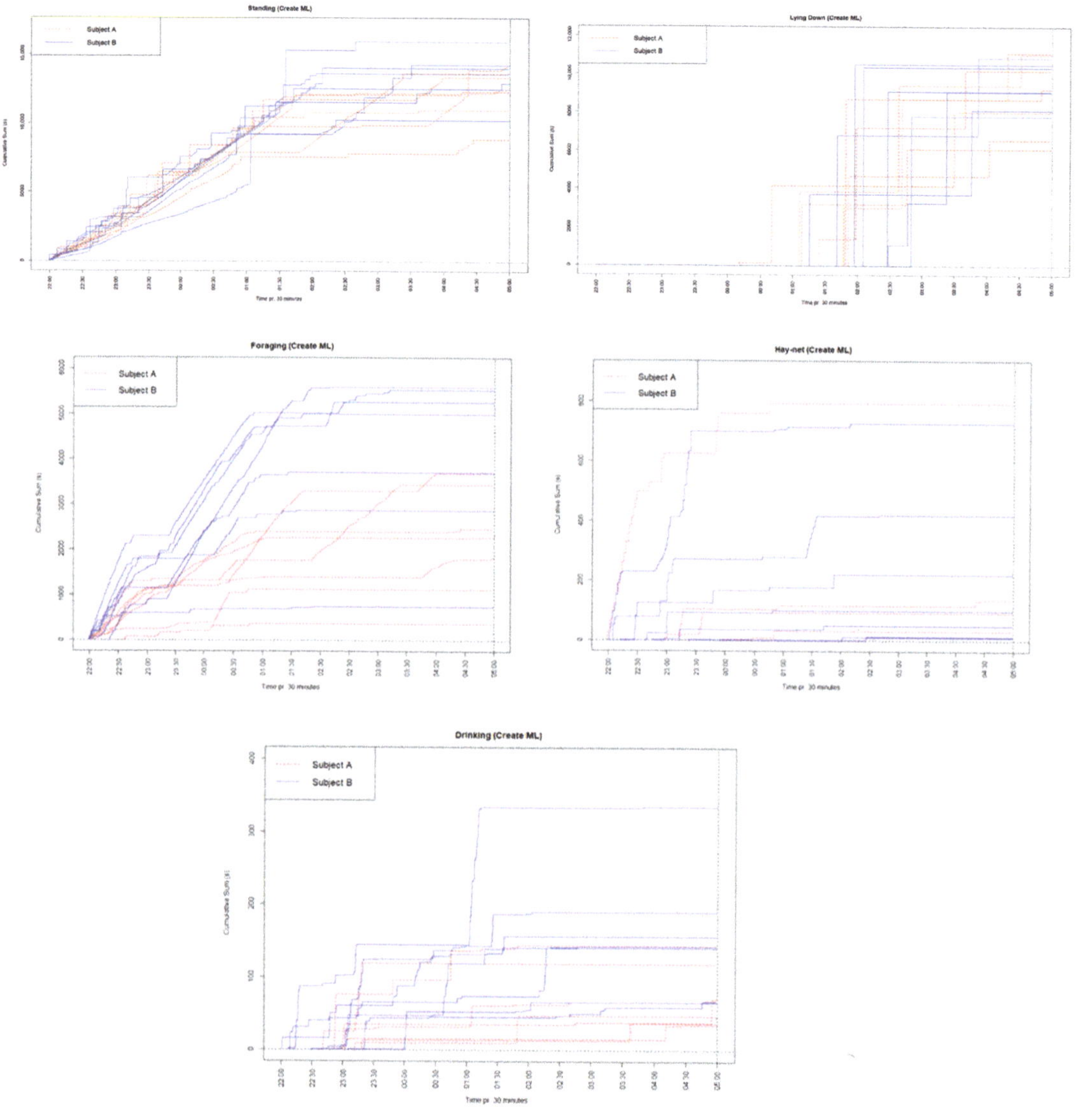

Figure A6. Cumulative graphs for every behavior for both subjects each observed night. Each color/line type signifies the subject, and every line signifies a different night.

The 'Standing' behavior was roughly similar in shape and sum for each subject and across nights, although Subject B had somewhat higher sums during several nights. 'Lying down' was also roughly similar for both subjects and between nights. 'Lying down' generally started occurring later at night but would then usually occur a lot for the rest of the night. 'Foraging' occurred mostly at the beginning of the night for both subjects, although Subject B tended to have higher sums than Subject A. 'Foraging' shapes and sums also varied from night to night. 'Hay-net' did not occur much for either individual but generally occurred most at the beginning of the night. 'Drinking' similarly did not occur much for either individual but mostly occurred at the beginning of the night, and furthermore displayed higher sums for Subject B.

Appendix F. Spearman Rank Correlation between Days

Table A5. DeepLabCut—Subject A. Spearman rank correlation matrix for DeepLabCut model of Subject A, based on data collected between 16 April 2024 and 22 April 2024. The matrix displays the pairwise correlation coefficients between the data sets corresponding to each date. The color intensity represents the strength of the correlation, with darker green indicating higher positive correlation and darker red indicating higher negative correlation.

	16-04-2024	17-04-2024	18-04-2024	19-04-2024	20-04-2024	21-04-2024	22-04-2024
16-04-2024	1.000						
17-04-2024	0.820	1.000					
18-04-2024	0.756	0.965	1.000				
19-04-2024	0.465	0.779	0.791	1.000			
20-04-2024	0.367	0.307	0.237	0.207	1.000		
21-04-2024	0.716	0.938	0.924	0.700	0.276	1.000	
22-04-2024	0.786	0.612	0.520	0.250	−0.053	0.450	1.000

Table A6. DeepLabCut—Subject B. Spearman rank correlation matrix for DeepLabCut model of Subject B, based on data collected between 16 April 2024 and 22 April 2024. The matrix displays the pairwise correlation coefficients between the data sets corresponding to each date. The color intensity represents the strength of the correlation, with darker green indicating higher positive correlation and darker red indicating higher negative correlation.

	16-04-2024	17-04-2024	18-04-2024	19-04-2024	20-04-2024	21-04-2024	22-04-2024
16-04-2024	1.000						
17-04-2024	0.930	1.000					
18-04-2024	0.860	0.777	1.000				
19-04-2024	0.299	0.115	0.573	1.000			
20-04-2024	0.297	0.401	0.357	0.019	1.000		
21-04-2024	0.997	0.924	0.872	0.315	0.314	1.000	
22-04-2024	0.843	0.760	0.999	0.601	0.372	0.856	1.000

Table A7. Create ML—Subject A. Spearman rank correlation matrix for Create ML model of Subject A, based on data collected between 16 April 2024 and 22 April 2024. The matrix displays the pairwise correlation coefficients between the data sets corresponding to each date. The color intensity represents the strength of the correlation, with darker green indicating higher positive correlation and darker red indicating higher negative correlation.

	16-04-2024	17-04-2024	18-04-2024	19-04-2024	20-04-2024	21-04-2024	22-04-2024
16-04-2024	1.000						
17-04-2024	0.902	1.000					
18-04-2024	0.872	0.918	1.000				
19-04-2024	0.759	0.841	0.859	1.000			
20-04-2024	0.222	0.043	0.102	−0.135	1.000		
21-04-2024	0.896	0.934	0.970	0.855	0.112	1.000	
22-04-2024	0.737	0.741	0.501	0.473	0.048	0.516	1.000

Table A8. Create ML—Subject B. Spearman rank correlation matrix for Create ML model of Subject B, based on data collected between 16 April 2024 and 22 April 2024. The matrix displays the pairwise correlation coefficients between the data sets corresponding to each date. The color intensity represents the strength of the correlation, with darker green indicating higher positive correlation and darker red indicating higher negative correlation.

	16-04-2024	17-04-2024	18-04-2024	19-04-2024	20-04-2024	21-04-2024	22-04-2024
16-04-2024	1.000						
17-04-2024	0.933	1.000					
18-04-2024	0.913	0.842	1.000				
19-04-2024	0.747	0.662	0.882	1.000			
20-04-2024	0.311	0.414	0.379	0.558	1.000		
21-04-2024	1.000	0.933	0.917	0.752	0.313	1.000	
22-04-2024	0.852	0.775	0.988	0.892	0.373	0.857	1.000

Appendix G. Spearman Rank Correlation between Individuals

Table A9. Spearman rank correlation matrix comparing the data between Subject A and Subject B for the DeepLabCut model, from 16 April 2024 to 22 April 2024. The matrix presents the correlation coefficients between corresponding dates for the two subjects. The color intensity represents the strength of the correlation, with darker green indicating higher positive correlation and darker red indicating higher negative correlation.

		Subject B						
		16-04-2024	17-04-2024	18-04-2024	19-04-2024	20-04-2024	21-04-2024	22-04-2024
Subject A	**16-04-2024**	0.476						
	17-04-2024	0.772	0.700					
	18-04-2024	0.839	0.762	0.975				
	19-04-2024	0.884	0.790	0.783	0.407			
	20-04-2024	0.372	0.508	0.194	−0.361	0.717		
	21-04-2024	0.654	0.622	0.881	0.551	0.672	0.672	
	22-04-2024	0.203	0.068	0.441	0.879	0.197	0.224	0.474

Table A10. Spearman rank correlation matrix comparing the data between Subject A and Subject B for the Create ML model, from 16 April 2024 to 22 April 2024. The matrix presents the correlation coefficients between corresponding dates for the two subjects. The color intensity represents the strength of the correlation, with darker green indicating higher positive correlation and darker red indicating higher negative correlation.

		Subject B						
		16-04-2024	17-04-2024	18-04-2024	19-04-2024	20-04-2024	21-04-2024	22-04-2024
Subject A	**16-04-2024**	0.690						
	17-04-2024	0.673	0.563					
	18-04-2024	0.855	0.785	0.977				
	19-04-2024	0.532	0.470	0.790	0.920			
	20-04-2024	0.206	0.314	0.112	0.052	0.602		
	21-04-2024	0.743	0.711	0.918	0.874	0.491	0.748	
	22-04-2024	0.347	0.194	0.419	0.512	0.066	0.349	0.472

References

1. Veasey, J.S. Differing animal welfare conceptions and what they mean for the future of zoos and aquariums, insights from an animal welfare audit. *Zoo Biol.* **2022**, *41*, 292–307. [CrossRef] [PubMed]
2. Barongi, R.; Fisken, F.A.; Parker, M.; Gusset, M. *Committing to Conservation: The World Zoo and Aquarium Conservation Strategy*; World Association of Zoos and Aquariums (WAZA) Executive Office: Gland, Switzerland, 2015.
3. European Association of Zoos and Aquaria. *Standards for the Accommodation and Care of Animals in Zoos and Aquaria*; EAZA: Amsterdam, The Netherlands, 2014.

4. Danish Association of Zoos and Aquaria. *DAZA Etiske Retningslinjer*; DAZA: Aalborg, Denmark, 2022. Available online: https://www.daza.dk/pdf/DAZA%20etiske%20retningslinjer%20-%20godkendt%20GF%2015.06.2022-underskrevet.pdf (accessed on 9 April 2023).
5. Sutherland, W.J. The importance of behavioural studies in conservation biology. *Anim. Behav.* **1998**, *56*, 801–809. [CrossRef] [PubMed]
6. Wolfensohn, S.; Shotton, J.; Bowley, H.; Davies, S.; Thompson, S.; Justice, W.S.M. Assessment of Welfare in Zoo Animals: Towards Optimum Quality of Life. *Animals* **2018**, *8*, 110. [CrossRef] [PubMed]
7. Perdue, B.M.; Sherwen, S.L.; Maple, T.L. Editorial: The Science and Practice of Captive Animal Welfare. *Front. Psychol.* **2020**, *11*, 1851. [CrossRef] [PubMed]
8. Sherwen, S.L.; Hemsworth, P.H. The Visitor Effect on Zoo Animals: Implications and Opportunities for Zoo Animal Welfare. *Animals* **2019**, *9*, 366. [CrossRef] [PubMed]
9. Hauenstein, S.; Jassoy, N.; Mupepele, A.; Carroll, T.; Kshatriya, M.; Beale, C.M.; Dormann, C.F. A systematic map of demographic data from elephant populations throughout Africa: Implications for poaching and population analyses. *Mammal Rev.* **2022**, *52*, 438–453. [CrossRef]
10. Rees, A.P. Activity budgets and the relationship between feeding and stereotypic behaviors in Asian elephants (*Elephas maximus*) in a Zoo. *Zoo Biol.* **2009**, *28*, 79–97. [CrossRef] [PubMed]
11. Yon, L.; Williams, E.; Harvey, N.D.; Asher, L. Development of a behavioural welfare assessment tool for routine use with captive elephants. *PLoS ONE* **2019**, *14*, e0210783. [CrossRef] [PubMed]
12. Greco, B.J.; Meehan, C.L.; Heinsius, J.L.; Mench, J.A. Why pace? The influence of social, housing, management, life history, and demographic characteristics on locomotor stereotypy in zoo elephants. *Appl. Anim. Behav. Sci.* **2017**, *194*, 104–111. [CrossRef]
13. Bansiddhi, P.; Nganvongpanit, K.; Brown, J.L.; Punyapornwithaya, V.; Pongsopawijit, P.; Thitaram, C. Management factors affecting physical health and welfare of tourist camp elephants in Thailand. *Biodivers. Conserv.* **2019**, *7*, e6756. [CrossRef]
14. Bertelsen, S.S.; Sørensen, A.S.; Pagh, S.; Pertoldi, C.; Jensen, T.H. Nocturnal Behaviour of Three Zoo Elephants (*Loxodonta africana*). *Genet. Biodivers. J.* **2020**, *4*, 92–113. [CrossRef]
15. Holdgate, M.R.; Meehan, C.L.; Hogan, J.N.; Miller, L.J.; Rushen, J.; de Passillé, A.M.; Soltis, J.; Andrews, J.; Shepherdson, D.J. Recumbence Behavior in Zoo Elephants: Determination of Patterns and Frequency of Recumbent Rest and Associated Environmental and Social Factors. *PLoS ONE* **2016**, *11*, e0153301. [CrossRef] [PubMed]
16. Boyle, S.A.; Roberts, B.; Pope, B.M.; Blake, M.R.; Leavelle, S.R.; Marshall, J.J.; Smith, A.; Hadicke, A.; Falcone, J.F.; Knott, K.; et al. Assessment of Flooring Renovations on African Elephant (*Loxodonta africana*) Behavior and Glucocorticoid Response. *PLoS ONE* **2015**, *10*, e0141009. [CrossRef]
17. Schiffmann, C.; Hellriegel, L.; Clauss, M.; Stefan, B.; Knibbs, K.; Wenker, C.; Hård, T.; Galeffi, C. From left to right all through the night: Characteristics of lying rest in zoo elephants. *Zoo Biol.* **2022**, *42*, 17–25. [CrossRef]
18. Greco, B.J.; Meehan, C.L.; Hogan, J.N.; Leighty, K.A.; Mellen, J.; Mason, G.J.; Mench, J.A. The Days and Nights of Zoo Elephants: Using Epidemiology to Better Understand Stereotypic Behvaior of African Elephants (*Loxodonta africana*) and Asian Elephants (*Elephas maximus*) in North American Zoos. *PLoS ONE* **2016**, *11*, e0144276. [CrossRef]
19. Broom, D.M. Welfare of Animals: Behavior as a Basis for Decisions. *Encycl. Anim. Behav.* **2010**, *3*, 580–584.
20. Jacobson, S.L.; Ross, S.R.; Bloomsmith, M.A. Characterizing abnormal behavior in a large population of zoo-housed chimpanzees: Prevalence and potential influencing factors. *PeerJ* **2016**, *4*, e2225. [CrossRef] [PubMed]
21. Bacon, H. Behaviour-Based Husbandry—A Holistic Approach to the Management of Abnormal Repetitive Behaviors. *Animals* **2018**, *8*, 103. [CrossRef]
22. Fuktong, S.; Yuttasaen, P.; Punyapornwithaya, V.; Brown, J.L.; Thitaram, C.; Luevitoonvechakij, N.; Bansiddhi, P. A survey of stereotypic behaviors in tourist camp elephants in Chiang Mai, Thailand. *Appl. Anim. Behav. Sci.* **2021**, *243*, 105456. [CrossRef]
23. Mason, G.J. Stereotypies and suffering. *Behav. Process.* **1991**, *25*, 103–115. [CrossRef] [PubMed]
24. Mason, G.J.; Rushen, J. *Stereotypic Animal Behaviour: Fundamentals and Applications to Welfare*; CABI Digital Library: Wallingford, UK, 2006.
25. Mostard, K.E. *General Understanding, Neuro-Endocrinologic and (Epi)Genetic Factors of Stereotypy*; Radboud University of Nijmegen: Nijmegen, The Netherlands, 2011.
26. Altmann, J. Observational Study of Behavior: Sampling Methods. *Behaviour* **1974**, *49*, 227–267. [CrossRef]
27. Vanitha, V.; Thiyagesan, K.; Baskaran, N. Prevalence of stereotypies and its possible causes among captive Asian elephants (*Elephas maximus*) in Tamil Nadu, India. *Appl. Anim. Behav. Sci.* **2016**, *174*, 137–146.
28. Adams, J.; Berg, J.K. Behavior of Female African Elephants (*Loxodonta africana*) in Captivity. *Appl. Anim. Ethol.* **1980**, *6*, 257–276. [CrossRef]
29. Dell, A.I.; Bender, J.A.; Branson, K.; Couzin, I.D.; de Polavieja, G.G.; Noldus, L.P.; Pérez-Escudero, A.; Perona, P.; Straw, A.D.; Wikelski, M.; et al. Automated image-based tracking and its application in ecology. *Trends Ecol. Evol.* **2014**, *29*, 417–428. [CrossRef] [PubMed]
30. Gomez-Marin, A.; Paton, J.J.; Kampff, A.R.; Costa, R.M.; Mainen, Z.F. Big Behavioral Data: Psychology, Ethology and the Foundations of Neuroscience. *Nat. Neurosci.* **2014**, *17*, 1455–1462. [CrossRef]
31. Anderson, D.J.; Perona, P. Toward a Science of Computational Ethology. *Neuron* **2014**, *84*, 18–31. [CrossRef] [PubMed]

32. Mathis, A.; Mamidanna, P.; Cury, K.M.; Abe, T.; Murthy, V.N.; Mathis, M.W.; Bethge, M. DeepLabCut: Markerless pose estimation of user-defined body parts with deep learning. *Nat. Neurosci.* **2018**, *21*, 1281–1289. [CrossRef]
33. Mirkó, E.; Dóka, A.; Miklósi, Á. Association between subjective rating and behaviour coding and the role of experience in making video assessments on the personality of the domestic dog (*Canis familiaris*). *Appl. Anim. Behav. Sci.* **2013**, *149*, 45–54. [CrossRef]
34. Zhao, Z.-Q.; Zheng, P.; Xu, S.-T.; Wu, X. Object Detection With Deep Learning: A Review. *IEEE Trans. Neural Netw. Learn. Syst.* **2019**, *30*, 3212–3232. [CrossRef] [PubMed]
35. Zhou, Z.-H.; Liu, S. *Machine Learning*; Springer Nature: Nanjing, China, 2021.
36. Lenzi, J.; Barnas, A.F.; ElSaid, A.A.; Desell, T.; Rockwell, R.F.; Ellis-Felege, S.N. Artificial intelligence for automated detection of large mammals creates path to upscale drone surveys. *Sci. Rep.* **2023**, *13*, 947. [CrossRef]
37. Bain, M.; Nagrani, A.; Schofield, D.; Berdugo, S.; Bessa, J.; Owen, J.; Hockings, K.J.; Matsuzawa, T.; Hayashi, M.; Biro, D.; et al. Automated audiovisual behavior recognition in wild primates. *Sci. Adv.* **2021**, *7*, eabi4883. [CrossRef] [PubMed]
38. Sakib, F.; Burghardt, T. Visual Recognition of Great Ape Behaviours in the Wild. *arXiv* **2020**, arXiv:2011.10759.
39. Hebbar, P.K.; Pullela, P.K. Deep Learning in Object Detection: Advancements in Machine Learning and AI. In Proceedings of the 2023 International Conference on the Confluence of Advancements in Robotics, Vision and Interdisciplinary Technology Management (IC-RVITM), Bangalore, India, 28–29 November 2023.
40. Hardin, A.; Schlupp, I. Using machine learning and DeepLabCut in animal behavior. *Acta Ethologica* **2022**, *25*, 125–133. [CrossRef]
41. Marks, M.; Jin, Q.; Sturman, O.; von Ziegler, L.; Kollmorgen, S.; von der Behrens, W.; Mante, V.; Bohacek, J.; Yanik, M.F. Deep-learning-based identification, tracking, pose estimation and behaviour classification of interacting primates and mice in complex environments. *Nat. Mach. Intell.* **2022**, *4*, 331–340. [CrossRef]
42. Nath, T.; Mathis, A.; Chen, A.C.; Patel, A.; Bethge, M.; Mathis, M.W. Using DeepLabCut for 3D markerless pose estimation across species and behaviors. *Nat. Protoc.* **2019**, *14*, 2152–2176. [CrossRef] [PubMed]
43. Marques, O. Machine Learning with Core ML. In *Image Processing and Computer Vision in IOS*; Springer International Publishing AG: Cham, Switzerland, 2020; pp. 29–40.
44. Andersen, T.A.; Herskind, C.; Maysfelt, J.; Rørbæk, R.W.; Schnoor, C.; Pertoldi, C. The nocturnal behaviour of African elephants (*Loxodonta africana*) in Aalborg Zoo and how changes in the environment affect them. *Genet. Biodivers. J.* **2020**, *4*, 114–130. [CrossRef]
45. Ruuska, S.; Hämäläinen, W.; Kajava, S.; Mughal, M.; Matilainen, P.; Mononen, J. Evaluation of the confusion matrix method in the validation of an automated system for measuring feeding behaviour of cattle. *Behav. Process.* **2018**, *148*, 56–62. [CrossRef] [PubMed]
46. Larsen, J.F.; Andersen, K.K.D.; Cuprys, J.; Fosgaard, T.B.; Jacobsen, J.H.; Krysztofiak, D.; Lund, S.M.; Nielsen, B.; Pedersen, M.E.B.; Pedersen, M.J.; et al. Behavioral analysis of a captive male Bornean orangutan (*Pongo pygmaeus*) when. *Arch. Biol. Sci.* **2023**, *75*, 443–458. [CrossRef]
47. Field, A.P. Kendall's Coefficient of Concordance. *Encycl. Stat. Behav. Sci.* **2005**, *2*, 1010–1011.
48. Brownlee, J. Machine Learning Mastery. Guiding Tech Media. 19 August 2020. Available online: https://machinelearningmastery.com/distance-measures-for-machine-learning/ (accessed on 22 May 2024).
49. Malviya, M.; Buswell, N.T.; Berdanier, C.G.P. Visual and Statistical Methods to Calculate Intercoder Reliability for Time-Resolved Observational Research. *Int. J. Qual. Methods* **2021**, *20*, 16094069211002418. [CrossRef]
50. Whiteway, M.R.; Biderman, D.; Friedman, Y.; Dipoppa, M.; Buchanan, E.K.; Wu, A.; Zhou, J.; Bonacchi, N.; Miska, N.J.; Noel, J.-P.; et al. Partitioning variability in animal behavioral videos using semi-supervised variational autoencoders. *PLoS Comput. Biol.* **2021**, *17*, e1009439. [CrossRef] [PubMed]
51. García-Mateos, G.; Hernández-Hernández, J.; Escarabajal-Henarejos, D.; Jaén-Terrones, S.; Molina-Martínez, J. Study and comparison of color models for automatic image analysis in irrigation management applications. *Agric. Water Manag.* **2015**, *151*, 158–166. [CrossRef]
52. Hernández-Hernández, J.L.; García-Mateos, G.; González-Esquiva, J.M.; Escarabajal-Henarejos, D.; Ruiz-Canales, A.; Molina-Martínez, J.M. Optimal color space selection method for plant/soil segmentation in agriculture. *Comput. Electron. Agric.* **2016**, *122*, 124–132. [CrossRef]
53. Tobler, I. Fundamental Research Behavioral Sleep in the Asian Elephant in Captivity. *Sleep* **1992**, *15*, 1–12. [PubMed]
54. Finch, K.; Sach, F.; Fitzpatrick, M.; Rowden, L.J. Insights into Activity of Zoo Housed Asian Elephants (*Elephas maximus*) during Periods of Limited Staff and Visitor Presence, a Focus on Resting Behaviour. *Zool. Bot. Gard.* **2021**, *2*, 101–114. [CrossRef]
55. Casares, M.; Silván, G.; Carbonell, M.D.; Gerique, C.; Martinez-Fernandez, L.; Cáceres, S.; Illera, J.C. Circadian rhythm of salivary cortisol secretion in female zoo-kept African elephants (*Loxodonta africana*). *Zoo Biol.* **2016**, *35*, 65–69. [CrossRef] [PubMed]
56. Yin, Z.; Zhao, Y.; Xu, Z.; Yu, Q. Automatic detection of stereotypical behaviors of captive wild animals based on surveillance videos of zoos and animal reserves. *Ecol. Inform.* **2024**, *79*, 102450. [CrossRef]

animals

MDPI

Review

Monitoring Pig Structural Soundness and Body Weight in Pork Production Systems Using Computer Vision Approaches

Ryan Jeon [1,*], Caleb Rykaczewski [2], Thomas Williams [3], William Harrington [4], James E. Kinder [2,3] and Mark Trotter [3]

[1] Integer Technologies LLC, 1556 Main Street, Suite 200, Columbia, SC 29201, USA
[2] Department of Animal Sciences, College of Food, Agricultural, and Environmental Sciences, The Ohio State University, Columbus, OH 43210, USA; rykaczewski.2@buckeyemail.osu.edu (C.R.); kinder.15@osu.edu (J.E.K.)
[3] Institute of Future Farming Systems, School of Health, Medical, and Applied Sciences, CQUniversity, Rockhampton, QLD 4701, Australia; t.m.williams@cqu.edu.au (T.W.); m.trotter@cqu.edu.au (M.T.)
[4] College of Business, Law, and Governance, James Cook University, Townsville QLD 4811, Australia; william.harrington@my.jcu.edu.au
* Correspondence: rjeon94@gmail.com

Simple Summary: Opportunities exist to integrate computer vision systems into pork production to enhance monitoring and decision-making, particularly in the areas of structural soundness, lameness, and body weight prediction. These aspects are important for maintaining herd health, optimizing productivity, and ensuring economic viability. Traditional methods for assessing pig body weight and leg structural soundness are labor-intensive, subjective, and often inaccurate. Computer vision offers an opportunity by providing automated, noninvasive, and precise assessments, making these tasks more efficient and reliable. With the capability to autonomously monitor important physical traits using deep learning models and 3D imaging techniques, computer vision systems are likely to be important in the future of pork production. In this review, we explore the current advancements, challenges, and future potential of computer vision in pork production systems.

Abstract: As the global demand for products from food-producing animals increases with greater household economic capacity, there is an increased emphasis on the development of precision technologies for monitoring the health, product production, and wellbeing of these animals. The present review focuses on pork production. Using these systems is advantageous for enhancing pork production efficiency when trained personnel utilize these technologies to full capacity and have objective, automated, and uninterrupted streams of data collection. While these systems have great potential for revolutionizing food animal production, the nascent stage of computer vision in precision technology has precluded its integration into traditional agricultural practices and systems. In this review paper, there is a focus on the need to (1) evaluate the performance and effective use of computer vision technologies to collect and evaluate reliable data from pork production enterprises; and (2) focus on the current state of sensor-based animal management using a data fusion approach to monitor pig health/performance. Many of these technologies are in various stages of development; therefore, these technologies have not been integrated into pork production or other food animal producing systems. Even though the focus of this review article is on the utilization of these technologies in pork production systems, these technologies are relevant in other food animal production systems, particularly dairy and poultry production. Therefore, we describe an approach that emphasizes the important need for computational capacity and speed, edge computing, data storage and transmission, and maintaining connectivity in rural settings.

 https://doi.org/10.3390/ani15050635

Keywords: precision technologies; computer vision; pork production; weight; skeletal structure

1. Conventional Approaches to Visual Assessment of Pigs in Breeding Herds

The focus of this review article is on body weight and lameness. These factors have a large effect on the cost of pork production [1] and are aspects where initial progress can be made using precision technologies. Feet and leg structural soundness and body weight are important pig characteristics that require regular monitoring and evaluation to ensure the economic viability of pork production systems; however, accurately assessing these features is difficult. The assessment of feet and leg skeletal characteristics associated with structural soundness is traditionally conducted by those trained in making these visual appraisals. In an industry with a relatively large workforce turnover, this reliance can lead to inconsistencies and inaccuracies in these evaluations [2]. Subjective visual evaluations, though requiring minimal time to conduct, lack consistency and accuracy [1]. Similarly, human visual appraisal of pigs for marketing estimation leads to inaccuracies [3].

1.1. Foot and Leg Structure in Sows

In pork production systems, structural soundness of the feet and legs is a major problem. Depictions of typical leg structural problems can be accessed via this link: https://i0.wp.com/porkgateway.org/wp-content/uploads/2015/07/2LegStructure.png (accessed on 20 December 2024). These physical structural deficiencies will often lead to premature culling, removing affected sows from the breeding population and causing a loss of potential revenue. Research results indicate that increasing the average parity at removal by one-tenth could increase profit in the USA swine industry by USD 15 million [2]. Premature culling is often labeled and classified as locomotion disorders, which encompass a range of other problems, such as lameness, injuries to the sow and piglets, multiple syndromes, and general ambulatory unsoundness [4]. Due to the subjective nature of diagnostic techniques for assessing these problems, however, there have been numerous studies conducted to evaluate the correlation between sow locomotion, productivity, and mortality. The individual animal effects of locomotor dysfunctions affect the financial viability of intensively managed food animal production systems due to the detrimental effects on product production. These problems can also be particularly evident in [4] large commercial poultry meat [5] and dairy milk product [6] production, where animals are housed in confinement buildings. As with all food-producing animal systems, there are "downstream" effects of locomotor dysfunctions on the sustainability of these enterprises from an economic viability perspective [7].

1.2. Association Between Sow Locomotion and Mortality

The results of many studies indicate a close association between sow locomotion and mortality rates. Sow mortalities attributed to locomotion problems vary considerably. Estimates in studies are these mortalities range from 9% [8] to 23% [9]. It was found that lameness or foot lesions accounted for 8.6% of sow culling from breeding herds [10]. Similar results were reported with locomotion problems being responsible for 11% and 13% of sow culling during the early production stages [11]. In some of the initial studies, there was a focus on sow locomotion and motility, and there were similar findings with specific physical structure indicators that could be used to predict the risk of locomotion disorder occurrences [12]. The results from more recent studies are consistent in that sow longevity is an important animal welfare and economic concern for commercial swine breeders [3,13,14]. Locomotor dysfunctions, therefore, can significantly reduce efficiency in pork production enterprises due to increased mortality and other production-related inefficiencies associated with locomotor problems [15,16].

1.3. Subjectivity of Structural Appraisal in Sows

Visual evaluation of the structure, frame, and gait of sows can provide important information on the likelihood of an individual to have leg structural problems. Evaluations are often conducted by trained farm staff, and these manual inspections are considered standard practice in pork production enterprises. The inherent subjectivity in these assessments, however, has historically been a point of contention. The importance of reliable measurements for leg structural integrity was evident, with as many as 25% of females being culled because of issues with their feet and legs [7]. Furthermore, farm staff experience had effects on the consistency and repeatability of the scores for sow leg structural integrity [17,18]. Similar findings were subsequently reported, confirming the results of these studies [19,20].

The current objective measurement procedures for evaluating sow and gilt leg integrity need to yield data that are more quantitative. There are efforts to develop an objective measurement method of joint angles for knee, hock, front, and rear pasterns and a rear stance position in swine using digital imaging technology and to assess the repeatability of the objective measurement process [21]. Based on the results from the intraclass correlation coefficient analyses, the repeatability of the objective method used in this study to evaluate feet and leg structural soundness ranged from 0.552 to 0.879. It was concluded that an objective feet and leg structural soundness trait measurement could be implemented as an alternative to subjective methods because of the repeatability of determinations and the accuracy of joint evaluations.

The application of computer vision algorithms for evaluating images and video recordings of sows/gilts provides an unbiased and objective approach to assessing various indicators of foot and leg integrity. The results from a recent study in which computer vision models were developed to identify ten key body landmarks of pigs from their side profile images and two from their rear profile images resulted in a mean average precision (mAP) of 0.94 across all areas of the body that were evaluated [22]. Trigonometric formulae were developed to calculate the hock and knee angles from these body landmarks. These automated angle measurements were validated with comparisons to manual measurements. There was an average root mean square error (RMSE) of 4.13 deg and correlation coefficients (average $r^2 = 0.84$) different from zero, confirming the consistency of the data when there were evaluations using the Bland–Altman procedure. This methodology, therefore, provided a reliable method for obtaining precise leg angle measurements, which can be valuable for refining gilt replacement criteria and ultimately enhancing sow breeding programs.

2. Body Weight Assessment of Pigs at Time of Marketing

Efficacious and efficient determinations of pig body weight is of great value in commercial swine production systems. The precise estimation of body weight influences economic profitability margins for swine producers and serves as an important indicator of health. For example, trends of decreasing body weight can be a potential indicator of disease and illness in food-producing animals. Detection can be enhanced using precision technologies [23]. Body weight is associated with dietary nutrient intake which, in turn, has biological implications for reproductive performance [24]. When body weight is routinely monitored, data can be plotted as growth curves, which is a useful production approach for evaluating feed efficiency utilization and pig growth in pork production enterprises. For these reasons, the accurate and precise estimation of animal body weight is of multifaceted significance and is particularly important at nearly every stage of production in pork-producing enterprises, but especially when determining optimal times for marketing pigs.

2.1. Current Techniques for Estimating Pig Body Weight

The approaches used in commercial pork production enterprises for estimating pig body weight are a direct measurement, using a scale, and visual estimates. The utilization of accurately calibrated scales will result in the collection of accurate body weight data, but this is labor-intensive and requires multiple individuals to effectively and efficiently collect data, which can result in injury and/or stress, leading to welfare issues for both pigs and those conducting the weighing task. Furthermore, a considerable time investment is required to obtain individual pig body weights using a scale, where employees could be conducting other duties instead [25]. The direct measurement technique, therefore, is impractical for large commercial pork production enterprises. The current alternative, using visual appraisal, has many other shortfalls. Visual estimations of pig body weight are not only highly subjective but also lack consistency, with accuracy depending on the experience of the farm staff [17].

2.2. Correlation Between Values for Pig Body Weight and Biometric Determinations

Historically, in numerous studies, there have been investigations of the relationship between pig body weight and quantifiable biometrics (such as heart girth, length, and height), with the consistent reporting of close correlations between body weight and these biometric variables. Three techniques have been compared for obtaining pig body biometric data: tape measure/caliper, livestock scales, and projections from 2D images of pigs [26]. The main findings were that calipers and measuring tapes were the most effective methods for estimating pig body weight. There are also close correlations between values for body weight and various biometric values such as height, length, width, flank, and girth, leading to the conclusion that the combination of biometric data can be used as independent variables to predict body weight [27]. Interestingly, the significant limitations reported for the utilization of biometric data in these studies were the same as those previously described when there were pig body weight determinations using scales: the need to immobilize the animals, inherent human measurement errors, subjectivity leading to non-reliable data, and time constraints. Considering the large number of pigs housed in present-day pork production enterprises, the manual collection of biometric data is not a feasible alternative to directly measuring pig body weight using scales.

2.3. Automated Electronic Weighing Systems

An emerging alternative to manual weighing is the use of automated weighing systems. While initially expensive, these systems are a viable alternative to manual weighing when combined with electronic identification tags, enabling individual animal tracking and automated data management. Without such tags, the benefits of automation are reduced. The initial findings were that a single employee could weigh as many as 100 pigs per hour by allowing pigs to pass through automated gates that led to the automatic collection of pig body weights [28]. There was a subsequent report that the automated approach was reliable with greater repeatability compared to manual weighing methods [29].

While both automated weighing systems and computer vision technologies offer promising alternatives to manual weighing in pig production, each alternative has unique implementation considerations. Automated weighing systems, while potentially resulting in less initial retrofitting costs than previously considered, require specific infrastructure modifications. Computer vision systems, even if there is reduced weight estimation accuracy and precision as compared with direct weighing using a scale, have advantages in terms of continuous, non-contact monitoring and the potential to collect additional data on animal behavior and body condition. Further research is needed to optimize the cost-effectiveness and practical application of both technologies, particularly in addressing challenges such as data interpretation for actionable management decisions and minimizing RFID tag loss when there is the utilization of automated weighing systems.

3. Computer Vision Techniques in Pig Assessment

Computer vision approaches are a promising solution for the objective collection of data, with there being the potential for automating the evaluation of feet and leg structural soundness and predicting pig body weight. The use of these technologies will enable continuous, noninvasive monitoring, providing consistent and accurate assessments that are important for effective pork production management. By automating the evaluation of key metrics such as structural soundness and body weight, the utilization of computer vision systems will lead to a reduction in the reliance for subjective visual assessments, thereby improving the reliability of the data used in decision-making processes. The application of computer vision, however, is not without challenges and requires a concerted research effort to fully realize its potential. In subsequent sections of this review article, the methods and application of computer vision techniques in making assessments in pork production enterprises are explored, focusing on the importance of enhancing the precision and efficiency of monitoring systems.

3.1. Introduction to Neural Networks

Neural networks are computational models that draw inspiration from the architecture and function of the human brain, enabling the processing of information and pattern recognition [30]. When well-trained neural networks are versatile, applications range from simple linear regression to categorizing animal behavior recognition. Details regarding what neural networks can be evaluated via information provided at this link: https://www.ibm.com/think/topics/neural-networks (accessed on 20 December 2024). The utilization of neural networks to detect and predict intricate patterns, particularly in complex data sources such as images, results in well-trained neural networks with great potential in various fields, including computer vision. By leveraging these capabilities, neural networks can break down images into fundamental components and progressively build to more complex representations, such as shapes and objects.

Neural networks consist of layers of interconnected nodes, referred to as neurons, that perform mathematical calculations. In computer vision applications, data in the form of images are passed through these layers and undergo a series of mathematical transformations. Each neuron applies a weighted sum to the inputs, which produces a "signal" that represents the neuron's output at this stage. This signal is then passed through a nonlinear activation function, with there being the resulting decision of whether the signal should continue to propagate through the network or be deactivated. Signals that are deactivated are zeroed out from influencing subsequent layers of the final output. This selective process enables the network to learn from important features that activate the function while ignoring features that do not, enabling the network to learn complex patterns in the data. Training a neural network involves minor adjustments to the weights of the connections to minimize the error between the predicted output and the actual output, a process known as backpropagation. Ultimately, the final output of the network is a prediction or classification result that aligns with the data for which the network was trained.

3.2. Object Detection and Feature Extraction in Computer Vision

Computer vision techniques can be utilized to extract important visual features from images and videos. This field initially gained traction with the Deformable Parts Model (DPM), a pioneering object detection algorithm that relies on a structured support vector machine (SVM) rather than a neural network. The DPM represents objects as a collection of parts, each with specific geometric relationships, and uses a "sliding window" approach to scan the image and detect potential objects based on the alignment of these parts [31]. As the need for more efficient and faster object detection grew, the region-based convolutional neural network (R-CNN) was developed. The R-CNN improved upon the DPM by strategically assessing the "interesting" regions of an image instead of evaluating every possible region [32]. Although the R-CNN was an improvement compared to the DPM, the speed was inadequate for real-time applications. Subsequently, the "You Only Look Once" (YOLO) algorithm emerged as a more practical solution.

3.3. Advent of the "You Only Look Once" (YOLO) Algorithm

The development of the YOLO algorithm marked an advancement in object detection due to the time efficiency of detection and real-time capabilities for detection [33]. Both YOLO and the R-CNN are object detection models. These are designed to locate and classify objects, such as pigs or their body parts, within images and videos. The YOLO algorithm has become a central feature when utilizing computer vision approaches, with uses ranging from agricultural to facial recognition. The YOLO model has three main components: the backbone, neck, and head. The backbone is responsible for extracting essential features from the input image, the neck connects these features into a feature pyramid that enhances the network's capacity to detect objects at different scales, and the head produces the final output, including the bounding boxes and class predictions.

What sets YOLO apart is "single-shot" detection capacities, which allows for evaluating the entire image in one pass, predicting both the objects that are present and their locations. This structure enables rapid object detection, resulting in YOLO being a more effective and efficient option as compared with previously utilized methods such as the R-CNN and DPM.

The application of the YOLO algorithm in pork production systems provides an important opportunity to advance precision agriculture. The use of this algorithm enables the precise collection of visual data, facilitating the objective and automated assessment of various aspects of pig anatomy. This capability has the potential to revolutionize animal husbandry by allowing producers to optimize feeding strategies, improve welfare conditions, and enhance productivity. By leveraging computer vision models such as YOLO, producers can make more informed decisions, ultimately increasing the efficiency and efficacy of pork production systems.

3.4. Localization in Computer Vision

With the utilization of YOLO, there is more precise localization, which leads to greater precision for the spatial location of objects within an image through boundary box and centroid coordinates. The utilization of this process begins with overlaying the image with randomly sized anchor boxes and the calculation of the Intersection of Union (IoU) for each anchor box against the "ground truth" (i.e., determined from data collected manually) annotation [33]. This allows for the determination of the confidence of each anchor box containing a class object. The utilization of this intricate procedure leads to the assurance that only the most statistically reliable boundary boxes are predicted, which leads to enhancing the accuracy of a trained YOLO model.

3.5. "Black Boxes" in Deep Learning

The utilization of CNNs has led to the efficacious estimating of pig body weight [34–36]. Convolutional neural networks (CNNs) have been proven to be highly effective in various applications, resulting in accurate predictions for conducting tasks involving complex visual data. A major barrier for the utilization of CNNs, however, is the "black box" nature, which refers to the lack of transparency in how the models make decisions. This opacity creates several challenges. First, it is difficult to pinpoint which specific features the model is using to make predictions, raising concerns about the reliability and consistency of these predictions across different datasets or conditions. For example, if a CNN model is trained on a particular dataset, it may rely on features that are not universally applicable, leading to potential inaccuracies when applied to a different dataset.

Furthermore, the inability to understand the internal workings of these models makes troubleshooting particularly challenging. When predictions are inaccurate or when the model fails to perform as expected, it can be difficult for scientists, clinicians, or producers to diagnose the root cause of the problem. This lack of interpretability can be a major barrier to the broader utilization of CNNs in practical applications, as end-users may be hesitant to trust a system they cannot fully understand or explain.

For scientists and researchers, the "black box" nature of CNNs limits the capacity to extract meaningful insights from the data, which is essential for advancing knowledge and refining models. To address these issues, there is a growing interest in developing more interpretable models or enhancing the transparency of existing CNNs using techniques like explainable AI (XAI). With these efforts, there is an attempt to bridge the gap between the useful predictive capabilities of CNNs and the need for definitive, understandable decision-making processes in fields like animal husbandry and precision agriculture.

3.6. Metrics of Assessing Computer Vision Models

The performance of computer vision models, particularly in object detection and classification tasks, is commonly evaluated using metrics that provide insight into accuracy and reliability. Important metrics include mean average precision (mAP) [37], Intersection of Union (IoU) [38], precision, and recall [39]. Mean average precision is a widely subscribed-to metric that combines precision (i.e., the ratio of true positive predictions to all positive predictions) and recall (i.e., the ratio of true positive predictions to all actual positives across various thresholds). This metric provides for an aggregated measure of a model's capacity to correctly identify objects, reflecting its overall performance when there are varying extenuating circumstances. Intersection of Union (IoU) assesses the accuracy of object localization by measuring the overlap between predicted bounding boxes and the actual bounding boxes where data were collected manually, which are often manually labeled. The IoU ranges from 0 to 1, where a larger IoU indicates more precise localization, making it particularly valuable in applications where accuracy in locating objects within an image is essential.

In traditional statistics, precision evaluations are a simpler determination of a model's predictive capabilities by indicating the likelihood that a positive prediction is correct. It, however, does not account for the model's capacity to detect all real positives (recall) or accuracy in localization (IoU).

Ultimately, mAP and IoU together offer a nuanced assessment tailored to the unique challenges of computer vision tasks such as object detection. While mAP reflects the model's overall detection performance, IoU provides a direct measure of localization accuracy. These metrics, alongside precision and recall, provide a comprehensive toolkit for quantitatively assessing the efficacy of computer vision models.

4. Application of Computer Vision in Pig Assessment

4.1. Estimating Pig Body Weight

Computer vision has emerged as a promising alternative for overcoming the limitations of conventional pig body weight estimation. Biometric methods for body weight estimation in pigs have been determined to be accurate and reliable [40]. Computer vision systems can be utilized to determine biometric measurements with precision, making these systems candidates for integration. These systems can be specifically modified for use in pork production enterprises and are noninvasive, automated, and allow for the collection of objective data for determining biologically important features of food-producing animals.

4.1.1. Traditional Image Processing Techniques for Pig Body Weight

Traditional image processing techniques differ significantly from deep learning-based methods, like YOLO and RCNN. These are designed to use manually crafted algorithms and features to process images. By design, these are suitable for simple tasks (such as thresholding) but are not very useful with complex "real-world" scenarios (thresholding in varying lighting). In an early study, a traditional computer vision technique was utilized to evaluate the dorsal pig surface, and from the data collected, there was the estimation of pig body weight with an accuracy of 5% [41]. Similarly, when contrasting lighting conditions were evaluated to estimate the dorsal surface area of pigs, there were predictions of pig body weight within an accuracy of 0.9 kg [42]. Utilization of an eclipse fitting model, and the Hough transform image extraction technique, resulted in an estimated pig body weight to an accuracy of 96.2% with an average error of 1.23 kg [43]. Even with these advancements, one significant limitation of these 2D image processing techniques is the lack of a capacity to obtain a comprehensive variety of relevant biological features. For example, in these previous studies, the dorsal images of pigs could be evaluated and utilized for effectively determining pig width and back length, but this approach is not effective for determining pig girth or flank measurements. Furthermore, with the use of these techniques, there is the utilization of Euclidian distances, which are not effective at determining geodesic features such as pig girth.

4.1.2. Three-Dimensional Computer Vision Techniques for Pig Body Weight

While 2D computer vision is important in the detection of live animals, the advent of 3D computer vision was a transformative advancement. This technology enables the capture of 3D geodesic data, which include important measurements such as the girth of an animal, ultimately expanding the scope of what is possible with the precision assessment of body weight in food-producing animals. With the use of depth imaging procedures, there is the utilization of 3D data, a compelling alternative to relying solely on 2D images. One pioneering utilization of depth imaging for swine-related research included mounting an RGB depth camera on the ceiling of a pig housing facility. In this study [44], an RGB-D computer vision system was developed for predicting the body weight of non-restrained pigs using top-view RGB and depth images. For 38 days, images of eight pigs were recorded via video daily using an Intel RealSense D435 camera for 3 min at six frames per second, while ground truth weights were determined directly using a scale. Pigs were of crossbred Yorkshire and Large White breeding and were 5 weeks of age at the initiation of the experiment. On average, the pigs weighed 23.5 kg (SD = 7.6 kg) at the start of the experiment and 46.7 kg (SD = 8.7 kg) at experimental cessation. Morphological features, such as length, width, and height, were extracted using Python's OpenCV library. By conducting quality control evaluations, there was the removal of frames with motion blur, distorted shapes, or non-standard positions. Linear mixed models were used to make predictions, achieving coefficients of determination ranging from 0.49 to 0.98, highlighting the robustness and potential for widespread application in pork production enterprises.

In a similar study, there was the validation of the use of depth images for predicting pig body weight [45]. A total of 772 depth images and corresponding mass measurements were collected from 234 pigs in a grow-finish pork production facility of pigs of Landrace, Duroc, and Yorkshire breeding. Within this group, there was an equal number of barrows and gilts. On average, piglets weighed 27 kg (SD = 4.4 kg) at the start and 40 kg (SD = 6.5) at the end of the experiment. Body weight was calculated from these depth images collected from a Kinect sensor (Microsoft, Milipitas, CA, USA) and mounted on the wall above the animal scale. Both color and depth images were acquired at approximately 1 s intervals. Metrics were collected using a program developed in MATLAB software (version R2015b). Using this software, a system of linear equations was developed to predict the weight from the volume. The global equation was determined to be $R^2 = 0.9905$. These findings indicate that depth sensors can provide an accurate and scalable solution to continuous body weight estimation.

There was another study where there was use of photogrammetry and artificial neural networks to estimate the body weight of Holstein cattle [46]. Body dimensions (wither height, hip height, body length, and hip width) were captured using Canon EOS 400D cameras (Canon, Oita, Japan) that were synchronized and calibrated to obtain 3D data. From the resulting data, metrics on body dimensions were collected. This dataset was divided into test and training subsets, with the most precise performing ANN having reliable accuracy in predicting pig body weight. The correlation coefficient was 0.995 when compared to weights obtained manually. These results indicate that photogrammetrically derived body metrics can be a reliable method for predicting pig body weight.

The use of 3D point clouds allows for three-dimensional representations of the animal, capturing spatial information that goes beyond two-dimensional imaging. Early research explored the potential of 3D point cloud technology for accurate pig body weight estimations. In a recent study [47], a hybrid approach was utilized, combining statistical filtering and DBSCAN clustering for denoising point clouds. This technique mitigated bias and improved feature extraction. The model incorporated pig dorsal body area parameters and a CNN, achieving a mean absolute error of 12.45 kg and a mean absolute percentage error of 5.36%. In a recent study [48], a 3D deep learning approach on point clouds collected in a pen environment called PointNet was utilized. PointNet is a model used to process point cloud data, which consist of points in 3D space, representing the shape and structure of objects. With the utilization of this model, there was a coefficient of determination of 0.94 and a root mean squared error of 6.88 kg. Comparing the PointNet model results to that of a volume-based method, there was an improved accuracy ($R^2 = 0.94$ compared with 0.75). Concurrently, ref. [49] there was the development of a non-contact model using point cloud data from the dorsal evaluation of the pig's anatomy, with there being an absolute error of 11.552 kg and a relative error of 4.812%. The findings in these studies highlight the potential of 3D point cloud technology to provide accurate and noninvasive techniques for estimating pig body weight. Challenges with point cloud technology remain, such as the computational requirements for processing large point cloud datasets, interference from environmental factors such as dust and airborne particles, and the need for standardization across different sensor systems. Nonetheless, advancements in hardware and software have promise to improve the potential of 3D point cloud technology in food-producing animal management.

While the findings from these results are promising, it is essential to assess the robustness of models when analyzing datasets that have not been previously evaluated using this model. Image data can vary in terms of environmental conditions, lighting, position of the animal, size, color, and age. For example, models trained on specific breeds or age groups will likely not be effective when evaluating animals of other ages, e.g., as in [20], because models are trained on specific key anatomical features. Depth data accuracy is also affected by hardware variability, camera angles, lighting conditions, and housing environment. These changes can potentially limit the robustness of a depth estimation model in less controlled settings. Depth imaging technology is highly susceptible to "noise" resulting from dust in the air, ambient lighting, and the occlusion of objects, impacting the precision of 3D measurements collected from these types of sensors. To address these challenges, improvements include the collection of more diverse datasets across different enterprises, breeds, age groups, and other conditions to improve the breadth of model utilization. Incorporating advanced filtering techniques or using more sophisticated sensors like a multi-camera array can help reduce noise and increase the resolution of captured data. The integration of CNN's successful 2D image tasks, with depth sensors to capture complex 3D data patterns, has recently been utilized to conduct studies with food-producing animals. The results from recent studies indicate the practical efficacy of these approaches. For example, in [50], the utilization of depth images and CNNs, enhanced with transfer learning and model ensembling, to estimate body condition scores in dairy cows, resulted in a relatively precise accuracy (82% within 0.25 BCS units; 97% within 0.50 units). Similarly, an automated cow body condition scoring system using multiple 3D cameras has been developed, along with the training of independent CNN models and combining the estimations through ensemble modeling for significantly improved accuracy [51]. These findings suggest that similar techniques can be effectively applied to other food-producing animals, including swine.

4.2. Scoring of Feet and Leg Structural Soundness

Before the advent of deep learning technology, there was the evaluation of objective and automated techniques to quantify animal leg soundness [52–55]. These studies were initially conducted with horses and cattle but were the foundational information for a similar assessment of pigs. Subsequently, pig-related research was conducted where there were evaluations of sow feet and leg soundness traits using video recordings [56]. There, however, was a small correlation between values for manually measured traits and those obtained using automated techniques, which is indicative of the need for more precise measurement approaches. There were also computer vision techniques evaluated to assess the feet and leg structural soundness of pigs [57]. The results from these studies are indicative of the fact that computer vision has potential to assess intricate details, such as joint angles, stride lengths, and gait patterns, ultimately allowing for a more precise assessment opportunity for physical traits, such as feet and leg soundness.

4.3. Challenges in Adopting Machine Learning Technologies

While many of the previously discussed technologies can be used without internet connectivity, the development, testing, and refinement of machine learning models involve significant computational complexity and large dataset management requirements. Training datasets for models can range from a few gigabytes to hundreds of gigabytes, as demonstrated by version 3.5 of the large language model (LLM) ChatGPT (OpenAI, San Francisco, CA, USA), where there was the utilization of 570 GB of training data [58]. Beyond connectivity, the computational infrastructure needed to train and utilize these models requires access to specialized hardware, such as GPUs or TPUs, and robust storage systems that are inaccessible on most rural farms. This makes deployment in these technologies challenging unless cloud-based solutions or edge computing devices are leveraged.

Edge computing, a technology that processes data locally on devices rather than relying on centralized infrastructure, offers a promising solution by enabling real-time data processing. This reduces dependency on high-speed internet and is particularly advantageous in rural settings, where local devices can perform important computations and continue to function independently from internet connectivity. By integrating cloud computing with local deployments, farmers can benefit from a hybrid system that enhances feasibility and functionality in rural farms. For example, the results from one study [59] are indicative of how cloud computing can be integrated with remote sensing technologies to optimize farming practices. This system was tested in rural agricultural settings and addressed network and resource limitations to enhance crop yield and resource efficiency. This integrated system highlights the feasibility of combining cloud-based and localized systems to improve the adoption of machine learning in resource-constrained settings.

Furthermore, adopting ML technologies requires technical expertise in data management, model training, and integration into farm workflows. Many farmers lack access to training or resources to effectively use this software and hardware-based tools, creating a barrier to machine learning adoption on the farm. Lightweight models, such as MobileNet [60], or techniques such as model pruning, are specifically designed for resource-constrained settings. These approaches make ML tools more accessible and practical for farmers in rural areas. For example, MobileNet employs depthwise separable convolutions to reduce computational complexity, resulting in less power consumption and faster inference times when compared to traditional convolutional neural networks. The results indicated that the quantized version of SSD-MobileNet-v2 has an inference time of ~68.96 ms with a COCO mAP of 60.99, while the default SSD model requires greater than 120 ms per inference with a similar mAP score [58]. This increase in speed makes lightweight models particularly advantageous for deployment in rural settings where computational resources are limited. Even when edge computing devices are used to process the data locally, there are still challenges, such as maintenance, software updates, and troubleshooting of the devices. These problems require skilled personnel, which adds to operational costs. Furthermore, ethical considerations such as data privacy and security remain pertinent, particularly when transmitting sensitive food animal production unit-derived data to centralized data centers.

High-speed internet connectivity is necessary for transmitting large datasets to data centers for model training and integration. Many food animal producers in rural areas in the USA still lack access to advanced broadband services (100 mb/s download, 20 mb/s upload) [61,62]. This lack of connectivity limits the adoption of technologies such as centralized cloud-based herd management and video-assisted evaluations of food-producing animals, which require robust broadband for real-time analysis. While these techniques have potential, the reliance on stable network connections creates barriers in rural areas without sufficient infrastructure. To address this, offline deployment strategies, such as preloading essential data or enabling models to inference offline [63], can ensure uninterrupted functionality. Lightweight ML models, such as MobileNet, further enhance this by allowing for efficient computations on resource-constrained devices. When combined with hybrid approaches that integrate edge and cloud computing, localized real-time processing can be achieved while still periodically syncing with the cloud for updates and more precise analytics. These layered solutions integrate the benefits of cloud-based systems, lightweight models, and offline strategies, making these effective for overcoming connectivity challenges in rural settings.

5. Conclusions

The evolution of the pork production industry has been marked by continuous changes and a quest for precision, objectivity, and optimization. While traditional techniques have served the industry for centuries, the limitations of manual techniques, particularly for pig body weight estimation and feet and leg structural soundness, have become evident. Computer vision approaches allow for objective, automated, and noninvasive techniques to be utilized for evaluating pigs. From trained deep learning detection models to complex algorithms for 3D feature extraction, the advancements in computer vision procedures are going to rapidly change the way producers assess pigs in pork production systems. As the industry continues to evolve, there is no doubt that computer vision approaches will be pivotal in shaping the future of pork production.

There is an opportunity to integrate computer vision technologies with walk-over-weighing technologies for evaluating structural characteristics (e.g., structural soundness integrity) and sow body weight changes in pork production systems. The further refinement of precision technologies and access to adequate internet connectivity will enhance technology implementation in pork production, similar to advancements in other food animal (e.g., dairy) and food grain production systems.

Author Contributions: Conceptualization, R.J., J.E.K. and C.R.; methodology, R.J., J.E.K., C.R. and T.W.; software, R.J. and C.R.; validation, R.J., C.R., T.W. and J.E.K.; formal analysis, R.J. and J.E.K.; investigation, R.J. and J.E.K.; resources, R.J. and J.E.K.; data curation, R.J. and C.R.; writing—original draft preparation, R.J.; writing—review and editing, R.J., C.R., T.W., W.H., J.E.K. and M.T.; visualization, R.J.; supervision, J.E.K.; project administration, J.E.K.; funding acquisition, not applicable. All authors have read and agreed to the published version of the manuscript.

Funding: This research received no external funding.

Institutional Review Board Statement: Not applicable.

Informed Consent Statement: Not applicable.

Data Availability Statement: No new data were created or analyzed in this study.

Conflicts of Interest: Author Ryan Jeon was employed by the company Integer Technologies LLC. The remaining authors declare that the research was conducted in the absence of any commercial or financial relationships that could be construed as a potential conflict of interest.

References

1. Black, N.J.; Arruda, A.G. Turnover events of animal caretakers and it impact on productivity in swine farms. *Prev. Vet. Med.* **2021**, *193*, 105418. [CrossRef] [PubMed]
2. Moeller, G.A.; Stalder, K.J. Sow Longevity. In *Advances in Pig Welfare*, 2nd ed.; Camerlink, I., Baxter, E.M., Eds.; Series in Food Science, Technology and Nutrition; Woodhead Publishing: Sawston, UK, 2024; pp. 163–184. [CrossRef]
3. Cabezon, F.; Schinckel, A.P.; Que, Y. Evaluation of statistics to be used to quantify the magnitude of errors in the sorting of pigs for market via simulation. *Prof. Anim. Sci.* **2016**, *32*, 495–506. [CrossRef]
4. Lucia, T.; Dial, G.D.; Marsh, W.E. Lifetime reproductive performance in female pigs having distinct reasons for removal. *Livest. Prod. Sci.* **2000**, *63*, 213–222. [CrossRef]
5. Korvar, D.R. Review: Current challenges in poultry nutrition, health, and welfare. *Animal* **2023**, *17*, 100755. [CrossRef] [PubMed]
6. De Vries, A.; Marcondes, M.I. Review: Overview of factors affecting productive lifespan of dairy cows. *Animal* **2020**, *14*, 155–164. [CrossRef] [PubMed]
7. Stalder, K.J.; Lacy, R.C.; Cross, T.L.; Conatser, G.E. Financial impact of average parity of culled females in a breed-to-wean swine operation using replacement gilt net present value analysis. *J. Swine Health Prod.* **2003**, *11*, 69–74. [CrossRef]
8. D'Allaire, S.; Stein, T.E.; Leman, A.D. Culling patterns in selected Minnesota swine breeding herds. *Can. J. Vet. Res.* **1987**, *51*, 506–512. Available online: https://pubmed.ncbi.nlm.nih.gov/3453273/ (accessed on 20 December 2024). [PubMed]
9. Zhao, Y.; Liu, X.; Mo, D.; Chen, Q.; Chen, Y. Analysis of reasons for sow culling and seasonal effects on reproductive disorders in Southern China. *Anim. Reprod. Sci.* **2015**, *159*, 191–197. [CrossRef] [PubMed]

10. Engblom, L.; Lundeheim, N.; Dalin, A.M.; Andersson, K. Sow removal in Swedish commercial herds. *Livest. Sci.* **2007**, *106*, 76–86. [CrossRef]
11. Stein, T.E.; Dijkhuizen, A.; D'Allaire, S.; Morris, R.S. Sow culling and mortality in commercial swine breeding herds. *Prev. Vet. Med.* **1990**, *9*, 85–94. [CrossRef]
12. Lisgara, M.; Skampardonis, V.; Kouroupides, S.; Leontides, L. Hoof lesions and lameness in sows in three Greek swine herds. *J. Swine Health Prod.* **2015**, *23*, 244–251. [CrossRef]
13. Authement, M.R.; Knauer, M.T. Associations between Sow Body Lesions with Body Condition and Subsequent Reproductive Performance. *Open J. Vet. Med.* **2023**, *13*, 111–121. [CrossRef]
14. Kikuta, M.; Preis, G.M.; Deen, J.; Pinilla, J.C.; Corzo, C.A. Sow mortality in a pig production system in the midwestern USA: Reasons for removal and factors associated with increased mortality. *Vet. Rec.* **2023**, *192*, 7. [CrossRef] [PubMed]
15. Heinonen, M.; Peltoniemi, O.; Valros, A. Impact of lameness and claw lesions in sows on welfare, health and production. *Livest. Sci.* **2013**, *156*, 2–9. [CrossRef]
16. Vargovic, L.; Athorn, R.Z.; Hermesch, S.; Bunter, K.L. Improving sow welfare and outcomes in the farrowing house by identifying early indicators from pre-farrowing assessment. *J. Anim. Sci.* **2022**, *100*, skac294. [CrossRef] [PubMed]
17. Van Steenbergen, E.J. Description and evaluation of a linear scoring system for exterior traits in pigs. *Livest. Prod. Sci.* **1989**, *23*, 163–181. [CrossRef]
18. Main, D.C.J.; Clegg, J.; Spatz, A.; Green, L.E. Repeatability of a lameness scoring system for finishing pigs. *Vet. Rec.* **2000**, *147*, 574–576. [CrossRef]
19. Van Nuffel, A.; Sprenger, M.; Tuyttens, F.A.M.; Maertens, W. Cow gait scores and kinematic gait data: Can people see gait irregularities? *Anim. Welf.* **2009**, *18*, 433–439. [CrossRef]
20. D'Eath, R.B. Repeated locomotion scoring of a sow herd to measure lameness: Consistency over time, the effect of sow characteristics and inter-observer reliability. *Anim. Welf.* **2012**, *21*, 219–231. [CrossRef]
21. Stock, J.D.; Calderón Díaz, J.A.; Abell, C.E.; Baas, T.J.; Mote, B.F.; Rothchild, M.F.; Stalder, K.J. Development of an Objective Feet and Leg Conformation Evaluation Method Using Digital Imagery in Swine. *J. Anim. Sci. Livest. Prod.* **2018**, *96*, 3549–3557. [CrossRef]
22. Jeon, R.L.; Peschel, J.M.; Ramirez, B.C.; Stock, J.D.; Stalder, K.J. Deep Learning Based Landmark Detection for Measuring Hock and Knee Angles in Sows. *Transl. Anim. Sci.* **2024**, *8*, txad033. [CrossRef] [PubMed]
23. Sadeghi, E.; Kappers, C.; Chiumento, A.; Derks, M.; Havinga, P. Improving piglets health and well-being: A review of piglets health indicators and related sensing technologies. *Smart Agric. Technol.* **2023**, *5*, 100246. [CrossRef]
24. Noblet, J.; van Milgen, J. Energy value of pig feeds: Effect of pig body weight and energy evaluation system. *J. Anim. Sci.* **2004**, *82*, E229–E238. Available online: https://pubmed.ncbi.nlm.nih.gov/15471802 (accessed on 20 December 2024).
25. Phillips, R.W.; Dawson, W.M. A study of methods for obtaining measurements of swine. *J. Anim. Sci.* **1936**, *1*, 93–99. [CrossRef]
26. Brandl, N.; Jørgensen, E. Determination of live weight of pigs from dimensions measured using image analysis. *Comp. Electron. Agric.* **1996**, *15*, 57–72. [CrossRef]
27. Walugembe, M.; Nadiope, G.; Stock, J.D.; Stalder, K.J.; Pezo, D.; Rothschild, M.F. Prediction of live body weight using various body measurements in Ugandan village pigs. *Livest. Res. Rural Dev.* **2014**, *26*, 5. Available online: https://www.researchgate.net/publication/288185380 (accessed on 20 December 2024).
28. Smith, R.A.; Turner, M.J.B. Electronic aids for use in fastock weighing. *J. Agric. Eng. Res.* **1974**, *19*, 299–311. [CrossRef]
29. Turner, M.J.B.; Smith, R.A. Recent developments in the handling and weighing of farm animals. *Livest. Prod. Sci.* **1975**, *2*, 289–295. [CrossRef]
30. LeCun, Y.; Bengio, Y.; Hinton, G. Deep learning. *Nature* **2015**, *521*, 436–444. [CrossRef]
31. Forsyth, A.D.; Ponce, J. *Computer Vision: A Modern Approach*; Prentice Hall: Hoboken, NJ, USA, 2002; p. 1.
32. Girshick, R. Fast R-CNN. *arXiv* **2015**, arXiv:1504.08083. Available online: https://arxiv.org/abs/1504.08083 (accessed on 20 December 2024).
33. Redmon, J.; Divvala, S.; Girshick, R.; Farhadi, A. You only look once: Unified, real-time object detection. In Proceedings of the IEEE Computer Society Conference on Computer Vision and Pattern Recognition, Las Vegas, NV, USA, 27–30 June 2016. [CrossRef]
34. Suwannakhun, S.; Daungmala, P. Estimating Pig Weight with Digital Image Processing using Deep Learning. In Proceedings of the 14th International Conference on Signal Image Technology and Internet Based Systems (SITIS), Las Palmas de Gran Canaria, Spain, 26–29 November 2018; pp. 320–326. [CrossRef]
35. Cang, Y.; He, H.; Qiao, Y. An Intelligent Pig Weights Estimate Method Based on Deep Learning in Sow Stall Environments. *IEEE Access* **2019**, *7*, 164867–164875. [CrossRef]
36. Schofield, C.P. Evaluation of image analysis as a means of estimating the weight of pigs. *J. Agric. Eng. Res.* **1990**, *47*, 287–296. [CrossRef]

37. Herlocker, J.L.; Konstan, J.A.; Terveen, L.G.; Riedl, J.T. Evaluating Collaborative Filtering Recommender Systems. *ACM Trans. Inf. Syst.* **2004**, *22*, 5–53. [CrossRef]
38. Rezatofighi, H.; Tsoi, N.; Gwak, J.; Sadeghian, A.; Reid, I.; Savarese, S. Generalized Intersection over Union: A Metric and A Loss for Bounding Box Regression. *arXiv* **2019**, arXiv:1902.09630. [CrossRef]
39. Powers, D.M.W. Evaluation: From precision, recall and F-measure to ROC, informedness, markedness and correlation. *arXiv* **2020**, arXiv:2010:16061. [CrossRef]
40. Minagawa, H.; Ichikawa, T. Determining the Weight of Pigs with Image Analysis. *Trans. ASAE* **1992**, *37*, 1011–1015. [CrossRef]
41. Marchant, A.; Schofield, C.P.; White, R.P. Pig growth and conformation monitoring using image analysis. *Anim. Sci.* **1999**, *68*, 141–150. [CrossRef]
42. Kashiha, M.A.; Bahr, C.; Ott, S.; Moons, C.P.H.; Niewold, T.A.; Tuyttens, F.; Berckmans, D. Automatic monitoring of pig locomotion using image analysis. *Livest. Sci.* **2014**, *159*, 141–148. [CrossRef]
43. Kongsro, J. Development of a computer vision system to monitor pig locomotion. *Open J. Anim. Sci.* **2013**, *3*, 254–260. [CrossRef]
44. Yu, H.; Lee, K.; Morota, G. Forecasting dynamic body weight of nonrestrained pigs from images using an RGB-D sensor camera txab006. *Trans. Anim. Sci.* **2021**, *5*, txab006. [CrossRef]
45. Condotta, I.C.F.S.; Brown-Brandl, T.M.; Silva-Miranda, K.O.; Stinn, J.P. Evaluation of a depth sensor for mass estimation of growing and finishing pigs. *Biosyst. Eng.* **2018**, *173*, 11–18. [CrossRef]
46. Tasdemir, S.; Ozkan, I.A. ANN approach for estimation of cow weight depending on photogrammetric body dimensions. Inter. *J. Eng. Geosci.* **2019**, *4*, 36–44. [CrossRef]
47. Liu, Z.; Hua, J.; Xue, H.; Tian, H.; Chen, Y.; Liu, H. Body Weight Estimation for Pigs Based on 3D Hybrid Filter and Convolutional Neural Network. *Sensors* **2023**, *23*, 7730. [CrossRef] [PubMed]
48. Paudel, S.; de Sousa, R.V.; Sharma, S.R.; Brown-Brandl, T. Deep Learning Models to Predict Finishing Pig Weight Using Point Clouds. *Animals* **2024**, *14*, 31. [CrossRef] [PubMed]
49. Liu, Y.; Zhou, J.; Bian, Y.; Wang, T.; Xue, H.; Liu, L. Estimation of Weight and Body Measurement Model for Pigs Based on Back Point Cloud Data. *Animals* **2024**, *14*, 1046. [CrossRef]
50. Rodríguez Alvarez, J.; Arroqui, M.; Mangudo, P.; Toloza, J.; Jatip, D.; Rodriguez, J.M.; Teyseyre, A.; Sanz, C.; Zunino, A.; Machado, C.; et al. Estimating Body Condition Score in Dairy Cows From Depth Images Using Convolutional Neural Networks, Transfer Learning and Model Ensembling Techniques. *Agronomy* **2019**, *9*, 90. [CrossRef]
51. Summerfield, G.I.; De Freitas, A.; van Marle-Koster, E.; Myburgh, H.C. Automated Cow Body Condition Scoring Using Multiple 3D Cameras and Convolutional Neural Networks. *Sensors* **2023**, *23*, 9051. [CrossRef] [PubMed]
52. Ratzlaff, M.H.; Hyde, M.L.; Hutton, D.V.; Rathgeber, R.A.; Balch, O.K. Interrelationships between moisture content of the track, dynamic properties of the track and the locomotor forces exerted by galloping horses. *J. Equine Vet. Sci.* **1997**, *17*, 35–42. [CrossRef]
53. Auer, J.A.; Fackelman, G.E.; Gingerich, D.A.; Fetter, A.W. Effect of hyaluronic acid in naturally occurring and experimentally induced osteoarthritis. *Am. J. Vet. Res.* **1980**, *41*, 568–574. Available online: https://pubmed.ncbi.nlm.nih.gov/7406275/ (accessed on 20 December 2024). [CrossRef]
54. Imamura, S.; Zin, T.T.; Kobayashi, I.; Horii, Y. Automatic Evaluation of Cow's Body-Condition-Score Using 3D Camera. In Proceedings of the 2017 IEEE 6th Global Conference on Consumer Electronics (GCCE 2017), Nagoya, Japan, 24–27 October 2017; pp. 1–2. [CrossRef]
55. Zin, T.T.; Seint, P.T.; Tin, P.; Horii, Y.; Kobayashi, I. Body Condition Score Estimation Based on Regression Analysis Using a 3D Camera. *Sensors* **2020**, *20*, 3705. [CrossRef] [PubMed]
56. Calabotta, D.F.; Kornegay, E.T.; Thomas, H.R.; Knight, J.W.; Notter, D.R.; Veit, H.P. Restricted energy intake and elevated calcium and phosphorus intake for gilts during growth. I. Feedlot performance and foot and leg measurements and scores during growth. *J. Anim. Sci.* **1982**, *54*, 565–575. [CrossRef]
57. Zhang, A.L.N.; Wu, B.P.; Jiang, C.X.H.; Xuan, D.C.Z.; Ma, E.Y.H.; Zhang, F.Y.A. Development and validation of a visual image analysis for monitoring the body size of sheep. *J. Appl. Anim. Res.* **2018**, *46*, 1004–1015. [CrossRef]
58. Brown, T.B.; Mann, B.; Ryder, N.; Subbiah, M.J.; Kaplan, J.; Dhariwal, P.; Neelakantan, A.; Shyam, P.; Sastry, G.; Askell, A.; et al. Language Models are Few-Shot Learners. *arXiv* **2020**, arXiv:2005.14165. [CrossRef]
59. Yadav, A.L.; Khare, S.; Talwandi, N.S. Cloud-Based Agricultural Monitoring System for Precision Farming. In Proceedings of the 2024 11th International Conference on Reliability, Infocom Technologies and Optimization (Trends and Future Directions) (ICRITO), Noida, India, 14–15 March 2024; pp. 1–6. [CrossRef]
60. Vasu, P.K.A.; Gabriel, J.; Zhu, J.; Tuzel, O.; Ranjan, A. MobileOne: An Improved One Millisecond Mobile Backbone. *arXiv* **2023**, arXiv:2206.04040. Available online: https://arxiv.org/abs/2206.04040 (accessed on 20 December 2024).

61. Federal Communications Commission. Broadband Progress Report. 2016. Available online: https://docs.fcc.gov/public/attachments/FCC-16-6A1.pdf (accessed on 20 December 2024).
62. Federal Communications Commission. FCC Fact Sheet. 2024. Available online: https://docs.fcc.gov/public/attachments/FCC-24-6A1.pdf (accessed on 20 December 2024).
63. Reis, D.; Kupec, J.; Hong, J.; Daoudi, A. Real-Time Flying Object Detection with YOLOv8. *arXiv* **2024**, arXiv:2305.09972. Available online: https://arxiv.org/abs/2305.09972 (accessed on 20 December 2024).

animals

Article

An Interactive Feeder to Induce and Assess Emotions from Vocalisations of Chickens

Antonis Golfidis, Buddhamas Pralle Kriengwatana *, Mina Mounir and Tomas Norton *

Faculty of Bioscience Engineering, Katholieke Universiteit Leuven (KU LEUVEN), Kasteelpark Arenberg 30, 3001 Leuven, Belgium; antonis.gkolfidis@kuleuven.be (A.G.); mina.mounir@kuleuven.be (M.M.)
* Correspondence: pralle.kriengwatana@kuleuven.be (B.P.K.); tomas.norton@kuleuven.be (T.N.)

Simple Summary: The vocalisations that an animal produces could provide a window into its emotional state. More knowledge on how emotional states are expressed in the vocalisations of birds could even be used to improve the welfare of farmed poultry. The present study introduced a novel device designed to trigger different emotional states in hens using different physical and chemical stimuli. The final device was able to elicit a broad range of vocalisations without any human interference. It was found that the hens not only actively used the device as a feeder but also responded vocally to both positive and negative stimuli. Preliminary findings indicate that the vocal responses of the hens vary as a function of the intensity of the emotional state they experience. This research contributes to understanding poultry emotions and has the potential to open new opportunities in understanding and improving their welfare in farming environments.

Abstract: Understanding the emotional states of animals is a long-standing research endeavour that has clear applications in animal welfare. Vocalisations are emerging as a promising way to assess both positive and negative emotional states. However, the vocal expression of emotions in birds is a relatively unexplored research area. The goal of this study was to develop an interactive feeding system that would elicit positive and negative emotional states, and collect recordings of the vocal expression of these emotions without human interference. In this paper, the mechatronic design and development of the feeder is described. Design choices were motivated by the desire for the hens to voluntarily interact with the feeder and experience the different stimuli that were designed to induce (1) positive low-arousal, (2) positive high-arousal, (3) negative low-arousal, and (4) negative high-arousal states. The results showed that hens were motivated to engage with the feeder despite the risk of receiving negative stimuli and that this motivation was sustained for at least 1 week. The potential of using the interactive feeder to analyse chicken vocalisations related to emotional valence and arousal is being explored, offering a novel proof of concept in animal welfare research. Preliminary findings suggest that hens vocalised in response to all four stimulus types, with the number of vocalisations, but not the probability of vocalising, distinguishing between low- and high-arousal states. Thus, the proposed animal–computer interaction design has potential to be used as an enrichment device and for future experiments on vocal emotions in birds.

Keywords: vocal emotions; animal–computer interaction; laying hens; chickens; vocalisations; birds

1. Introduction

The welfare of captive non-human animals (hereafter, animals) under human care is a growing public concern, with both direct and indirect impacts on human health and well-being and environmental sustainability. Recent scientific interest in the role of emotional states in animal welfare is motivated by the idea that good welfare should not only be characterised as a lack of negative states but also include the opportunity to experience positive emotional states [1–3]. In short, animals should not only survive, but thrive in environments that provide a balance between what an animal likes to do and what is healthy for it to do. Thus, knowing when an animal is in a positive emotional state remains a critical and ongoing research endeavour in animal welfare sciences.

The scientific study of animal emotions is flourishing, being supported by different conceptual frameworks. These frameworks can enable cross-species comparisons of emotional states and are complemented by technological and methodological advancements that facilitate the observation and quantification of these states [4,5]. Emotions can be defined as internal, short-lived psychobiological states that are reactions to specific internal or external objects or events that are biologically or ecologically relevant to an individual. Emotions are conceptually different from moods, which are more persistent internal states that are not necessarily triggered by a specific object or event [5]. Animal emotions can be viewed as consisting of two dimensions: arousal and valence [4]. Arousal refers to the intensity of the emotion while valence refers to the hedonic value of the emotion (positive or negative). Although there is no access to animals' subjective emotional experiences, they are likely accompanied by changes in arousal and valence that can be measured as changes in behaviour, cognitive processing, physiology, and neural activity [4,5].

Vocalisations are emerging as promising non-invasive measures of animal emotional states [6]. Vocalisations are produced by almost all vertebrates and may be reliable markers of emotions because the midbrain and limbic systems that process emotional stimuli also play a role in vocal production [7–9]. In vertebrates, the arousal response includes elevation of sympathetic nervous system activity, which leads to a higher respiratory rate and muscle tension, both of which modulate acoustic features of vocalisations [10].

Research into acoustic features that distinguish vocalisations produced in positive and negative emotional states shows that across several mammalian species, vocalisations produced in positive emotional states tend to be shorter and have a lower fundamental frequency (although species differences exist [6,7,11,12]). For instance, horse whinnies produced during positive situations (social reunion) were lower pitched and shorter in duration than whinnies produced during negative situations (social isolation; [13]). Similarly, grunts, screams, and squeals of wild boars during positive situations (food anticipation and affiliative interactions) were shorter in duration with less amplitude modulation and lower frequency than in negative situations (agonistic interactions) [14]. Compared to mammals, little progress has been made on the identification of vocal indicators of positive emotions in birds. Research on vocal emotions in birds is highly concentrated on negative emotions, with a large bias towards high-arousal calls that signal current levels of pain, fear, or distress [6,7,15].

The aim of this study was to design a device capable of stimulating domestic layer hens to vocalise across different levels of emotional arousal and valence. To the best of the authors' knowledge, no existing device has been able to induce vocalisations across all four quadrants of the dimensional model of emotions, i.e., positive low-arousal, positive high-arousal, negative low-arousal, and negative high-arousal emotional states. The design was inspired by findings of two studies on vocalisations during positive emotional states in domestic chickens, which observed variations in call types—namely food calls, fast clucks, and gakels—based on the anticipation of rewards [16,17]. While [17] found that the peak frequency increased with the arousal level in positively valenced situations, neither study analysed whether the acoustic features of these calls varied between the reward and no-reward conditions. Addressing this gap, the device is designed to explore how emotional valence and arousal are encoded in vocalisations, offering new insights for poultry welfare monitoring by assessing the birds' emotional states through their calls.

The ability to measure valence and arousal is essential for linking vocalisations to specific emotional states. Given the challenge of assessing an animal's emotional valence due to the absence of direct non-invasive methods, lack of verbal communication, and the inherent subjectivity in labelling emotional states, a feeder was developed to provide a range of stimuli known to impact approach or avoid behaviours. To quantify the intensity of an emotion, behavioural and physiological indicators that correlate with levels of arousal can be measured by devices such as heart rate sensors, thermal cameras, and RGB cameras. Therefore, the setup should accommodate the requirements these devices need to capture data.

Positioned within a broader research framework, the objective is to distinguish between vocalisations based on emotional valence and arousal, necessitating a system that prompts vocal responses to both positive and negative stimuli. This paper is structured around two main objectives: the development and testing of an interactive feeder as a novel method for eliciting and analysing chicken vocalisations, serving as a proof of concept in the intersection of animal welfare and technology. To this end, various automated feeder designs are discussed while focusing on refining the optimal design to facilitate this study's goals. Key aspects of the feeder's usability are investigated, such as the birds' ability to learn how to activate the feeder, their engagement frequency, and the consistency of daily usage patterns. Furthermore, audio data from two trials are analysed to explore the likelihood and frequency of vocalisations produced by hens while interacting with the feeder. The findings presented offer preliminary insights into the potential of such interactive devices in enhancing our understanding of animal emotional states, setting a foundation for further exploration in this novel intersection of fields.

2. Materials and Methods

2.1. Feeder

2.1.1. Stimuli

The objective was to create a device that hens could voluntarily activate to deliver stimuli inducing positive and negative emotional states of different arousal intensities (low and high). Therefore, the design needed to deliver the selected stimuli in a suitable manner. Stimuli were chosen to induce low- and high-arousal states, with each assigned as positive or negative based on the previous literature linking these stimuli with approach or avoid behaviours [18–23]. The rationale for using approach and avoid behaviours as proxies of emotional valence was based on the assumption that, in general, animals will move towards stimuli that are positive and away from stimuli that are negative. Hens are required to initially approach the device to activate it. Upon activation, stimuli—positive or negative—are randomly dispensed. Thus, the birds do not know whether they will receive a positive or negative stimulus until they approach and activate the feeder. Positive stimuli such as rice and mealworms generally encourage the birds to continue approaching the corresponding container, whereas negative stimuli lead to avoidance behaviour.

The selected positive stimuli were food items, including mealworms and rice [18]. These choices were based on preliminary observations indicating that these foods elicited stronger engagement compared to other options like corn and peanuts. Rice was chosen instead of corn (which was used in [18]) because pilot data showed that when given a choice between rice, corn, mealworms, and peanuts simultaneously, hens chose to consume mealworms and rice first and in greater quantities when compared to corn or peanuts. Negative stimuli included two puffs of air to the face (one every 30 s for 1 min), four puffs of air to the face (one every 15 s for 1 min) [20,23], and quinine-coated rice derived from 1% and 4% suspensions of quinine (a bitter tasting substance) dissolved in water that was mixed with rice [21,22]. Thus, it was assumed that mealworms and rice would trigger a positive valence of varying arousal levels, and rice + 1% quinine and rice + 4% quinine would trigger negative valence of varying arousal levels. Food colouring was used to make the rice with 1% quinine green and the rice with 4% quinine blue. This was based on the hypothesis that hens could differentiate between these conditions and learn to associate the colours with the different levels of bitterness. To motivate consistent engagement, the device was first intended to act as the hens' primary source of food. The device was designed as a feeder with the capacity to dispense 7 different types of stimuli to the hens (Table 1).

Table 1. List of experimental stimuli and their assigned valence and arousal conditions.

Stimulus	Valence	Arousal
Regular Feed	-	-
Mealworms	Positive	High
Rice	Positive	Low
1% Quinine Suspension	Negative	Low
4% Quinine Suspension	Negative	High
Two air puffs	Negative	Low
Four air puffs	Negative	High

In trial 1, regular feed was given 80% of the time the device was activated, and positive and negative stimuli could each be given 10% of the time (5% for mealworms, 5% for rice, 2.5% for rice + 1% quinine, 2.5% for rice + 4% quinine, 2.5% for two air puffs, and 2.5% for four air puffs). Hens were provided with ad libitum access to regular feed in a separate container to ensure that they received enough food during the first few days before they learned how to use the feeder. However, visual observations during the first few days of trial 1 indicated that hens frequently interacted with the device even though regular feed was available. Thus, the regular feed was left in a separate container throughout the trial. In trial 2, the separate container with regular feed was maintained for the duration of the trial, but the probability of positive stimuli was increased. Positive stimuli were provided 95% of the time (85% for mealworms, 10% for rice), and negative stimuli 5% of the time (only air puffs, with 2.5% for two air puffs and 2.5% for four air puffs), to enhance the sample size for vocal responses to positive stimuli.

2.1.2. Design Ideas

In addition to dispensing food and air puffs, the design of the device should allow for the measurement of arousal levels to validate the arousal level of animals at the time of vocalisation. To non-invasively measure arousal, a thermal camera was proposed to record the surface temperature of the head (face, eye, comb, and wattle) from a profile view, as this has been shown to change in response to both positive and negative events and different magnitudes of stress [23–26]. Considering all the specifications, three potential feeder designs were conceptualised. Each design has advantages and disadvantages that are summarised in Table 2.

Table 2. Advantages and disadvantages of the proposed design ideas.

Design	Advantages	Disadvantages
Tower	– Good thermal camera view – Low risk of food spilling	– Complex mechanism for food weight – Requires six motors – Mixed uneaten food leading to waste – Questionable accessibility by hens
Tube	– No need to remove uneaten food – Partially tested by farmers – Good thermal camera view	– Challenging motorised lid construction – Risk of disruption – Potential blockage of air puff
Floor Rotator	– No need to remove uneaten food – Mimics natural foraging – Single motor required – Hard to disrupt	– Potentially poor thermal camera view – Complex elevated platform construction

Design 1: The tower design (Figure 1) features a central container with compartments for different food stimuli, equipped with a sensor and air puff outlet at hen height. Upon

activation, food is delivered from individual compartments into a central feeding area controlled by motorised panels. After the programmed amount of food is dispensed, the birds can eat in the feeding area for a set amount of time. Once this time has expired, a motorised trap door on the floor of the feeding area opens, driving uneaten food in the waste disposal container.

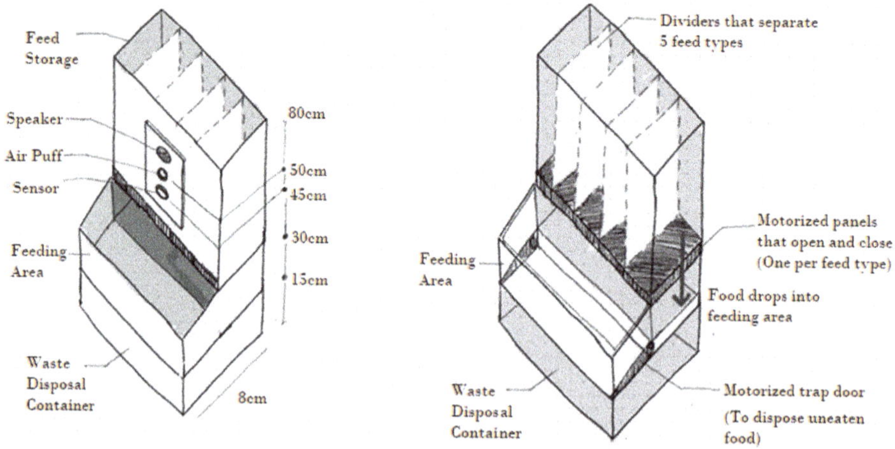

Figure 1. The tower design.

The tower design has an advantage that hens remain within a limited area, which would increase the probability of capturing usable images with a thermal camera, while minimising the risk of spilling food. However, some practical concerns arise with this design as it necessitates a complex mechanism to support the stored food's weight, only dispensing a controlled portion at a time. Additionally, it requires six motors in total: five for various food options and one dedicated to food disposal. Lastly, the potential mixing of uneaten food could lead to waste, and the placement of the sensor might require hens to stretch their necks.

Design 2: The tube design (Figure 2) consists of five independent 4″ PVC tubes, each containing a specific type of food, covered by a lid. On the lid of the central tube, there is the proximity sensor and the air puff outlet, with the speaker mounted in the central tube itself. Each tube has its own motor—connected to the central electronics box—which lifts the lid for the specified activation period, providing access to a distinct food stimulus.

The tube design has been partially used by farmers to feed chickens while addressing the challenge of removing uneaten food and has been partially used by farmers to feed chickens. It also facilitates the use of a thermal camera. The construction of the motorised lids presents its own set of challenges, both in the intricacy of the mechanism and the need for five distinct motors. A potential concern is the hens disrupting the stimulus delivery process, either by intentionally spilling food or interfering with the lids' operations. Lastly, the air puff outlet could be blocked by food.

Design 3: The floor rotator design (Figure 3) emerged from discussions with peers at the London Metropolitan University who explore ACI concepts [27]. The proposed system—ideally positioned on a raised platform—is designed with a multi-component structure. The upper section features a plastic surface with five holes along with a designated area for the interaction components, including the sensor, air puff outlet, and speaker. The lower section is intended to be at ground level to replicate hen foraging behaviour. This part houses five food containers for storing various food types and includes openings for wiring that connects the top section's components with the rest of the electronics. Between these two sections is a rotatable disc, controlled by a stepper motor, with a single hole that aligns with both the upper and lower sections' openings. To access food, hens trigger the sensor in the upper section, causing the disc to rotate and align its opening with one of the food containers, allowing the birds to feed.

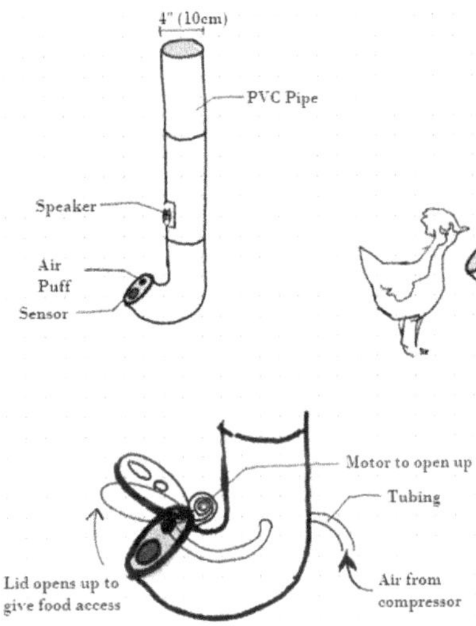

Figure 2. The tube design.

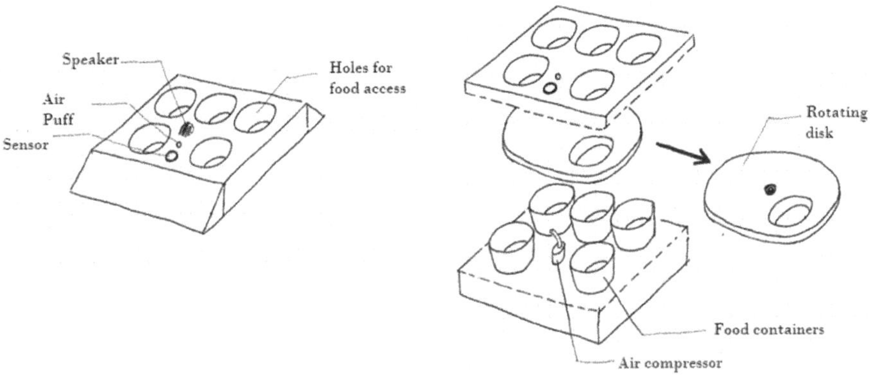

Figure 3. The floor rotator design.

The floor rotator design also bypasses the issue of uneaten food removal, while ensuring that there will be a low risk of spilling food. It also resembles the natural ground foraging behaviour, making it appealing to hens. A significant benefit lies in its electronic simplicity, anchored by a single stepper motor with some additional calibration required to fine-tune the rotation angles. Importantly, this option guarantees that the food delivery process would be hard to disrupt. Nevertheless, a couple of disadvantages remain, such as potential inconsistencies in thermal recording due to varying head angles of the hens in relation to the camera's position, and the complexity of the elevated platform construction, especially when considering the pen's dimensions and flooring.

2.1.3. Final Design

A modified version of the floor rotator design was implemented, believed to offer compelling advantages, while also addressing some of its challenges. The device is composed of two primary components: the rotatable disc and the containers. Figure 4 shows the feeder design.

Figure 4. The proposed interactive feeder. (**A**) The food containers, (**B**) the rotating disc, (**C**) the ultrasonic sensor on the side of the device, (**D**) metallic wheels aiding the rotation, (**E**) the box containing the electronics, positioned on the bottom, and (**F**) the final form of the feeder with pigeon spikes taped on the disc.

The feeder incorporates six distinct plastic containers (Figure 4A). The baseline container, which is empty of food, is the place where the hens have access while no stimulus is provided. As hens are anticipated to position their heads over this container when initiating the system, the air puff outlet is located in the baseline container. The remaining containers house the various food types. Besides these containers, the inner section of the feeder houses the stepper motor, the speaker, and the ultrasonic sensor. The sensor was fixed on the wall of the feeder (Figure 4B) so that system activation was triggered when a hen approached this particular side of the feeder.

The top layer of the device is a rotatable plastic disc, with an opening that matches the containers' dimensions. This disc is secured to the stepper motor via a screw, ensuring synchronised rotation between the motor and the disc. Metal wheels were added to reduce friction (Figure 4D). The entire apparatus stands elevated on plastic legs, ensuring hens can access the container contents easily, instead of an elevated platform (due to pen size constraints).

The core electronic components are contained in a plastic compartment under the feeder (Figure 4E). This compartment connects the USB, power cables, wires, and tubes that are necessary for the air puff mechanism to the feeder. The air puff system comprises an air compressor, a solenoid valve, and the requisite plastic tubing connecting each component. Pigeon sticks were added to the rotating disc to prevent hens from stepping on it. (refer to Discussion for more details).

2.1.4. Hardware Description

The embedded system comprises several essential components, including an Arduino Mega 2560 microcontroller, a Real-Time Clock (RTC) module, an SD card reader, an ultrasonic sensor, a miniature speaker, a pneumatic solenoid valve, an air compressor, a uStepper S microcontroller, and a Nema Stepper Motor. Figure 5 provides a visual representation of the system's components and their interconnections. Below is a brief description of each component:

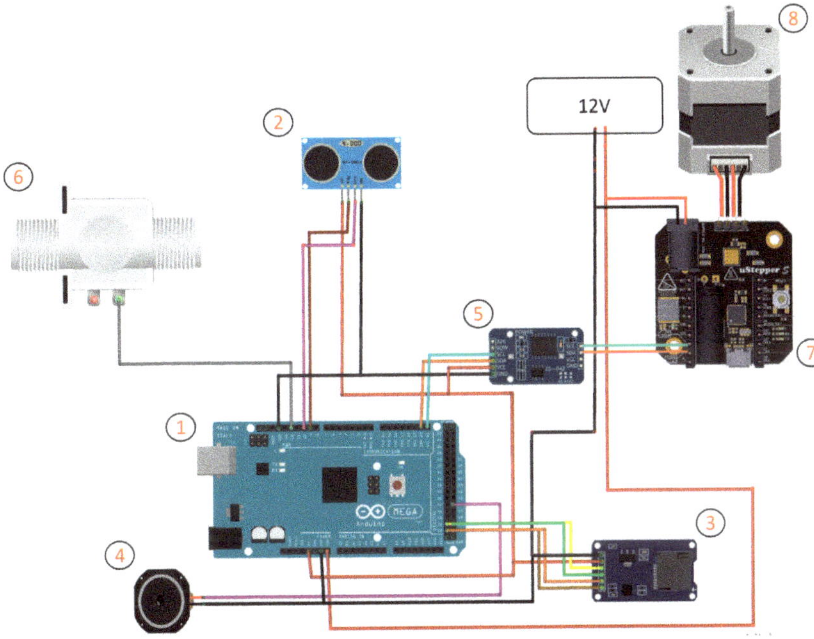

Figure 5. System's block diagram. Hardware components: (1) Arduino Mega, (2) Ultrasonic Sensor, (3) Real-Time Clock, (4) miniature speaker, (5) Micro-SD card reader, (6) pneumatic solenoid valve, (7) uStepper S stepper driver, and (8) Nema 23 stepper motor.

(1) Arduino Mega 2560: An open-source microcontroller board serving as the central control unit responsible for coordinating the operation of the entire system.

(2) HC-SR04 Ultrasonic sensor: A proximity sensor that measures distance using sound waves. It is set to detect when a hen is within 10 cm.

(3) DS3231 Real-Time Clock module: The Real-Time Clock (RTC) module reports feeder activation times, measures the duration of stimuli delivery, and determines whether the feeder should be active during the daytime or inactive during the night.

(4) Miniature speaker: A 0.5 W miniature speaker was used to emit a pure tone sinewave right before the offset of the stimulus presentation. It also serves to attract the attention of the hens, encouraging them to approach the feeder, and startles their reflexes, prompting them to withdraw their heads from the container as the disc returns to its baseline position.

(5) Micro-SD card reader SPI interface + Micro-SD card: Essential for data storage, it logs the activation times of the feeder, including details like feeder ID, stimulus ID, date, and time. Data retrieval can be achieved in various ways, such as removing the SD card after the experiment, connecting a laptop to the Arduino via USB to view activation logs in the Serial Monitor, or using a Bluetooth module for remote data acquisition to minimise human presence.

(6) Pneumatic solenoid valve: This component was integrated into the experimental apparatus to deliver controlled air puffs. Operational control of this valve is managed by the Arduino microcontroller, enabling the release of air stored in an air compressor.

(7) uStepper S stepper driver: This is a microcontroller stepper driver which is compatible with Arduino boards. The two controllers, namely the uStepper S and the Arduino Mega, communicate using the I2C protocol, with the Arduino acting as the leader. This bidirectional interaction allows for the exchange of messages between the controllers. Specifically, the Arduino Mega sends instructions specifying the desired angle for the stepper motor, while the uStepper S responds by reporting the current angle at which the stepper motor is positioned. This dynamic information exchange achieves accurate and real-time control over the rotatable disc's movements.
(8) Nema 23 stepper motor: A Nema stepper motor is an electric motor which converts digital input pulses from the uStepper S into precise mechanical shaft rotation in a series of equally spaced steps. The rotation parameters such as speed, acceleration, deceleration, and the exact angles were fine-tuned to ensure the smooth operation of the feeder.

2.1.5. Description of Software

The software, developed in the Arduino environment, consists of two distinct files: one tailored for the Arduino Mega board and the other for the uStepper S board. The programme's flowchart is illustrated in Figure 6. The basic steps of the algorithm are described below:

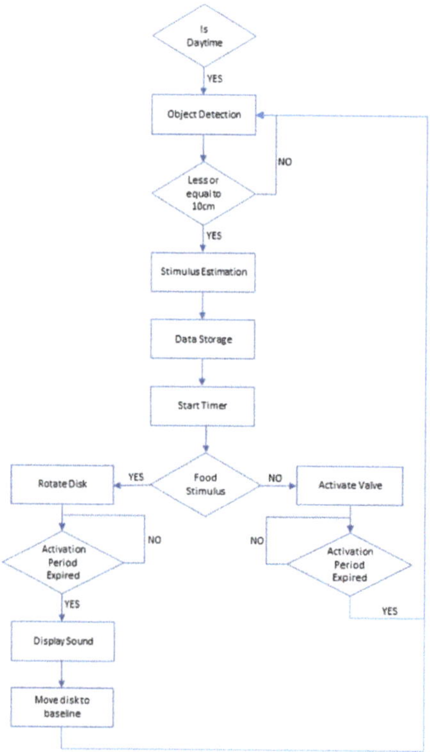

Figure 6. The flowchart of the proposed algorithm.

- Time check: The current time is verified to ensure that feeder operations occur only during the daytime (8:00 AM to 10:00 PM).
- Object detection: The system continuously assesses the distance measured by the ultrasonic sensor. If an object is detected within a 10 cm range, it is interpreted as an intention to activate the feeder.

- Stimulus estimation: Each stimulus type has an assigned probability. The system generates a random number (ranging from 0.0 to 100.0) to determine the exact stimulus to be delivered.
- Data storage: Details about the chosen stimulus, along with the timestamp of its presentation, are recorded on the SD card.

Stimulus delivery:

- Air puff: For low- or high-intensity air puffs, the Arduino activates the solenoid valve.
- Food stimulus: The Arduino sends an encoded message to the uStepper S board indicating the desired angle of rotation. The stepper motor moves the rotatable disc to the defined angle, providing access to the specified food type.
- Sound display: After the expiration of the opening period, a simple tone is produced via the miniature speaker.
- System reset: Once the stimulus delivery concludes, the stepper motor returns to its initial position. Concurrently, the ultrasonic sensor initiates a new scanning cycle to detect objects.

2.2. *Animals and Setup*

Ethical approval was obtained prior to experiments from the Ethical Committee for Animal Experimentation at KU Leuven (project number 082/2023).

2.2.1. Animals and Housing

Four ISA brown laying hens were obtained from TRANSfarm, KU Leuven. The hens were tested in pairs, in two separate trials. The hens were 30 weeks old at the start of trial 1 (from 7 July 2023 to 31 July 2023) and 39 weeks old at the start of trial 2 (from 12 September 2023 to 22 September 2023). For both trial 1 and trial 2, the experiments were concluded at midday on their respective final days. Due to unforeseen technical difficulties, the device did not deliver stimuli reliably until 24/07; thus, for trial 1, the hens' data were analysed only between 24th and 31st July, which was 15 days after the feeder was first introduced to the hens. The hens were housed in pens 2.3 × 2.3 × 0.8 m (L × W × H), with a net covering the top of the pen, in a climate-controlled room at the Department of Biosystems, KU Leuven animal facility. The ambient temperature was kept at 21 °C and humidity between 60 and 70%. The daily light/dark cycle was 14 h/10 h light/dark, with lights on between 08:00 and 22:00. The hens received ad libitum food and water in conventional poultry feeders, access to a dust-bathing substrate, a pecking stone, an elevated perching area for roosting, and nest boxes (2 per pair). Daily checks were conducted to ensure that the birds had sufficient food and water and were in good health.

2.2.2. Recording Setup

In each trial, two devices were placed in the pen (Figure 7). A directional microphone (AKG C 391 B) was set up above each device, and another omnidirectional one (AKG SE300 B) was used to record ambient sounds within the enclosure. All microphones were connected to a soundcard (Focusrite Clarett + 4Pre) which was then connected to a laptop that stored audio recordings. Recordings were made continuously (24 h/day). To capture video data, a camera was installed above the pen (Dahua DH-SD1A203T-GN). This camera was connected to a Network Video Recorder (Dahua DHI-NVR4208-8P-4KS2) that stored the video recordings. Recordings were made only when the lights were on (from 08:00 to 22:00).

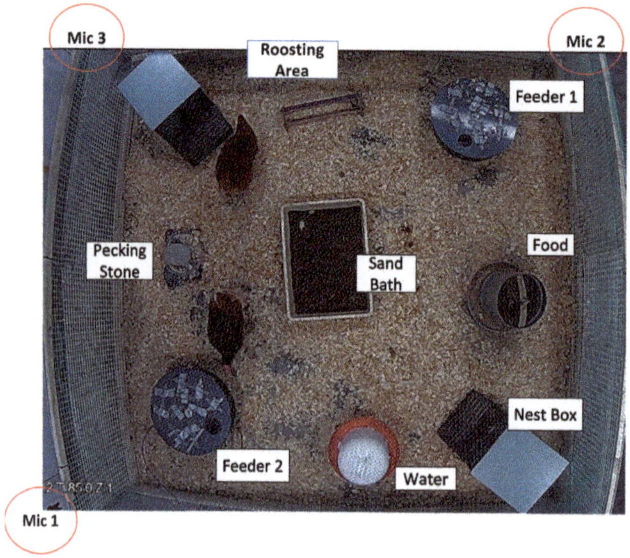

Figure 7. The pen and placement of recording devices.

2.3. Data Analysis

2.3.1. Feeder Usage Statistics

To investigate the number of feeder activations across experimental days and to identify potential hourly patterns, the activation logs retrieved from the SD card of each feeder were utilised. Following each experimental trial, feeder activation data were compiled in the form of text files. These logs contained details regarding which feeder was activated, the stimulus delivered, and the exact date and time of the activation. The text files were transformed into CSV format to enable easier further analysis steps. Subsequently, activation logs were processed using R software, version 4.3.2.

2.3.2. Vocalisations

One of the primary objectives of this study was to provide a proof of concept by analysing vocalisations emitted across different experimental stimuli. While audio data were continuously recorded, feeder activations were irregular. Thus, the initial step of the analysis involved extracting audio that corresponded to the feeder activation logs. A Python script facilitated the extraction of 1 min segments that include feeder activations—spanning 5 s before to 55 s after the activation (Figure 8). For each feeder, the audio channel from the directional microphone aimed at that feeder was utilised.

Subsequently, each segment was processed using Audacity 3.3.3 to generate its spectrogram. These extracted segments were then annotated to determine the onset and offset of hen vocalisations. The times when the feeder opened and closed were also marked. This allowed any variances in the timing of vocalisations to be explored across different stimuli. Every segment was saved along with its respective Audacity project file and text annotation file. A subsequent Python script was employed to generate a database, registering the beginning and end of each vocalisation, its clarity grade, and the file in which it was located, setting the stage for in-depth statistical analysis.

Statistical analyses on feeder usage and vocalisations were conducted using the R Statistical language (version 4.3.0; R Core Team, 2023) on macOS 14.1, using the packages MuMIn [28], glmmTMB [29], lubridate [30], DHARMa [31], chron [32], report [33], patchwork [34], ggplot2 [35], dplyr [36], and tidyr [37].

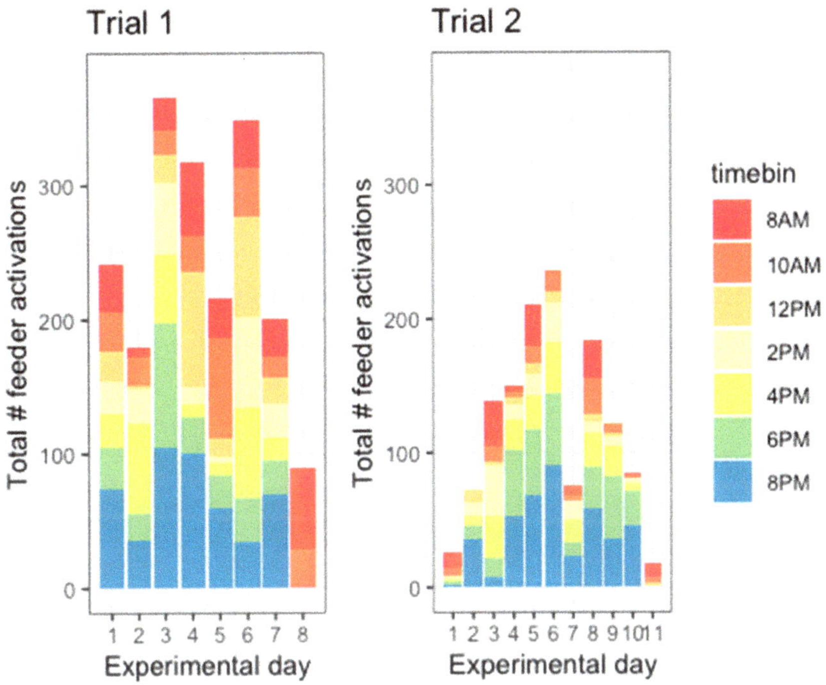

Figure 8. Number of times feeders were activated in trial 1 and trial 2 across all experimental days. Colours indicate time of day, divided into 2 h bins.

3. Results

3.1. Feeder Engagement

3.1.1. Engagement across Days

Hens activated the feeder a total of 1958 times in trial 1 (mean ± SD = 244.75 ± 93.94 times per day) and 1311 times in trial 2 (mean ± SD = 119.18 ± 72.11 times per day). Across both trials, the total number of times the birds were exposed to normal feed (Neutral) = 1571, rice (Positive + Low Arousal) = 181, mealworms (Positive + High Arousal) = 1127, rice + 1% quinine (Negative + Low Arousal) = 86, rice + 4% quinine (Negative + High Arousal) = 105, two air puffs (Negative + Low Arousal) = 117, and four air puffs (Negative + High Arousal) = 63. Feeder engagement varied between the two trials and across days (Figure 8). A generalised linear model with a negative binomial distribution was used to ask whether there was a trend for feeder engagement to increase or decrease over time. As feeder engagement and experimental day appeared to have a nonlinear relationship in trial 2, trial 1 and trial 2 were modelled separately. For both trials, a linear regression model was compared with a polynomial regression model, selecting the most parsimonious model as the model with the lowest Akaike information criterion (AIC). For trial 1, it was observed that feeder engagement did not show significant variation across days. The model's explanatory power was very weak (Nagelkerke's $R^2 = 3.15 \times 10^{-3}$). The model's intercept, corresponding to day = 0, was at 2.90 (95% CI [2.48, 3.35], $p < 0.001$). The effect of day was statistically non-significant and positive (beta = 0.02, 95% CI [−0.07, 0.12], $p = 0.640$). On the other hand, feeder engagement varied by experimental day in trial 2. The model's explanatory power was moderate (Nagelkerke's $R^2 = 0.19$). The model's intercept, corresponding to day = 0, was at 1.46 (95% CI [0.61, 2.38], $p < 0.001$). The effect of day [first degree] was statistically significant and positive (beta = 0.53, 95% CI [0.20, 0.84],

$p < 0.001$) while the effect of day [second degree] was also statistically significant and negative (beta = −0.05, 95% CI [−0.07, −0.02], $p < 0.001$).

3.1.2. Engagement within a Day

Hens interacted with the feeder throughout the day, possibly engaging more consistently in the evening (6 PM–10 PM) than in the morning (8 AM–10 AM, Figure 9). To check whether hens showed a preference for engaging with the feeder at specific times of the day, separate negative binomial models were fitted for trial 1 and trial 2 to predict the number of feeder activations by time (divided into 2 h bins). Model fit was evaluated using linear or polynomial regressions with AIC. For trial 1, the model's explanatory power was weak (Nagelkerke's $R^2 = 0.04$). The model's intercept, corresponding to timebin = 0, was at 2.70 (95% CI [2.31, 3.11], $p < 0.001$). The effect of the timebin was statistically non-significant and positive (beta = 0.07, 95% CI [−0.02, 0.16], $p = 0.135$). For trial 2, the model's explanatory power was substantial (Nagelkerke's $R^2 = 0.40$). The model's intercept, corresponding to timebin = 8 AM, was at 2.45 (95% CI [2.28, 2.63], $p < 0.001$). Within this model the effect of timebin [first degree] is statistically significant and positive (beta = 4.63, 95% CI [2.90, 6.37], $p < 0.001$). The effect of timebin [second degree] is statistically significant and positive (beta = 2.31, 95% CI [0.64, 4.01], $p = 0.009$). Thus, the hens in trial 1 did not have a preferred time to interact with the feeder, while the hens in trial 2 preferred to interact with it more in the evening.

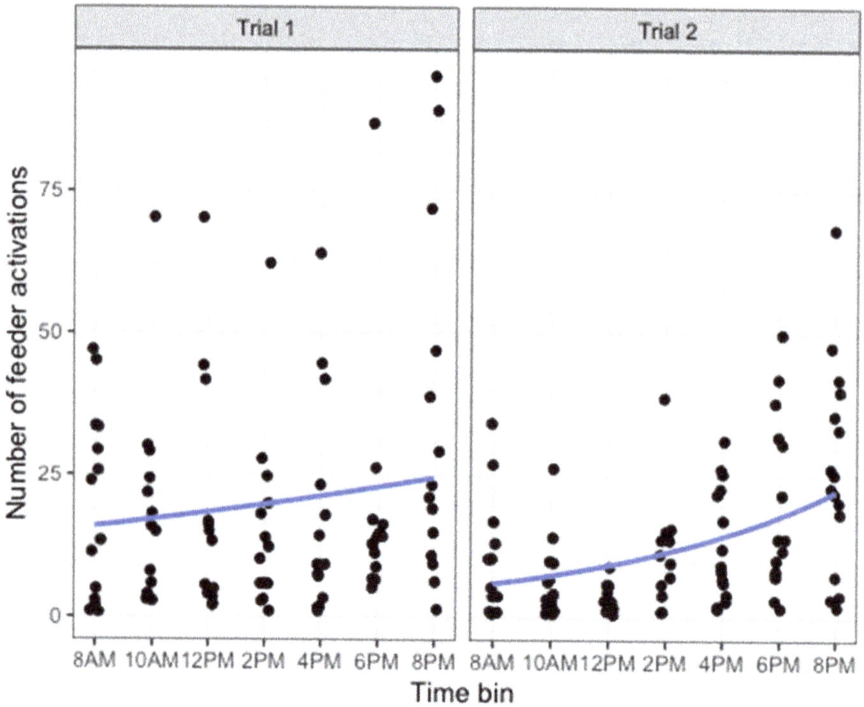

Figure 9. Number of times feeder activated at different times of the day. Time of day was divided into 2 h bins. Dots represent an activation. Blue lines show the trend across time of day.

3.2. *Vocal Activity*

The results presented herein should be considered preliminary, highlighting the interactive feeder's potential rather than providing definitive conclusions on emotional valence and arousal in chicken vocalisations.

3.2.1. Probability of Vocalising

Manual labelling of vocalisations was conducted on 930 sound files that recorded 1 min of audio after the feeder was activated. The number of files analysed for each stimulus type were normal feed (Neutral) = 200, rice (Positive + Low Arousal) = 161, meal-worms (Positive + High Arousal) = 277, rice + 1% quinine (Negative + Low Arousal) = 76, rice + 4% quinine (Negative + High Arousal) = 97, two air puffs (Negative + Low Arousal) = 77, and four air puffs (Negative + High Arousal) = 42.

Overall, it was found that the probability that hens would vocalise after activating the feeder was 51%. This value varied depending on the type of stimulus they experienced when activating the feeder (Figure 10). Hens were most likely to vocalise after receiving rice (Positive + Low Arousal) and rice coated with 4% quinine (Negative + High Arousal).

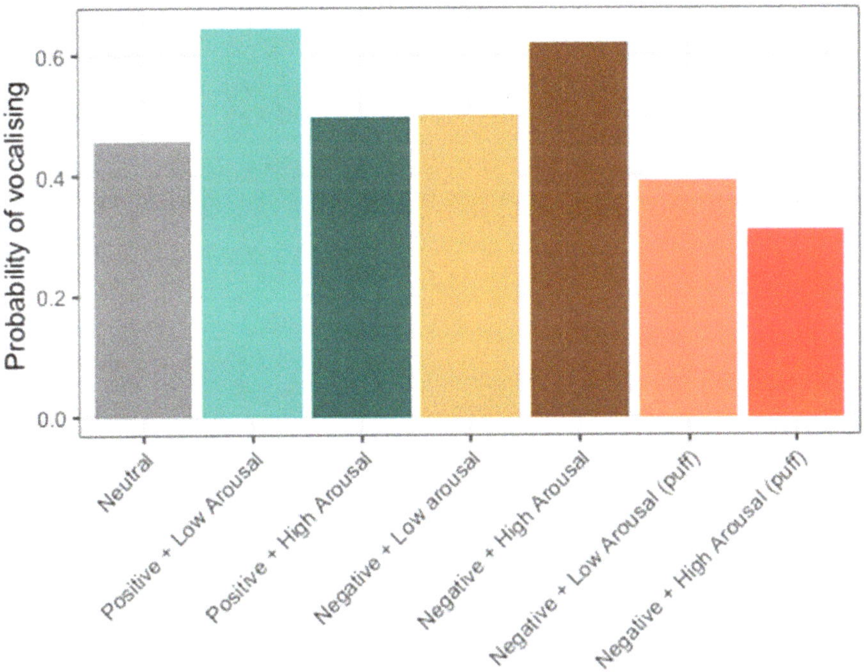

Figure 10. Probability of hens vocalising after receiving each stimulus type.

To explore whether the probability of vocalising could indicate valence or arousal, vocal activity was pooled from the negative stimulus types quinine and air puff, and we ran a generalised linear model with a binomial distribution. The stimulus' arousal and valence values (dummy coded) and the interaction of arousal and valence were used as predictors. It was found that the valence of stimuli and the interaction between arousal and valence of stimuli influenced the probability of vocalising (Figure 11); however, the model's overall explanatory power with only valence, arousal, and valence ×x arousal as predictors was very weak (Tjur's $R^2 = 0.02$). The model's intercept, corresponding to negative valence and low arousal, was at 0.12 (95% CI [−0.03, 0.27], $p = 0.130$). Within this model, the effect of valence was statistically significant and positive (beta = 0.18, 95% CI [0.03, 0.33], $p = 0.021$). The effect of arousal was statistically non-significant and negative (beta = −0.07, 95% CI [−0.22, 0.08], $p = 0.361$). The valence × arousal interaction was statistically significant and negative (beta = −0.23, 95% CI [−0.39, −0.08], $p = 0.003$). Post hoc comparisons of the interaction showed that hens were 0.56 times less likely to vocalise in low-arousal negative compared to low-arousal positive conditions ($p = 0.0021$, Table 3).

Hens were also 1.83 times more likely to vocalise in high-arousal positive than low-arousal positive conditions ($p = 0.015$, Table 3).

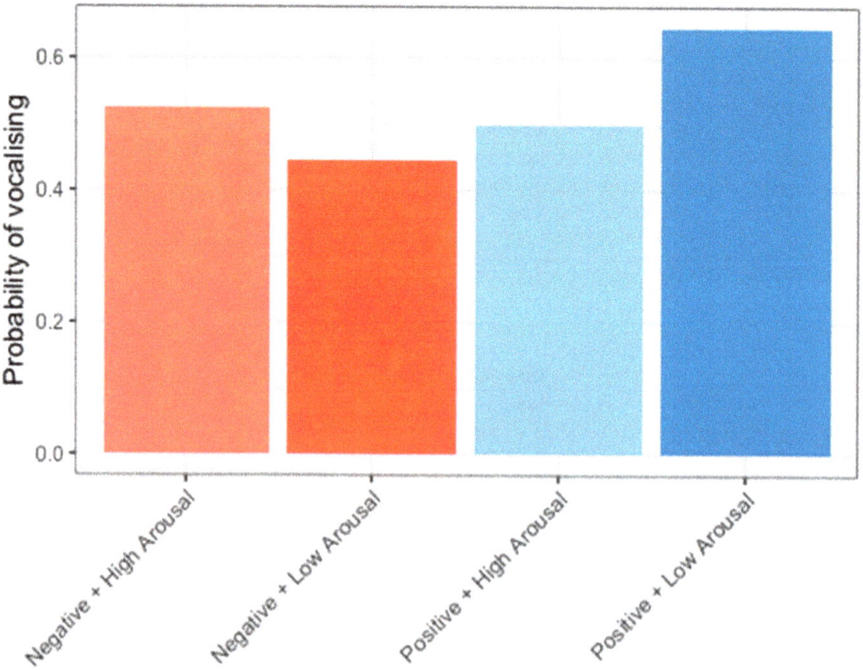

Figure 11. Probability of vocalising in response to stimuli varying in valence and arousal. The odds of hens vocalising was greater in positive low-arousal conditions than negative low-arousal and positive high-arousal conditions.

Table 3. Contingency table showing the number of times hens vocalised or did not vocalise after activating the feeder.

Condition	Negative + Low Arousal	Positive + Low Arousal	Negative + High Arousal	Positive + High Arousal
Vocalise_Yes	85	57	66	139
Vocalise_No	68	104	73	138

The probability of vocalisations exhibiting variation across the time of day and over experimental days was also evaluated. A generalised linear model with a binomial distribution was fitted to predict the probability of vocalising with experimental day, time (divided into 2 h bins from 08:00–22:00), and trial number. The model's explanatory power was weak (Tjur's $R^2 = 0.08$) but showed that the probability of vocalising varied depending on the experimental day, time of day, and trial. The model's intercept, corresponding to the first day, timebin 8 AM, and trial 1, was at 0.36 (95% CI [0.17, 0.55], $p < 0.001$). Within this model, the effect of day [first degree] was statistically significant and positive (beta = 10.66, 95% CI [6.14, 15.25], $p < 0.001$). The effect of day [second degree] was statistically significant and positive (beta = 6.69, 95% CI [2.50, 10.97], $p = 0.002$). The effect of timebin [first degree] was statistically significant and negative (beta = −10.38, 95% CI [−14.58, −6.23], $p < 0.001$). The effect of timebin [second degree] was statistically significant and negative (beta = −8.90, 95% CI [−13.05, −4.78], $p < 0.001$). The effect of trial was statistically significant and negative (beta = −0.71, 95% CI [−1.01, −0.41], $p < 0.001$). Thus, the probability of vocalising increased over experimental days and was higher in the morning and early afternoon than in the evening (Figure 12). This indicates a possible inverse relationship between feeder engagement and vocal activity. During the day, the hens used the feeder less but had a greater probability of vocalising. Similarly, over experimental days, the hens used the feeder less often but were more likely to vocalise when they did interact with it. The probability of vocalising also varied by trial, with the hens in trial 1 having 2.13 greater odds of vocalising than the hens in trial 2.

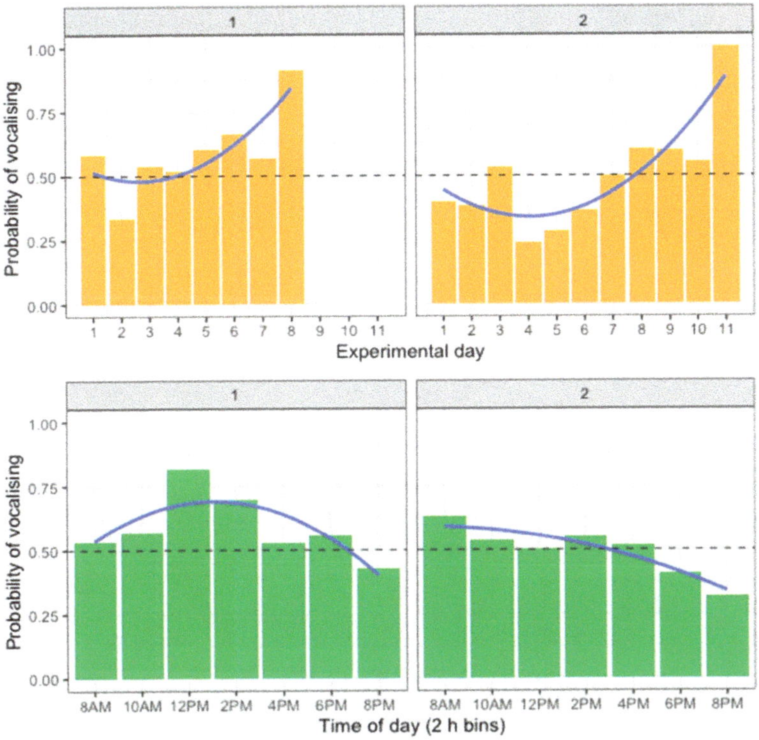

Figure 12. Change in probability of vocalising over experimental days and time of day, separated for trial 1 and trial 2. Lines show the trend across days and time of day.

3.2.2. Amount of Vocalisations

This study further investigated whether the amount of vocalisations differed depending on the stimulus that the hens received. Based on visual inspection, hens vocalised more in response to positively valenced stimuli; however, this may have been caused by the larger number of files in the positive conditions compared to the negative conditions (Figure 13A). To account for the unbalanced sample sizes, a random selection of 50 times the feeder delivered each valence and arousal combination (Positive + Low Arousal, Positive + High Arousal, Negative + Low Arousal, Negative + High Arousal, total 200 files) was made. The number of vocalisations made in each 1 min recording was then counted. Consecutive vocalisations were counted as a single vocalisation if they were separated by less than 2 s of silence. Plotting the number of vocalisations for these 50 selected files shows that hens seem to vocalise more to negative than positive stimuli (Figure 13B).

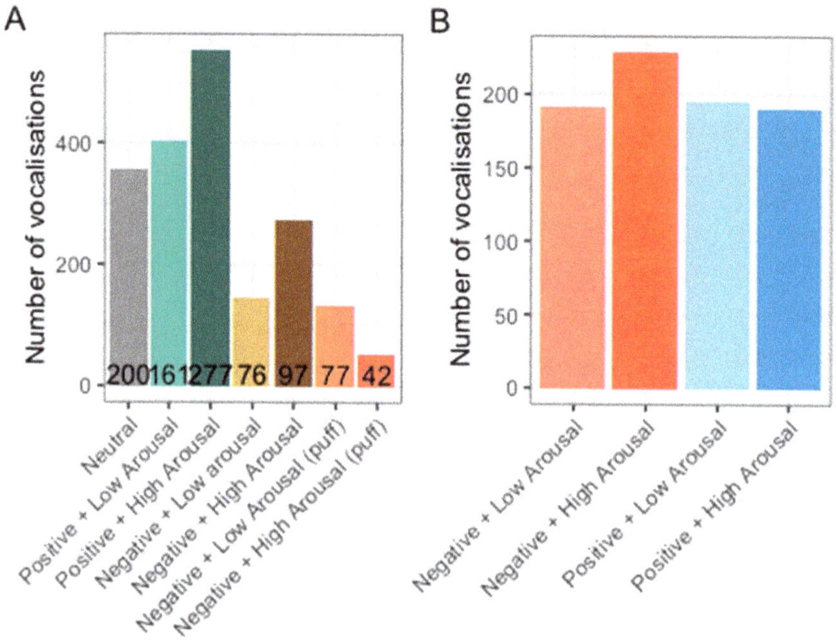

Figure 13. Number of vocalisations in response to the different stimuli types (**A**) and in response to 50 deliveries of the four different valence and arousal combinations (**B**). Numeric values in (**A**) indicate the number of files analysed per stimulus type.

To assess this difference statistically, the glmmTMB package was used to run a generalised linear mixed model with a Generalised Poisson distribution on the number of vocalisations. Stimulus valence, arousal, and valence × arousal interaction were included as fixed predictors in the model. The model's explanatory power was weak (pseudo-R-squared calculated with MuMIn package = 0.042). The model's intercept, corresponding to valence = negative and arousal = low, was at 1.37 (95%CI [1.33, 1.42, $p < 0.001$). Within this model, arousal was the only significant predictor, with hens vocalising significantly more to high- than low-arousal stimuli (beta = 0.06, 95% CI [0.02, 0.11], $p = 0.001$, Figure 14). The effect of valence was statistically non-significant and negative (beta = −0.04, 95% CI [−0.08, 0.002], $p = 0.065$). The effect of valence x arousal was also statistically non-significant and positive (beta = −0.02, 95% CI [−0.06, 0.02], $p = 0.271$).

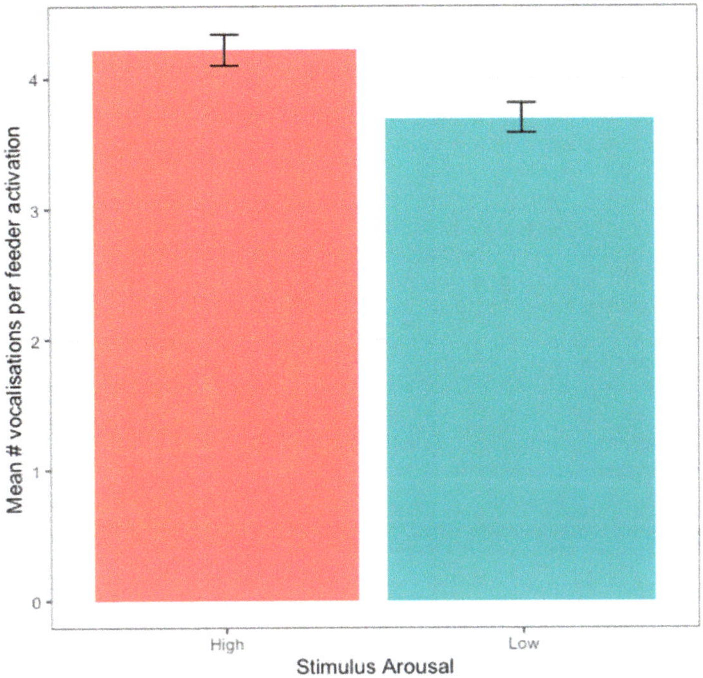

Figure 14. Number of vocalisations in response to stimuli with positive or negative arousal.

4. Discussion

To the best of the authors' knowledge, the present study is the first to explore hens' voluntary interaction with an ACI device and their vocal activity under evoked emotional arousal and valence states. The results show that the proposed design motivated hens to engage with the feeder frequently despite receiving both positive and negative stimuli, and they only showed a gradual decline in interest after more than a week in one of the trials. In trial 1, regular feed was provided upon the vast majority of activations, with the birds engaging with the device even in those cases, suggesting that the hens' interest was not entirely due to the presence of novel food rewards (mealworms and rice). This finding is aligned with previous suggestions that animals have an intrinsic drive to explore that goes beyond foraging for resources [38,39]. The engagement levels observed in both trials highlight the ACI device's potential as an effective environmental enrichment tool for hens. Enrichment strategies that enhance environmental complexity and provide opportunities for voluntary exploration have been shown to stimulate investigation and reduce neophobia [40]. The nature of the device, which encourages hens to approach and interact with it, is aligned with such strategies and might have beneficial effects on hens' welfare by reducing excessive fear, enhancing cognitive capacity, and improving adaptation skills [41].

The hens' feathers absorbed the ultrasonic waves, leading to a situation in which a bird could be less than 10 cm from the sensor and not trigger it. The system only triggered when the ultrasound hit featherless areas like the hens' legs or head. Nonetheless, the hens quickly learned to activate the system with their head in just a day (multiple activations were detected; Supplementary Video S1). One possible explanation for this behaviour was that the sensor itself produces sound within the hens' audible range. Some hens even started pecking on the sensor. However, this meant that the hens were not positioning their head directly above the feeder's baseline container upon feeder activation. Thus, hens did not receive an air puff to the face, but were startled by the loud and sudden noise of the puff.

The movement or sound of the disc's rotation seemed to attract the hens' interest. Thus, the speaker was mostly useful for defining the onset and offset of feeder activations in sound analysis and labelling (see Section 2.3.2), rather than attracting hens to the feeder. Disc movement also posed a problem, as hens started stepping on the disc, which disrupted the rotating process. Remarkably, the hens even managed to intentionally turn the disc to reach containers with mealworms (Supplementary Video S3). Pigeon sticks taped on the surface of the disc successfully prevented such behaviours.

The hens quickly learned how to activate the sensor to obtain preferred treats, although there was a noticeable difference between the two trials (averaging 244 activations daily in trial 1 and 72 in trial 2). In both trials, their usage persisted over a week, demonstrating peaks and troughs despite the introduction of mild punishments such as air puffs and distasteful food. The following discussion will explore how variations between the two trials could be due to differences in the probability of reward, the inclusion of mild negative stressors, and individual differences.

Stimuli were dispensed randomly, but with varying probabilities across trials. Trial 2, in which mealworms were provided 85% of the time, had 33% less activations than trial 1, in which mealworms were presented 5% of the time, and the hens' interest seemed to be shorter. Such results suggest faster habituation when the reward probability is high. This aligns with the observations in [42], where it was noted that because animals have adapted to the unpredictability inherent in their natural environments, the most effective enrichment strategies are likely to involve rewards that are varied and unexpected. The lower reward probability of mealworms and rice in trial 1 could have increased the hens' effort to obtain them, thus increasing the time it took for hens to habituate to the feeder. When that effort was minimal and mealworms were often provided in trial 2, the hens showed decreased engagement after a week. Longer trial durations would be needed, with higher stimulus uncertainty, to fully understand the scope of this phenomenon.

While the negative effects of distress have been widely recognised, the authors of [43,44] introduced the concept of 'eustress' to describe the enhancement of the body's natural responses to stressors that are not disruptive to an animal's homeostasis. Reference [45] also suggests that introducing positive stressors, which also alter the arousal of animals but in a beneficial manner, could enhance animals' cognitive functioning. This is reflected in our findings, where the integration of mild, non-harmful punishments in trial 1 led to increased feeder engagement. In contrast, trial 2, without negative food stimuli, had fewer activations. The probability of encountering negative stimuli was 10% in the first trial and 5% in the second. Despite the startling effect of the air puffs, the hens continued to actively use the feeders while showing high interest in consuming the mealworms, suggesting that they were willing to risk being startled to acquire preferred foods. Future trials could eliminate negative stimuli to investigate how these mild punishments contribute to the overall interest in the device.

Distinct differences in behaviour among individual birds were observed, particularly in terms of curiosity, inventiveness, and food preferences, which likely contributed to the varying levels of engagement. All hens were highly motivated to consume mealworms, but those in the first trial demonstrated a remarkable determination to access them. For example, a hen manipulated the feeder by manually rotating the disc with her feet to access the mealworms. This behaviour points to varying degrees of exploration and curiosity among the birds, with some showing a heightened eagerness to interact with the feeder. Such differences can be explained with respect to the proactive–reactive axis [46], with proactive or bold birds showing a higher tendency to explore their environment, while reactive or shy hens are more cautious with external objects [47].

Regarding this study's long-term objective, data from the first two trials indicate the device's efficacy in eliciting vocalisations, although the probability of vocalising did not reliably distinguish between emotional valence and arousal. Birds vocalised slightly more than half of the time (51%) during feeder activation. Interestingly, the daily and diurnal patterns in feeder usage did not match the probability of vocalisations or the number of vocalisations. The amount of vocalisations was positively related to emotional arousal, aligning with expectations, and suggests that stimuli were effective at inducing the intended level of arousal. Nonetheless, the measured indices, i.e., the probability of vocalising and the number of vocalisations, were not reliable indicators of emotional valence. Further analysis of the vocalisations' properties would be beneficial to investigate differences across valence states. It has been proposed [7] that the duration and the frequency distribution of the spectrum of the calls, as well the type of the call, can help in distinguishing the valence of animals. Other behavioural and physiological data would help to further validate that hens' arousal levels were indeed higher or lower as anticipated by the stimuli. Moreover, expanding this study to include a larger sample of birds would be essential for making more definitive statements about emotional valence and arousal in chicken vocalisations.

Another unexpected finding was the inverse correlation between feeder activations and the likelihood of vocalisations across experimental days. Staring with a relatively high probability, perhaps due to the novelty of the environment, the hens' engagement increased while the probability decreased. Over the last experimental days of each trial, the probability reached the maximum levels, while engagement was either constant or slightly decreased. Our speculation on this interesting result is that as the hens became familiar with the provided stimuli, they increasingly produced anticipatory calls when facing a reward [48] and frustration-related calls when negative stimuli were provided or when not receiving rewards [49]. Future studies should identify the call types made in various reward and non-reward conditions and examine whether they increase over time. The same inverse relationship was observed within experimental days, with birds being slightly more likely to activate the feeder towards the evening, while the probability of vocalising was at the lowest level during the same time. Natural behaviours could explain this phenomenon, with birds being less vocally active during the evening and more active during the day. Morning hours often include egg-laying and feeding activities [50,51], during which hens are more likely to produce vocalisations. Analysing the spontaneous vocal activity of hens could support this interpretation.

One potential issue with the existing setup involves the competition of birds over access to the provided stimulus, mainly the mealworms and rice. As observed in [40], hens can be highly competitive even when their nutritional needs are fulfilled. This was the case mainly in trial 1 and to some extent for trial 2, where dominant hens often monopolised access to both feeders, using them even when the subordinate hen was the one that activated the apparatus (Supplementary Video S2). Moreover, in the absence of direct competitive behaviours, it was observed that birds often used the feeder simultaneously. Such events can lead to ambiguous emotional states and, consequently, vocalisations that cannot be assigned to a single context, since one bird might enjoy the provided mealworms while the other bird produces sounds related to frustration and negative valence. This study did not specifically focus on identifying which bird activated the feeder or produced a vocalisation; however, this could further enhance our understanding of individuals' emotional states. The integration of RFID sensors, coupled with a proximity sensor to allow only one individual to activate a feeder, might be one way to overcome this problem. This could reduce the undesired effect of simultaneous usage but not entirely eliminate it, since the other hen could easily try to reach the feeder's content after the activation. Another possible way to bypass this could be the assignment of activations and vocalisations to individual birds, either algorithmically or by inspecting the video data.

5. Conclusions

This study introduced an interactive feeder designed to elicit vocalisations from laying hens in different emotional valence and arousal states. The active engagement of the hens with the feeder over several days highlights its potential as an environmental enrichment tool. The birds showed motivation to interact with the device, which dispensed both positive and negative stimuli, leading to a substantial collection of vocalisations. A preliminary vocalisation analysis indicated that birds vocalised more in states of high arousal compared to low arousal. However, the number of vocalisations did not significantly differentiate between positive and negative valence states. This study lays the groundwork for a deeper exploration into the emotional states of poultry using vocal expression and underscores the potential of the presented ACI device in animal welfare research.

Supplementary Materials: The following supporting information can be downloaded at: https://www.mdpi.com/article/10.3390/ani14091386/s1.

Author Contributions: Conceptualization, B.P.K. and T.N.; methodology, B.P.K. and A.G.; software, A.G. and M.M.; validation, A.G., B.P.K. and M.M.; formal analysis, B.P.K.; investigation, A.G.; resources, T.N.; data curation, A.G., B.P.K. and M.M.; writing—original draft preparation, A.G. and B.P.K.; writing—review and editing, A.G., B.P.K., M.M. and T.N.; visualization, A.G. and B.P.K.; supervision, T.N. and B.P.K.; project administration, T.N.; funding acquisition, T.N., B.P.K. and M.M. All authors have read and agreed to the published version of the manuscript.

Funding: The authors acknowledge funding from KU Leuven internal funds: C24M/22/022 (3E220566).

Institutional Review Board Statement: The animal study protocol was approved by the Institutional Review Board (or Ethics Committee) of KU Leuven (protocol code 082/2023 and date of approval 31 May 2023).

Informed Consent Statement: Not applicable.

Data Availability Statement: The original contributions presented in the study are included in the article and Supplementary Materials, further inquiries can be directed to the corresponding authors.

Acknowledgments: We would like to thank Ludo Happaerts, Bertram Van Soom, Panason Manorost, and Frank Mathijs for their valuable technical guidance. The construction of the feeders would not have been possible without their expertise.

Conflicts of Interest: The authors declare no conflict of interest.

References

1. Mellor, D. Updating Animal Welfare Thinking: Moving beyond the "Five Freedoms" towards "A Life Worth Living". *Animals* **2016**, *6*, 21. [CrossRef] [PubMed]
2. Webb, L.E.; Veenhoven, R.; Harfeld, J.L.; Jensen, M.B. What Is Animal Happiness? *Ann. N. Y. Acad. Sci.* **2019**, *1438*, 62–76. [CrossRef] [PubMed]
3. Rault, J.L.; Hintze, S.; Camerlink, I.; Yee, J.R. Positive Welfare and the Like: Distinct Views and a Proposed Framework. *Front. Vet. Sci.* **2020**, *7*, 370. [CrossRef] [PubMed]
4. Mendl, M.; Burman, O.H.P.; Paul, E.S. An Integrative and Functional Framework for the Study of Animal Emotion and Mood. *Proc. R. Soc. B Biol. Sci.* **2010**, *277*, 2895–2904. [CrossRef] [PubMed]
5. Kremer, L.; Klein Holkenborg, S.E.J.; Reimert, I.; Bolhuis, J.E.; Webb, L.E. The Nuts and Bolts of Animal Emotion. *Neurosci. Biobehav. Rev.* **2020**, *113*, 273–286. [CrossRef]
6. Laurijs, K.A.; Briefer, E.F.; Reimert, I.; Webb, L.E. Vocalisations in Farm Animals: A Step towards Positive Welfare Assessment. *Appl. Anim. Behav. Sci.* **2021**, *236*, 105264. [CrossRef]
7. Briefer, E.F. Coding for 'Dynamic' Information: Vocal Expression of Emotional Arousal and Valence in Non-Human Animals. In *Coding Strategies in Vertebrate Acoustic Communication*; Springer: Cham, Switzerland, 2020; pp. 137–162. [CrossRef]
8. Burgdorf, J.; Panksepp, J. The Neurobiology of Positive Emotions. *Neurosci. Biobehav. Rev.* **2006**, *30*, 173–187. [CrossRef] [PubMed]
9. Panksepp, J. The Basic Emotional Circuits of Mammalian Brains: Do Animals Have Affective Lives? *Neurosci. Biobehav. Rev.* **2011**, *35*, 1791–1804. [CrossRef]
10. Briefer, E.F. Vocal Expression of Emotions in Mammals: Mechanisms of Production and Evidence. *J. Zool.* **2012**, *288*, 1–20. [CrossRef]
11. Maigrot, A.-L.; Hillmann, E.; Anne, C.; Briefer, E.F. Vocal Expression of Emotional Valence in Przewalski's Horses (*Equus Przewalskii*). *Sci. Rep.* **2017**, *7*, 8779. [CrossRef]
12. Wöhr, M.; Schwarting, R.K.W. Activation of Limbic System Structures by Reply of Ultrasonic Vocalization in Rats. In *Handbook of Mammalian Vocalization. An Integrative Neuroscience Approach*; Brudzynski, S.M., Ed.; Elsevier: Amsterdam, The Netherlands, 2010; pp. 113–124, ISBN 978-0-12-374593-4.
13. Briefer, E.F.; Maigrot, A.-L.; Mandel, R.; Freymond, S.B.; Bachmann, I.; Hillmann, E. Segregation of Information about Emotional Arousal and Valence in Horse Whinnies. *Sci. Rep.* **2015**, *5*, 9989. [CrossRef]

14. Maigrot, A.-L.; Hillmann, E.; Briefer, E. Encoding of Emotional Valence in Wild Boar (Sus Scrofa) Calls. *Animals* **2018**, *8*, 85. [CrossRef]
15. Rose, P.; O'brien, M. Welfare Assessment for Captive Anseriformes: A Guide for Practitioners and Animal Keepers. *Animals* **2020**, *10*, 1132. [CrossRef]
16. Zimmerman, P.H.; Buijs, S.A.F.; Bolhuis, J.E.; Keeling, L.J. Behaviour of Domestic Fowl in Anticipation of Positive and Negative Stimuli. *Anim. Behav.* **2011**, *81*, 569–577. [CrossRef]
17. McGrath, N.; Dunlop, R.; Dwyer, C.; Burman, O.; Phillips, C.J.C. Hens Vary Their Vocal Repertoire and Struc-ture When Anticipating Different Types of Reward. *Anim. Behav.* **2017**, *130*, 79–96. [CrossRef]
18. Davies, A.C.; Radford, A.N.; Nicol, C.J. Behavioural and Physiological Expression of Arousal during Deci-sion-Making in Laying Hens. *Physiol. Behav.* **2014**, *123*, 93–99. [CrossRef] [PubMed]
19. Davies, A.C.; Radford, A.N.; Pettersson, I.C.; Yang, F.P.; Nicol, C.J. Elevated Arousal at Time of Decision-Making Is Not the Arbiter of Risk Avoidance in Chickens. *Sci. Rep.* **2015**, *5*, 8200. [CrossRef]
20. Paul, E.S.; Edgar, J.L.; Caplen, G.; Nicol, C.J. Examining Affective Structure in Chickens: Valence, Intensity, Persistence and Generalization Measured Using a Conditioned Place Preference Test. *Appl. Anim. Behav. Sci.* **2018**, *207*, 39–48. [CrossRef] [PubMed]
21. Skelhorn, J.; Rowe, C. Birds Learn to Use Distastefulness as a Signal of Toxicity. *Proc. R. Soc. B Biol. Sci.* **2010**, *277*, 1729–1734. [CrossRef]
22. Ganchrow, J.R.; Steiner, J.E.; Bartana, A. Behavioral Reactions to Gustatory Stimuli in Young Chicks (Gallus Gallus Domesticus). *Dev. Psychobiol.* **1990**, *23*, 103–117. [CrossRef]
23. Edgar, J.L.; Lowe, J.C.; Paul, E.S.; Nicol, C.J. Avian Maternal Response to Chick Distress. *Proc. R. Soc. B Biol. Sci.* **2011**, *278*, 3129–3134. [CrossRef] [PubMed]
24. Herborn, K.A.; Graves, J.L.; Jerem, P.; Evans, N.P.; Nager, R.; McCafferty, D.J.; McKeegan, D.E.F. Skin Temperature Reveals the Intensity of Acute Stress. *Physiol. Behav.* **2015**, *152*, 225–230. [CrossRef]
25. Edgar, J.L.; Nicol, C.J.; Pugh, C.A.; Paul, E.S. Surface Temperature Changes in Response to Handling in Domestic Chickens. *Physiol. Behav.* **2013**, *119*, 195–200. [CrossRef] [PubMed]
26. Moe, R.O.; Stubsjøen, S.M.; Bohlin, J.; Flø, A.; Bakken, M. Peripheral Temperature Drop in Response to Antici-pation and Consumption of a Signaled Palatable Reward in Laying Hens (Gallus Domesticus). *Physiol. Behav.* **2012**, *106*, 527–533. [CrossRef] [PubMed]
27. French Fiona (London Metropolitan University, London, UK). Personal communication, May 2023.
28. Bartoń, K. MuMIn: Multi-Model Inference. R Package Version 1.47.5. 2023. Available online: https://CRAN.R-project.org/package=MuMIn (accessed on 15 January 2024).
29. Brooks, M.E.; Kristensen, K.; van Benthem, K.J.; Magnusson, A.; Berg, C.W.; Nielsen, A.; Skaug, H.J.; Maechler, M.; Bolker, B.M. glmmTMB Balances Speed and Flexibility Among Packages for Zero-inflated Generalized Linear Mixed Modeling. *R J.* **2017**, *9*, 378–400. [CrossRef]
30. Grolemund, G.; Wickham, H. Dates and Times Made Easy with lubridate. *J. Stat. Softw.* **2011**, *40*, 1–25. Available online: https://www.jstatsoft.org/v40/i03/ (accessed on 15 January 2024). [CrossRef]
31. Hartig, F. DHARMa: Residual Diagnostics for Hierarchical (Multi-Level/Mixed) Regression Models. R Package Version 0.4.6. 2022. Available online: https://CRAN.R-project.org/package=DHARMa (accessed on 15 January 2024).
32. James, D.; Hornik, K. Chron: Chronological Objects which Can Handle Dates and Times. R Package Version 2.3-61. 2023. Available online: https://CRAN.R-project.org/package=chron (accessed on 15 January 2024).
33. Makowski, D.; Lüdecke, D.; Patil, I.; Thériault, R.; Ben-Shachar, M.; Wiernik, B. Automated Results Reporting as a Practical Tool to Improve Reproducibility and Methodological Best Practices Adoption. R Package Version 0.5.8. 2023. Available online: https://easystats.github.io/report/ (accessed on 15 January 2024).
34. Pedersen, T. Patchwork: The Composer of Plots. R Package Version 1.1.2. 2022. Available online: https://CRAN.R-project.org/package=patchwork (accessed on 15 January 2024).
35. Wickham, H. *Ggplot2: Elegant Graphics for Data Analysis. R Package Version 3.5.1*; Springer: New York, NY, USA, 2016; ISBN 978-3-319-24277-4. Available online: https://ggplot2.tidyverse.org (accessed on 15 January 2024).
36. Wickham, H.; François, R.; Henry, L.; Müller, K.; Vaughan, D. Dplyr: A Grammar of Data Manipulation. R Package Version 1.1.2. 2023. Available online: https://CRAN.R-project.org/package=dplyr (accessed on 15 January 2024).
37. Wickham, H.; Averick, M.; Bryan, J.; Chang, W.; McGowan, L.D.; François, R.; Grolemund, G.; Hayes, A.; Henry, L.; Hester, J.; et al. Welcome to the tidyverse. *J. Open Source Softw.* **2019**, *4*, 1686. [CrossRef]
38. Berlyne, D.E. *Conflict, Arousal, and Curiosity*; Martino Fine Books: Eastford, CT, USA, 1960. [CrossRef]
39. Wood-Gush, D.G.M.; Vestergaard, K. The seeking of novelty and its relation to play. *Anim. Behav.* **1991**, *42*, 599–606. [CrossRef]
40. Newberry, R.C. Exploratory behaviour of young domestic fowl. *Appl. Anim. Behav. Sci.* **1999**, *63*, 311–321. [CrossRef]
41. Campbell, D.L.M.; de Haas, E.N.; Lee, C. A review of environmental enrichment for laying hens during rearing in relation to their behavioral and physiological development. *Poult. Sci.* **2019**, *98*, 9–28. [CrossRef]
42. Watters, J.V. Toward a predictive theory for environmental enrichment. *Zoo Biol.* **2009**, *28*, 609–622. [CrossRef] [PubMed]
43. Moberg, G.P.; Mench, J.A. (Eds.) *The Biology of Animal Stress. Basic Principles and Implications for Animal Welfare*; CABI Publishing: Wallingford, UK, 2000; p. 377. [CrossRef]

44. Selye, H. Stress and distress. *Comprehens. Ther.* **1975**, *1*, 9–13.
45. Meehan, C.L.; Mench, J.A. The challenge of challenge: Can problem solving opportunities enhance animal welfare? *Appl. Anim. Behav. Sci.* **2007**, *102*, 246–261. [CrossRef]
46. Koolhaas, J.M.; Korte, S.M.; De Boer, S.F.; Van Der Vegt, B.J.; Van Reenen, C.G.; Hopster, H.; De Jong, I.C.; Ruis, M.A.; Blokhuis, H.J. Coping styles in animals: Current status in behavior and stress-physiology. *Neurosci. Biobehav. Rev.* **1999**, *23*, 925–935. [CrossRef] [PubMed]
47. Sih, A.; Bell, A.; Johnson, J.C. Behavioral syndromes: An ecological and evolutionary overview. *Trends Ecol. Evol.* **2004**, *19*, 372–378. [CrossRef] [PubMed]
48. McGrath, N. The Vocalisations and Behaviour of Chickens in Anticipation of Rewards. Ph.D. Thesis, University of Queensland, Brisbane, QLD, Australia, 2019. [CrossRef]
49. Zimmerman, P.H.; Koene, P.; van Hooff, J.A. Thwarting of behaviour in different contexts and the gakel-call in the laying hen. *Appl. Anim. Behav. Sci.* **2000**, *69*, 255–264. [CrossRef] [PubMed]
50. Villanueva, S.; Ali, A.B.A.; Campbell, D.L.M.; Siegford, J.M. Nest use and patterns of egg laying and damage by 4 strains of laying hens in an aviary system. *Poult. Sci.* **2017**, *96*, 3011–3020. [CrossRef]
51. Du, X.; Lao, F.; Teng, G. A Sound Source Localisation Analytical Method for Monitoring the Abnormal Night Vocalisations of Poultry. *Sensors* **2018**, *18*, 2906. [CrossRef]

animals

MDPI

Article

Developing a Preference Scale for a Bear: From "Bearly Like" to "Like Beary Much"

Jennifer Vonk

Department of Psychology, Oakland University, 654 Pioneer Drive, Rochester, MI 48309, USA; vonk@oakland.edu

Simple Summary: I trained an American black bear in human care to choose different response buttons when presented with an image of either a highly preferred or a less preferred food item. The bear learned to choose the appropriate response button when presented with the preferred food item at above chance levels and differentiated between the use of the buttons appropriately. However, she did not reach a high level of performance with the less preferred food item even after over 1000 trials, suggesting that performing a conditional discrimination on the basis of preferences may be challenging for black bears. However, the work presented here represents the first attempt to train a bear to indicate her relative preferences using something like a Likert scale commonly used with humans to indicate their preferences and could be a valuable welfare tool for animals in human care. Similar work with gorillas suggests that bears are as capable as great apes in learning such tasks and would also benefit from this type of technical enrichment.

Abstract: A preference scale for use by nonhuman animals would allow them to communicate their degree of liking for individual items rather than just relative preferences between pairs of items. It would also allow animals to report liking for images of objects that would be difficult to directly interact with (e.g., potential mates and habitat modifications). Such scales can easily be presented using touchscreen technology. Few zoos have used touchscreen technology for species other than nonhuman primates. I present a description of efforts taken to create such a scale for use with a single zoo-housed American black bear (*Ursus americanus*). Although the bear did not reach a high level of proficiency with assigning preferred and non-preferred food items to categorical responses of "like" and "dislike," she was able to learn how to use the like and dislike buttons differentially for a single preferred and less preferred food item and she selected the correct response button for the preferred item at above chance levels. These data contribute to our limited understanding of black bear cognition and suggest that conditional discriminations may be difficult for black bears. This finding can inform continued efforts to create a simpler tool for nonhumans to communicate their preferences to human caregivers in a more nuanced way than is currently possible. More generally, the current study contributes to the growing body of work supporting the use of touchscreen technology for providing enrichment to less studied species like bears.

Keywords: black bear; *Ursus americanus*; conditional discrimination; welfare; rating; ranking

1. Introduction

While touchscreens are becoming increasingly common for enrichment or research purposes in zoo-housed nonhuman primates, the number of other species provided with this level of technical enrichment remains extremely small. Bears are widely recognized to be highly intelligent and curious animals that could benefit from more complex and dynamic enrichment. Although there is a paucity of work describing bears' visual abilities, early work suggested that black bears discriminated various hues from grey, having difficulty with only red–green discriminations [1]. In addition, bears have been successfully trained to make categorical discriminations between stimuli presented on touchscreens (e.g., brown bears, *Ursus arctos*, Bernstein-Kurtycz et al., personal communication; American black bears, [2–6]; Malayan sun bears, *Helarctos malayanus* [7], and polar bears, *Ursus maritimus* (Jeremiasse et al., personal communication). Not only do bears appear to enjoy the stimulation provided from interacting with trainers through touchscreen training, the use of the computer presents researchers with novel ways to communicate with the bears. Computers have often been used in zoological settings to provide enrichment—most typically for nonhuman primates [8–10] but also for other species like parrots [11]—in the form of games, puzzles, or auditory enrichment. Computer interfaces have also been used to conduct assessments of animal well-being [9,12]. Computer interfaces can also be used to present images of foods and other objects, which subjects can then indicate their preferences for. For example, researchers recently presented a tablet to a Goffin's cockatoo (*Cacatua goffiana*) so that the bird could select symbols representing various items, activities, or interactions. Their results suggested that the single cockatoo subject could use the tablet effectively to request objects and interactions that presumably had positive effects on her well-being [13]. The current study aimed to provide a means for a bear to symbolically communicate preferences for the first time.

Understanding individual preferences is critical for optimizing an animal's environment and ensuring positive welfare. Preferences can inform habitat planning, husbandry, enrichment, and food provisioning [14]. Traditionally, preferences have been assessed indirectly by measuring degree of engagement with different enrichment items, foods, and environmental features etc. (e.g., [15,16]); objects that trigger different events like sounds [10], approach, and avoidance behavior (e.g., [17]); or efforts exerted to obtain access to space, social companions, or objects [18–20]. Preferences have been assessed more directly with forced choice tests between pairs or groups of real objects (e.g., [21–25]) or choices of symbols representing options like sounds [26]. These methods assess relative preferences among pairs or groups of choices, but fail to provide a more nuanced assessment of amount of liking (e.g., this object is liked to some degree compared to this item that is liked very much). Importantly, only objects that can be safely presented for investigation can be used in assessments involving real objects. In addition, paired-choice tests require the repetition of multiple pairings across items, which can be time consuming and can result in satiation when assessing preferences for foods or rewarding the individual for their choices. I attempted to develop a novel method to assess the degree of liking for various elements of the environment presented in pictorial form in an American black bear. This scale would ultimately allow her, and other nonhumans, to indicate preferences for food, enrichment, care staff, environments, sounds, and other stimuli that are not physically present at the time of assessment. Ultimately, I wished to be able to assess preferences for unfamiliar and previously inexperienced stimuli, such as planned habitat changes, possible mates, or even images representing more abstract concepts such as natural environments. Notably, preferences can be assessed in a single trial using this method once the animal understands the meaning of the end-points. I began training the bear to use such a nonverbal animal preference scale (NAPS) using images of foods for which her relative preferences could be determined.

In humans, preference scales are commonly encountered in product assessment, customer satisfaction surveys, and research into attitudes, beliefs, and personality traits. Such measures typically take the form of Likert scales [27], which allow respondents to indicate relative preference for items or agreement with ideas. One of the advantages of this type of scale is that respondents are able to indicate when they do not like an item at all rather than being forced to choose between equally preferred or non-preferred items. Rather, items are presented one at a time with a rating scale that has end-points representing a spectrum of agreement (e.g., from "strongly dislike" to "strongly like"). Using paired-choice tasks, an item might never be selected because it is less preferred than the other options, but it would not be possible to determine whether this item may also be liked rather than disliked. Preference scales have been widely adopted for research in multiple disciplines due to their flexibility [28]. Although there has been only one other known attempt to use such a scale with nonhumans, which I conducted concurrently with a bachelor group of gorillas (*Gorilla gorilla gorilla*, [29]), nonverbal versions have been used with human children [30,31] and clinical patients [32]. In these cases, verbal scale endpoints are replaced with intuitive images such as facial expressions to denote degree of liking [33], level of pain [34,35], and mood [36]. Therefore, although admittedly more complex and abstract relative to existing methods for assessing preferences, it seemed desirable and feasible to adopt similar methods to train nonhumans to use such a scale.

It should be noted that even pictorial Likert scales require verbal instruction, and a recent meta-analysis reveals that children below the age of five years cannot reliably use self-report measures of health outcomes. Furthermore, children below eight years of age may not be able to use a scale with more than two response options [37]. Therefore, training nonhumans to understand the construct of a sliding scale of preferences posed several challenges, not the least of which was deciding upon scale end points that might intuitively reflect "dislike" and "like." Training animals to understand task requirements without verbal instructions is not a novel challenge, but does necessitate a prolonged period of training prior to administering the test in contrast to the type of one-off assessments conducted with human respondents. Second, because constructs of liking or preferences might rely on explicit self-knowledge, use of a preference scale may depend upon a degree of abstraction beyond the grasp of most nonhuman animals. There is no existing work that suggests that nonhumans can accurately report on their own (or others') preferences when directly asked, and indeed, it has been noted that it would be difficult to learn signals of others' preferences when they are discordant from our own, especially inconsistently so [38]. However, as a starting point, effective use of the scale could be acquired through a simpler process of association between items that evoke a particular visceral response (e.g., disgust, excitement) and different operational responses (use of the different response buttons). A process of generalization might support the appropriate use of response buttons associated with negative and positive feelings toward novel stimuli. Thus, use of the NAPS could be assumed to measure relative preferences for categories or objects that can be represented physically regardless of whether the subject explicitly represents the items as "things I dislike and things I like".

Although stimuli to be rated could be presented in any modality perceived by the organism, use of a touchscreen system is most suitable for visual or auditory stimuli. Successful implementation of a visual NAPS requires that subjects understand the correspondence between pictures and their real-life referents. Many species have demonstrated picture–object correspondence (for review see [39]; e.g., in pigeons, *columbidae* [40]; in kea, *Nestor notabilis* [41]; macaques, *Macaca silenus* [42]), including the black bear that is the subject of the current study [2]. This apparently widespread ability supports the computer touchscreen methodology used here. However, another challenge with the NAPS is that stimuli, both in training and testing, must be subject-specific. To train subjects to use the NAPS, it is necessary to train them to understand what the different response buttons represent using stimuli for which the researchers already know the subject's preference. These buttons must be presented at the extreme ends of a spectrum (i.e., spatially) so that responses representing intermediate levels of preference can be added later to allow a more nuanced scale of preference. Once appropriate use of the most extreme response buttons is established for items for which preferences are known, researchers can introduce the use of intermediate buttons, and finally, begin assessment of preferences for novel items. Here, I presented the bear with images of food items based on her preferences as indicated by her care staff to train her on the use of the scale.

I conducted a simple validation of the food preferences indicated by the care staff by presenting the bear with a set of images of preferred versus less-preferred food items on a touchscreen in a two-alternative forced-choice task. As with gorillas tested previously [29], it was expected that the bear would spontaneously select images of the preferred items. As expected, the bear selected the images of preferred over less preferred foods at above chance levels even when items belonging to the preferred and less preferred categories were continuously changed, which was done to ensure the generalizability of the concept. Others have provided a similar validation of the use of pictorial stimuli to assess food preferences in other species with many of these subjects showing generalization of choices to novel food images within the same categories of preference (e.g., sloth bears, *Melursus ursinus*, [43]; lion-tailed macaques, *Macaca silenus*, [42]; gorillas, [18,44,45]; Japanese macaques, *Macaca fuscata*, and chimpanzees, *Pan troglodytes*, [45]). Early work with black bears showing their stable food preferences [15] and my own previous work with black bears making natural category discriminations using a touch-screen (e.g., [4,6,46]) made me optimistic that this new black bear subject would become proficient in communicating her preferences for items presented visually using the NAPS. Furthermore, there is some existing evidence that great apes—at least those that have received some symbolic/language training—can appropriately use symbols representing "bad" and "good" [47], and can use pictures to communicate desires [48] and my previous work with bears and apes suggested that they were capable of representing similar levels of abstraction compared to great apes [3–6,46,49,50]. Ultimately, I wished to present the task with a 5-point scale, but given the difficulties of gorillas trained previously [29] with the use of a neutral response button, I began training the bear with only the two extreme (non-preferred and preferred) response buttons. Had she demonstrated proficiency with these two end-points, I would have gradually added in additional response buttons along the spatial continuum.

2. General Method

The studies reported here were approved by the IACUC of Oakland University (Protocol #12082) and the Animal Welfare and Management Committee of the Detroit Zoological Society.

2.1. Subject

One wild-born female American black bear (Migwan, Basel, Switzerland), age 11 years at the beginning of the study, participated in this study when she resided at the Detroit Zoo, Royal Oak, MI, USA. Migwan was rescued from the wild at a very young age and rehabilitated due to injuries. She was housed individually. Although experimentally naïve, Migwan had participated in husbandry training. For example, she was target trained using positive reinforcement, including clicker training.

2.2. Materials

All experiments were programmed in Real Practice or Inquisit 3.0 (millisecond.com) and presented on a Panasonic CF-19 Toughbook or an Asus Aspire One Laptop projected to a 19" VarTech Armorall capacitive touch-screen monitor. The touchscreen monitor was affixed to the front of a rolling LCD cart. The touchscreen monitor was secured flush to the front of the steel mesh with bungee cords to secure the screen in place so that Migwan could touch the screen with her tongue through the gaps in the mesh. The care staff member and researcher always tested the touchscreen from the bear's side of the mesh prior to letting the bear into the indoor testing habitat. The laptop sat on a shelf on the cart behind the touchscreen (Figure 1). The experimenter stood against the back wall of the indoor area behind the cart and did not interact with the bear during trials. The care staff member placed the food rewards into a PVC tube affixed to the steel mesh to deliver food rewards for correct responses without any direct contact. This staff member always stood to the same side of the touchscreen during trials and did not direct attention to the bear or the laptop.

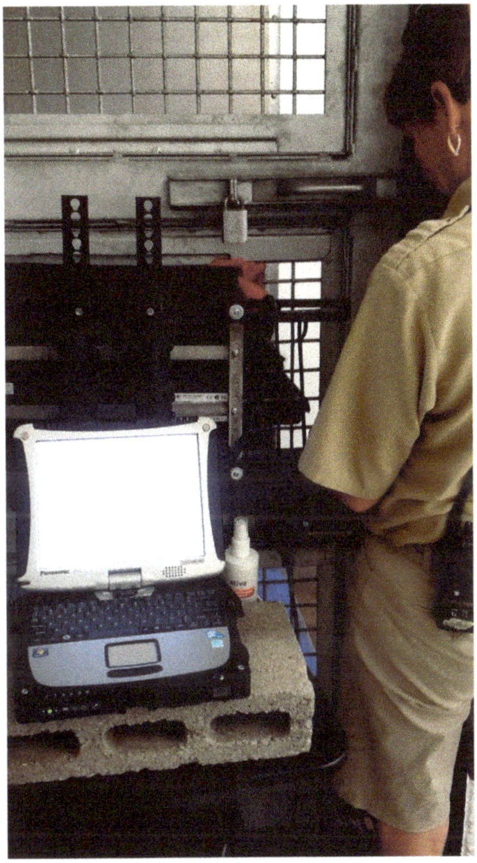

Figure 1. Experimental set-up showing Migwan peering from around the edge of the touchscreen.

Stimuli used in the experiments were non-copyrighted photographs downloaded from various websites or images drawn in Microsoft Paint. These stimuli included images of various foods, and two-dimensional shapes such as circles, squares, triangles, and misshapen objects drawn in blue, yellow, red, and green. Food items used to reward correct responses composed a minimal proportion of the bear's daily diet (e.g., almonds, biscuits, raisins, grapes).

2.3. General Procedure

The research took place in a non-public area of Migwan's indoor habitat. She participated in testing three afternoons a week at around 13:00 h between April to September in 2014 and 2016. Migwan did not participate in testing from October to March as she was in a state of torpor during the colder months. During the spring and summer months of 2015, Migwan participated in other tasks including a picture–object correspondence test [2] and an ambiguous cue affective bias task [3]. She also participated in a novel judgement bias task that was conducted simultaneously from April to September 2016 [5]. Testing took about 10–15 min each test day, and Migwan completed 4–5 sessions of testing each day. Participation in the tasks was entirely voluntary. Testing for the day ended when Migwan had consumed an appropriate number of rewards as determined by the care staff. A flowchart of the experimental phases is presented in Figure 2.

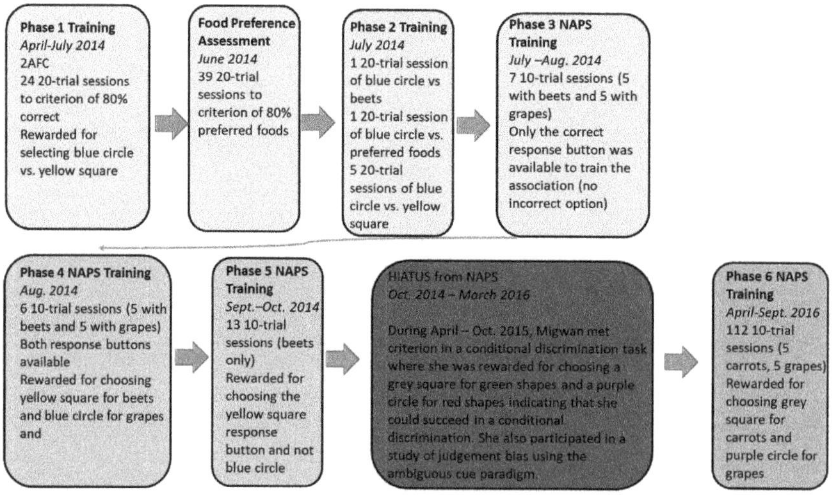

Figure 2. Flowchart of the phases of the experiment.

If Migwan selected the correct stimulus, a pleasant auditory beep was emitted, the touchscreen turned white, and the care staff member assisting with the trials placed a small food reward down a PVC chute affixed to the mesh. If Migwan selected an incorrect stimulus, there was no audio feedback, the touchscreen turned black, and there was a 500 ms inter-trial interval.

2.4. Phase 1 Training

Migwan had already been trained to target and to station by her caretakers using positive reinforcement. Prior to beginning the study, she was trained by her care staff to station in front of the touchscreen without it being turned on. She was rewarded for targeting to a familiar target by touching it with her nose. Once she was reliably touching the target positioned right in front of the screen, the care staff removed the target and rewarded Migwan for touching the blank screen with her nose. This training took a period of approximately one week.

To train Migwan to use the touchscreen, I first presented her with a two-alternative forced-choice task where she was presented with two stimuli drawn in Microsoft Paint: a yellow square on a white background and a blue circle on a black background. She was reinforced for selecting the blue circle and not reinforced for selecting the yellow square. The idea was to create a positive association with the blue circle and not with the yellow square so that these would be intuitive response buttons for the end-points of the scale, with the blue circle representing "like or preferred" and the yellow square representing "dislike or less preferred." The two stimuli filled most of the screen. The response button covered 80% of the stimulus so that Migwan had to touch the center of the stimulus and could not activate it just by nudging the edge of the stimulus. She was reinforced only if she used her tongue or nose to contact the touchscreen, not for using her paw. Migwan participated in, on average, four sessions a day, three days a week between April and July, 2014. The stimuli were presented in 20-trial sessions with the side of the correct stimulus (the blue circle) counterbalanced within the session.

In each trial, the stimuli appeared simultaneously and disappeared when one of them was selected. If Migwan selected the blue circle, a tone sounded, the screen turned white, the care staff member placed a food reward in the PVC tube affixed to the mesh, and the next trial commenced after 500 ms. If she selected the yellow circle, there was no sound, the screen turned black and she received no food reward. The inter-trial interval was the same. The criterion was set to four consecutive sessions at 80% correct or better (i.e.,

16/20 correct responses) or two consecutive sessions at 90% correct or better (i.e., 18/20 correct responses).

2.5. Food Preference Assessments

To assess a spontaneous preference for images of preferred foods, I again used a two-alternative forced-choice procedure. Sessions included 20 trials and were identical to the training task described above except for the stimuli used. Migwan completed 39 sessions of this task. On each trial, a food indicated by the care staff to be preferred by Migwan was randomly paired with beets, lettuce, or carrots (on sessions 5 and 6), which were foods identified by care staff as being least preferred by Migwan. An image from the preferred category was randomly paired with an image from the non-preferred category on each trial and presented in random order. Table 1 indicates which food items were presented in each category on each session along with the number of trials on which each food image was presented within a session. One photo was used for each of these food types. Changes in the food items presented were made to test the generalizability of Migwan's preferences. The side the non-preferred foods were presented on was counterbalanced within sessions with the constraint that they could not appear more than three times consecutively on the same side of the screen. Migwan was rewarded if she selected one of the presumed preferred foods and not if she selected the presumed non-preferred foods.

Table 1. Foods presented for each session of Food Preference Assessment.

Sessions	Preferred (# of Trials in Parentheses)	Non-Preferred
1–4, 7	Apples (4), Sweet Potatoes (4), Grapes (4), Kiwi (4), Oranges (4)	Beets (20)
5–6	Apples (4), Sweet Potatoes (4), Grapes (4), Kiwi (4), Pear (4)	Beets (8) Lettuce (8), Carrots (4)
8–10	Apples (4), Carrots (4), Grapes (4), Kiwi (4), Oranges (4)	Beets (20)
11–13	Sweet Potatoes (5), Grapes (5), Kiwi (5), Pear (5)	Beets (20)
14–15	Apples (5), Pineapple (5), Pear (5) Oranges (5)	Beets (20)
16	Pear (5), Apples (10), Oranges (5)	Beets (20)
17–18	Apples (4), Sweet Potatoes (4), Grapes (2), Strawberries (2) Pear (8)	Beets (20)
19–21	Apples (5), Sweet Potatoes (5), Pear (5), Kiwi (5)	Lettuce (20)
22–24	Apples (7), Sweet Potatoes (6), Pear (7)	Beets (20)
25–27	Apples (5), Sweet Potatoes (5), Grapes (5), Kiwi (5)	Beets (20)
28–31	Apples (4), Carrots (4), Grapes (4), Pear (4), Cantaloupe (4)	Beets (20)
32	Apples (5), Orange (5), Grapes (5), Cantaloupe (5)	Beets (20)
33	Grapes (7), Pineapple (6), Sweet Potato (7)	Lettuce (20)
34	Grapes (4), Pineapple (4), Sweet Potato (4), Carrots (4), Kiwi (4)	Beets (20)
35–36	Grapes (5), Pineapple (5), Sweet Potato (5), Kiwi (5)	Beets (20)
37–39	Apples (5), Carrots (5), Pear (5) Oranges (5)	Beets (20)

2.6. Phase 2 Training Continuation

I next presented Migwan with a session of 20 trials of photographs of beets paired with the image of the blue circle (with side counterbalanced), where she was rewarded only for touching the blue circle to reinforce the idea of the blue circle as something positive and the beets as something not positive.

I then presented a single 20-trial session where the blue circle was paired with images of the preferred foods. As expected, she performed at chance, choosing the blue circle only 10 times, suggesting that she perceived the blue circle as equally positive, or likely to lead to reward, as the images of the preferred foods.

I then presented Migwan with five additional sessions of the blue circle paired with the yellow square to ensure she was still performing at criterion with the training stimuli.

2.7. Phase 3 NAPS Training

I created a computer program in Inquisit v. 3 that presented an image of a less preferred, or a preferred food in the center of the screen that, once touched, prompted the appearance of a response button in the top left (yellow square) or top right (blue circle) of the screen. The food image remained on the screen, centered in the bottom half of the screen once the response buttons appeared (Figure 3). Each response button took up about 30% of the top half of the screen. These sessions consisted of 10 trials (5 with beets and 5 with grapes). Only one image was used to represent each food type (beets for the less preferred item and grapes for the most preferred item) and it was the same image used during the food preference assessments described above. Migwan was trained to associate images of beets with the yellow square response button and images of grapes with the blue circle response button. On each trial, there was only one available response button. When Migwan selected that button, a beep sounded, the screen turned white and the care staff member placed a food reward in the PVC chute. The next trial began once Migwan had touched the image of the food and the subsequent response button with her nose or tongue. Migwan completed 7 sessions of this phase.

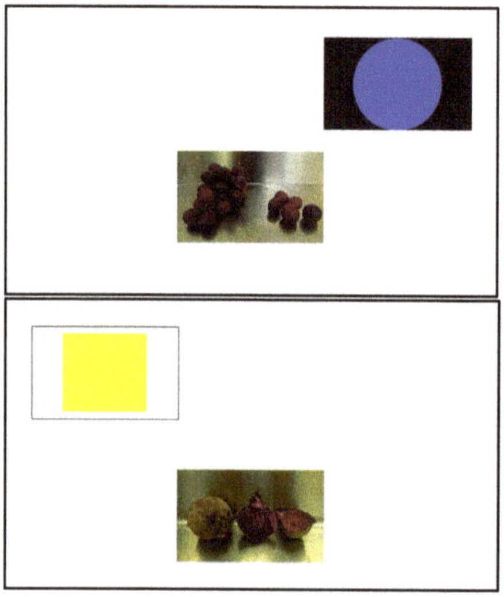

Figure 3. Sample trials of Phase 3 NAPS with the preferred food (**top**) and less preferred food (**bottom**).

2.8. *Phase 4 NAPS Training*

In Phase 4 of Training, sessions consisted of 10 trials, which were the same as above except that both response buttons appeared simultaneously on every trial and Migwan was rewarded only if she chose the correct one (Figure 4). Migwan completed six sessions of this phase.

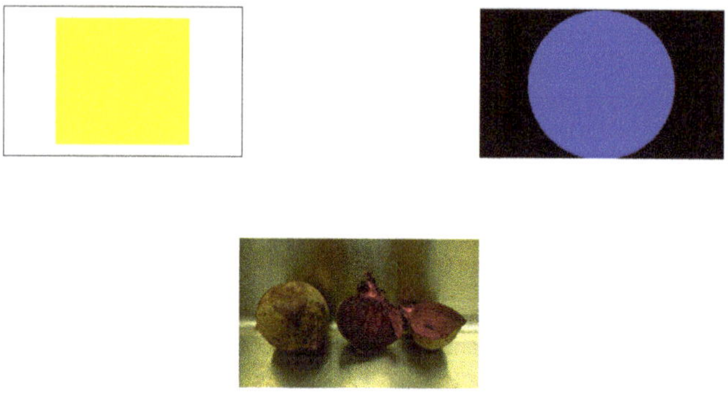

Figure 4. Trial of Phase 5 NAPS.

2.9. Phase 5 NAPS Training

Because I had successfully trained Migwan to associate the blue circle with reward, she was understandably reluctant to choose the yellow square even on trials when it would have been the correct response (i.e., when beets were presented). Therefore, I trained her to select the yellow square when beets were presented in this phase. This phase consisted of 10-trial sessions in which beets were always the food stimulus (always the same image as used previously) and both response buttons were presented simultaneously after she had touched the image of the beets (Figure 4). She was rewarded only for touching the yellow square/dislike button. Criterion was set to 80% correct responding for four consecutive sessions. Migwan received 13 sessions of this phase and then testing went on hiatus for fall torpor.

2.10. Testing Hiatus

When I resumed testing in April 2015, I focused instead on a picture–object correspondence task, which validated the use of two-dimensional images to represent objects for Migwan to rate [2]. I also presented her with a judgement bias test to assess affect changes across seasons [3]. In the spring of 2016, I returned to training Migwan in the current task. I trained her in several simpler conditional discrimination tasks using simple shapes (green triangle, red oval) rather than non-preferred and preferred foods to validate her ability to perform a conditional discrimination, before returning to the version of the task involving foods in the fall of 2016. Migwan learned to select a novel grey square response button conditional on being presented with a green triangle and to select a novel purple circle response button in response to being presented with a red oval. It took her 49 sessions (490 trials) to reach criterion and she successfully transferred at above chance levels to different images of the same shapes and colors as the original training stimuli. However, it took her 38 and 46 sessions to reach criterion again with the transfer shape and color stimuli, respectively.

2.11. Phase 6 NAPS Training

Having established that Migwan could learn this conditional discrimination task with less abstract decision rules, I returned to the task of training her to respond differentially to preferred and less preferred foods almost two years later. I presented Migwan with a version of the NAPS in which carrots were presented as the less preferred food and grapes were presented as the preferred food. I switched from beets to carrots as the less preferred foods to maximize the difference in appearance of the two presented foods as both beets and grapes were of a similar purplish color, and upon the suggestion of her care staff who noted that Migwan no longer preferred carrots relative to other foods from her daily diet. I verified that Migwan did not select carrots until all other foods were selected when presented with a handful of foods from her regular diet in her water trough. I also used the newly trained response buttons so that the dislike button was a gray square and the like button was a purple circle within a black background to mitigate against Migwan's retained preference for the blue circle as the like button. The locations of the stimuli remained the same with the food appearing in the center of the bottom half of the screen and the dislike button appearing on the top left and the like button appearing on the top right.

Each session consisted of 10 trials: 5 in which carrots were presented and 5 in which grapes were presented, in random order. Both response buttons appeared simultaneously on the screen after the food item was selected by Migwan. She was rewarded for selecting the dislike button if carrots were shown and the like button if grapes were shown. Migwan completed 112 sessions of this phase between 4 August and 30 September 2016 before testing was halted. Testing took place three times a week at 13:00 h. Migwan simultaneously participated in a novel test of judgement bias during this time [5]. Migwan moved to another facility in 2017 and could no longer be tested.

3. Results

Analyses were conducted using SPSS v. 28. Alpha was always set to $p = 0.05$.

3.1. *Phase 1 Training*

Migwan reached criterion in 35 sessions (approximately 700 trials with some sessions missing some trials).

3.2. *Food Preference Assessment*

Initially, Migwan had a strong left side bias. She eventually met criterion with two consecutive sessions at 90% correct by session 33 but I continued testing her with additional minor changes to the composition of the food items and she reached criterion again with four consecutive sessions at 90% or better by her 39th session (780 trials). Overall, she chose the preferred foods at levels above chance determined by a one sample Wilcoxon signed rank test ($Z = 4.434$, $p < 0.001$). Her performance improved across sessions, as can be seen in Figure 5. There was a significant difference in performance between the first and last halves of the testing sessions, Wilcoxon, $Z = -3.732$, $p < 0.001$.

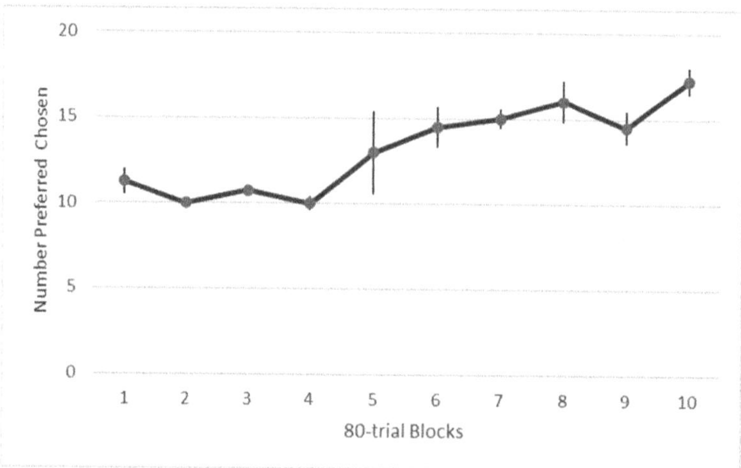

Figure 5. Average number of trials in which preferred foods were chosen during the food preference assessment.

3.3. *Phase 2 Training Continuation*

Migwan chose the blue circle 13 times on the single session in which it was paired with beets. In the single session, in which it was paired with preferred food items, she selected it 50% of the time. Across the five sessions in which it was paired with the yellow square, Migwan chose it on 80% or more trials on all but a single session, where she chose it 11 times.

3.4. *Phases 3 and 4 NAPS Training*

In Phase 3, only the correct response button appeared on each trial so Migwan was 100% correct on all 7 sessions. In Phase 4, Migwan chose the dislike response button only once across six sessions, so her performance was at chance.

3.5. *Phase 5 NAPS Training*

On the first four sessions, Migwan chose the dislike button correctly 50% of the time. However, by the end of 12 sessions, she had met the criterion, responding at 80% or more for four consecutive sessions. She was accidentally given a 13th session, on which she also scored 80% correct. Testing went on hiatus for torpor after this phase.

3.6. Phase 6 NAPS Training

Migwan completed 112 sessions. She responded equally quickly to touch photos of carrots (M = 2672.87, SD = 15,881.30) and grapes (M = 2596.05, SD = 18,051.841, $p = 0.93$ with a Wilcoxon signed ranks test).

Her average performance across all sessions was 56.61% correct, which was significantly above chance, (binomial test, N = 64, $p < 0.001$). She did not reach criterion; however, she had a run of three sessions at 80% correct between sessions 75 and 77 and she missed meeting criterion by a single trial by the 107th session. When comparing her performance on the first half of sessions (M = 55.36, SD = 1.33) to performance on the last half of sessions (M = 57.86, SD = 1.67), she showed little improvement. A Wilcoxon signed rank test confirmed no significant difference in performance between the first half and last half of sessions, $Z = -0.123$, $p = 0.092$. Figure 6 shows her performance across blocks of 4 sessions (40 trials).

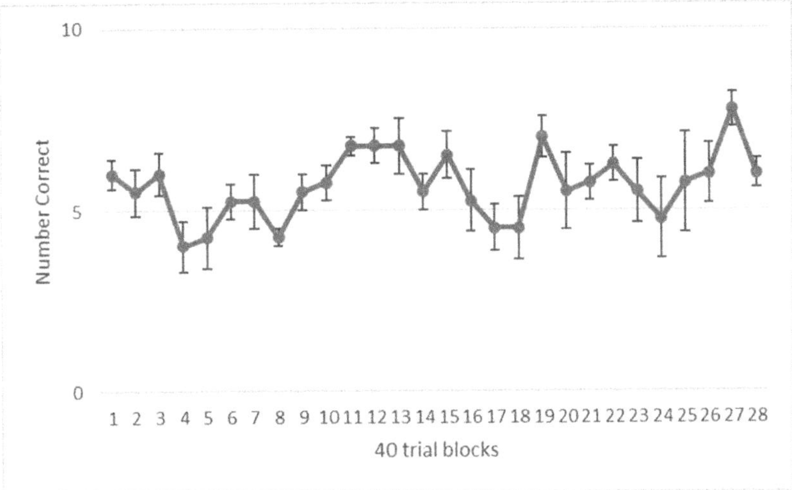

Figure 6. Performance on the Final NAPS Training Task.

Migwan's performance on trials where carrots were shown was not significantly different from chance, M = 0.48, SD = 0.500, binomial $p = 0.488$, but her performance was significantly above chance when grapes were shown, M = 0.67, SD = 0.471, $p < 0.001$.

I also conducted Chi square tests of independence to test whether the food item presented on that trial was significantly associated with selection of the different response buttons. The likelihood of choosing a particular response button was significantly associated with the food that was shown, $X^2 = 26.066$, $p < 0.001$. Migwan was more likely to choose the dislike button for carrots and the like button for grapes, as can be seen in Figure 7.

To test whether the latencies to respond were a function of the response button chosen (dislike, like, referred to henceforth as "response") and correctness of the response (henceforth "correct"), I used a generalized linear model (GLM) with a gamma distribution and a log link function. I included response, correct, and their interaction as fixed effects in the model. Response significantly predicted response time, $X^2 = 34.092$, $p < 0.001$, but correct did not, $X^2 = 0.139$, $p = 0.709$. However, response interacted with correct to predict response latencies, $X^2 = 6.918$, $p = 0.009$. The difference in response latencies for correct and incorrect responses was more pronounced if the dislike button was selected. In this case, Migwan was quicker to select dislike when it was the correct response (M = 952.99, SEM = 36.920) compared to when it was the incorrect response (M = 1082.40, SEM = 50.204). She showed the opposite pattern when choosing the like response. With the like response, she was faster to respond incorrectly (M = 1221.11, SEM = 45.816) than correctly (M = 1334.46, SEM = 43.864). These data appear in Figure 8.

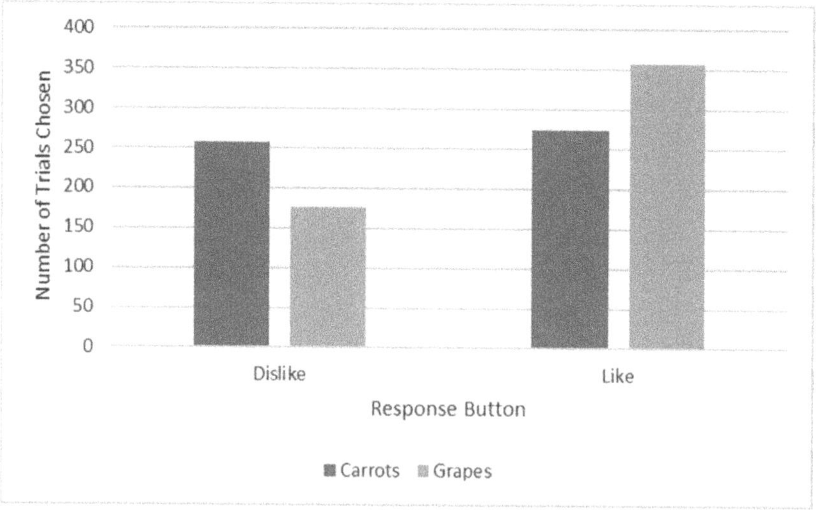

Figure 7. Number of Trials in which Responses were Selected for each Food in Phase Six.

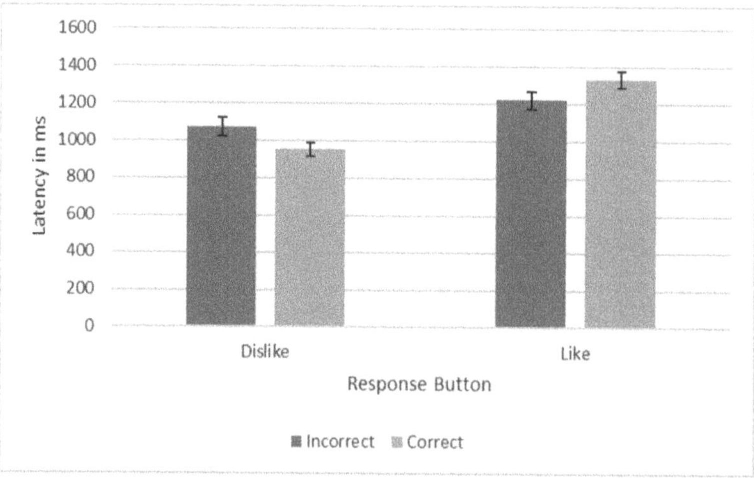

Figure 8. Response Latencies as a Function of Response and Correctness in Phase 6.

4. Discussion

I report on the first attempt to train a bear to use symbols to communicate her preferences. Members of many other species, including nonhuman primates [47,51], domestic dogs [52], dolphins [53] parrots [54], and a cockatoo [13] have shown the ability to use symbols to communicate to varying degrees. Despite lofty intentions of training a black bear to use a touchscreen to communicate her preferences for two-dimensional stimuli, the task, which depended upon a conditional discrimination, proved very difficult to train, as it had been for three gorillas trained in parallel [29]. This was somewhat surprising as the bear had previously outperformed the gorillas in two conditional discrimination tasks used to assess judgement bias [3,5,49,50]. Furthermore, conditional discrimination tasks have been mastered by individuals of various species, including pigeons [55], rats [56], octopuses, cuttlefishes [57], squirrel monkeys [58], and chimpanzees [59], so the task should not have been beyond Migwan's capability. Notably, the previous tasks involved associations between stimuli defined merely by shape and color and different response outcomes, whereas the current study aimed to test associations between broad abstract categories of preferred and less preferred foods. A construct concerning preferences is highly abstract and there is no existing evidence that nonhumans can represent a concept of their own or others' preferences. Because the preferences of others can inconsistently match or differ from our own, it is likely a challenging construct for nonverbal organisms to represent [38]. However, there is some evidence that at least three language-trained apes appropriately used lexigrams representing "good" and "bad" and applied them in a manner that was appropriate and consistent with their human caretakers' notions of good and bad behaviors [47]. There is also growing evidence that nonhuman mammals are capable of internally generating hedonic experiences in the absence of an external stimulus [60], making the use of a Likert scale for reporting preferences for symbolically represented aspects of their environment feasible. Furthermore, the current study progressed only to the point of training Migwan to associate a particular response with one preferred and one less preferred food, so should not have been conceptually more abstract than the previous studies. Unfortunately, due to Migwan's move to another facility, I was unable to continue testing her. I had initially aspired to train Migwan to use a five-point scale indicating a more nuanced sliding scale of preferences for items for which I did not already know her preferences, but I was unable to reach this goal.

However, there are some promising data from this project. First, Migwan did learn to perform at above chance levels in the NAPS presenting only a single preferred and less-preferred food item, and she missed our somewhat arbitrary criterion level of performance by only a single trial. Thus, one could conclude that she acquired the discrimination. However, she performed with greater accuracy when presented with an image of the preferred food—grapes. This may not be surprising given that, in this final phase of training, I had replaced the previously trained less preferred beets photo with a photo of carrots. Carrots were less consistently presented as a member of the "less preferred food" category across all of the training presented here. One of the major limitations of this study is that I do not have data from systematic preference tests verifying the care staff's indication of Migwan's preferences, although I did conduct informal assessments by presenting multiple food items in the water trough and I observed that Migwan did not eat beets when presented as rewards and did not choose carrots when presented alongside other options.

When presented with forced-choice tests of preferred versus non-preferred foods, Migwan did not spontaneously select the images of preferred foods at above chance rates unlike two of three gorillas [44], lion-tailed macaques [42], and two sloth bears [43] tested in food preference assessments with images of foods. However, Migwan did choose the preferred foods at above chance levels across all 39 sessions, and did learn to select them to a criterion of 80% over the course of testing, suggesting that she may have formed categories for "preferred foods" over "less preferred foods." It is less likely that she merely memorized which food photographs were associated with reward because I changed the food photographs periodically. However, it is true that beets, carrots, and lettuce were the only foods used as non-preferred foods so she may have simply learned not to select those images. I did not have the opportunity to test generalization to other photographs or to other preferred and less preferred foods to verify that she had formed such categories.

To further corroborate the conclusion that Migwan could learn to use the NAPS, she showed differential use of the response buttons dependent upon which food item had been presented on that trial, as did one of the gorillas tested in a similar procedure [29]. However, she used the buttons more accurately when the preferred grapes were shown, and did not clearly differentiate her use of the dislike and like buttons when the less preferred carrots were shown. It is possible that our initial training, in which I selectively rewarded Migwan for choosing the "like" button in order to reinforce its association with something positive, biased Migwan to the like button even when I changed its image in the final phase of training (the spatial location remained the same). It should be noted, though, that I did not train the gorillas selectively with the like icons used in their training, and they also struggled to learn this task with two of the three gorillas receiving many more trials in the initial training phase than the 1120 trials Migwan received [29]. Furthermore, although Migwan chose the like button more often, she did not choose it as often when it was incorrect as she did when it was correct. It was also not the case that every error involved inappropriate selection of the like button. Migwan also mistakenly chose the dislike button sometimes when grapes were presented. That she was above chance overall but did not show marked improvement across trials suggests that she had some spontaneous understanding of the task when I resumed testing in the fall of 2016, but did not develop an abstract conceptualization of the conditional discrimination nature of the task.

As with the gorillas who also struggled with this version of the task [29], I interrupted training on the NAPS with preferred and less preferred foods to present what I imagine to be a simplified conditional discrimination task using two-dimensional shapes of two colors (a green triangle and a red oval) associated with two new response buttons. Eventually, Migwan met a learning criterion. She met criterion more quickly with shape cues compared to color cues, which was in contrast to the gorillas who matched the colors more accurately. This is interesting because human children show a bias to attend to shape over color whereas chimpanzees tested in the same relational matching task showed the opposite bias and performed better on color matching trials [61]. These differences aside, both Migwan and two of the gorillas were able to learn the conditional discrimination task when presented with arbitrary shapes rather than images of food that were linked to their own preferences, suggesting that the mechanics of the task itself are not beyond their abilities, but that responding on the basis of their own preferences may be too abstract and require too much training to become a practical tool for use with animals in human care. This suggests that other procedures like token exchange [62] or the use of a progressive ratio reward schedule [63,64] to assess motivation to obtain rewards might have greater potential as a tool to assess animal preferences. In particular, although I had ultimately planned to present a 5-point Likert scale with five response buttons, adding more than two response buttons may be too challenging for nonhuman subjects [29], as human children cannot reliably use a 3-point scale until the age of eight years according to a recent meta-analysis [37].

Another limitation of the present study was the limited number of images used to represent the categories of preferred and non-preferred foods. Subjects show more robust transfer following training with a large number of exemplars representing categories, although training with multiple exemplars may slow acquisition of a category [65]. Training Migwan that she would not be rewarded for selecting the yellow square that represented the dislike button slowed her acquisition of the NAPS. In future, I would instead use images of items that held differential appeal for the like and dislike buttons. Although Migwan did quickly (i.e., within 120 trials) reach criterion in a task where selecting this button was always correct, she remained slightly biased toward the use of the like button even when the less preferred food image was presented and even when the images for the response buttons were replaced. That she did learn to use the dislike button, and to use it more often when it was rewarded (i.e., when the less preferred carrots were presented) indicates her flexibility in updating prior learned reward contingencies.

It is possible that presenting Migwan with other tasks in the intervening periods and breaking from training during torpor may have interfered with her reaching criterion levels of performance in this task. However, Migwan came very close to passing the admittedly somewhat arbitrary criterion. Furthermore, it should be noted that Migwan's performance was quite exceptional in other tasks presented to her over the same period. In fact, she outperformed gorillas on several similar cognitive tasks [3,5,49,50]. Therefore, her ability to perform accurately was not generally hampered by the presentation of multiple tasks during the same period of testing. In fact, she demonstrated remarkable flexibility in switching between tasks.

5. Conclusions

Although this particular attempt to develop a NAPS suffered from several limitations, it is my hope that other researchers are inspired to improve on these methods. Developing such a tool would provide a valuable new method in which nonhumans could communicate their degree of liking for various items in a single trial, including for things that cannot be physically presented. The use of a NAPS, once trained, allows an assessment of stability of preferences over time without numerous repetitions of pairs of items over many trials. This may be especially appealing for assessing food preferences when foods cannot be presented repeatedly due to satiety or other factors. Notably, the basic idea for the NAPS can be extended to the presentation of auditory, olfactory, or tactile stimuli and response buttons can be presented in other forms other than touchscreen buttons. Thus, the basic paradigm could be easily modified to suit various species and modalities of presentation. However, use of the touchscreen allows for random presentation of stimulus items and the recording of both responses and latencies to respond. Both measures provided some indication here as to how Migwan was understanding the task. Although the present study was motivated by the desire to develop a novel welfare tool, it also provides some insight into black bear cognition. Bears are still quite understudied with regard to their cognition. Migwan's performance in this study suggests that she can learn conditional discriminations, but that black bears, similar to other nonhumans, may not represent categories that are defined by unobservable features, such as relative preferences. Understanding such fundamental differences in how humans and nonhumans conceptualize their worlds will allow us to fully appreciate the uniqueness of other intelligent species and improve our abilities to provide them with the most appropriate stimulation while they are in our care.

Funding: This research received no external funding.

Institutional Review Board Statement: The study was conducted in accordance with the Declaration of Helsinki, and approved by the IACUC of Oakland University (Protocol #12082) and the Animal Welfare and Management Committee of the Detroit Zoological Society.

Informed Consent Statement: Not applicable.

Data Availability Statement: The data are available upon request from the author.

Acknowledgments: I thank the animal caretaker team at the Detroit Zoo, especially Marilynn Crowley, Flo Yates, Ela Wojtkowski, and Sarah Baker, for taking such excellent care of Migwan, and for participating in Migwan's sessions and providing crucial insight to her training. I also thank Mary Humbyrd and the zoo's Center for Zoo and Aquarium Animal Welfare and Ethics, Cynthia Bennett and Lauri Torgerson-White who encouraged and facilitated this project. Thank you to Zoe Johnson-Ulrich who assisted with data collection.

Conflicts of Interest: The author declares no conflict of interest.

References

1. Bacon, E.S.; Burghardt, G.M. Learning and color discrimination in the American black bear. *Bears Biol. Manag.* **1976**, *3*, 27–36. [CrossRef]
2. Johnson-Ulrich, Z.; Vonk, J.; Humbyrd, M.; Crowley, M.; Wojtkowski, E.; Yates, F.; Allard, S. Picture object recognition in an American black bear (*Ursus americanus*). *Anim. Cogn.* **2016**, *19*, 1237–1242. [CrossRef] [PubMed]
3. McGuire, M.C.; Vonk, J.; Johnson-Ulrich, Z. Ambiguous results when using the ambiguous-cue paradigm to assess learning and cognitive bias in gorillas and a black bear. *Behav. Sci.* **2017**, *7*, 51. [CrossRef] [PubMed]
4. Vonk, J.; Johnson-Ulrich, Z. Social and nonsocial category discriminations in a chimpanzee (*Pan troglodytes*) and American black bears (*Ursus americanus*). *Learn Behav.* **2014**, *42*, 231–245. [CrossRef]
5. Vonk, J.; McGuire, M.C.; Johnson-Ulrich, Z. Bearing fruit: Piloting a novel judgement bias task in an American black bear. *Zoo Biol.* **2021**, *40*, 89–97. [CrossRef] [PubMed]
6. Vonk, J.; Jett, S.E.; Mosteller, K.W. Concept formation in American black bears (*Ursus americanus*). *Anim. Behav.* **2012**, *84*, 953–964. [CrossRef]

7. Perdue, B.M. The effect of computerized testing on sun bear behavior and enrichment preferences. *Behav. Sci.* **2016**, *6*, 19. [CrossRef]
8. Gray, S.; Clark, F.; Burgess, K.; Metcalfe, T.; Kadijevic, A.; Cater, K.; Bennett, P. Gorilla Game Lab: Exploring modularity, tangibility and playful engagement in cognitive enrichment design. In Proceedings of the Fifth International Conference on Animal-Computer Interaction, Atlanta, GA, USA, 4–6 December 2018; pp. 1–13. [CrossRef]
9. Washburn, D.A. The four Cs of psychological wellbeing: Lessons from three decades of computer-based environmental enrichment. *Anim. Behav. Cogn.* **2015**, *2*, 218–232. [CrossRef]
10. Piitulainen, R.; Hirskyj-Douglas, I. Music for monkeys: Building methods to design with white-faced sakis for animal-driven audio enrichment devices. *Animals* **2020**, *10*, 1768. [CrossRef]
11. Gupfinger, R.; Kaltenbrunner, M. Animal-Centered sonic interaction design: Musical instruments and interfaces for Grey parrots. In Proceedings of the 6th International Conference on Animal-Computer Interaction (ACI 2019), Haifa, Israel, 12–14 November 2019.
12. Bethell, E.J.; Pfefferle, D. Cognitive bias tasks: A new set of approaches to assess welfare in nonhuman primates. In *Nonhuman Primate Welfare: From History, Science, and Ethics to Practice*; Springer International Publishing: Cham, Germany, 2023; pp. 207–230.
13. Cunha, J.; Rhoads, C. Use of a tablet-based communication board and subsequent choice and behavioral correspondences in a Goffin's Cockatoo (*Cacatua goffiana*). In Proceedings of the Seventh International Conference on Animal-Computer Interaction (ACI'2020), Milton Keynes, UK, 10–12 November 2020. [CrossRef]
14. Siegford, J.M. Multidisciplinary approaches and assessment techniques to better understand and enhance zoo nonhuman animal welfare. *J. Appl. Anim. Welf. Sci.* **2013**, *16*, 300–318. [CrossRef]
15. Bloomfield, R.C.; Gillespie, G.R.; Kerswell, K.J.; Butler, K.L.; Hemsworth, P.H. Effect of partial covering of the visitor viewing area window on positioning and orientation of zoo orangutans: A preference test. *Zoo Biol.* **2015**, *34*, 223–229. [CrossRef]
16. Dawkins, M.S.; Cook, P.A.; Whittingham, M.J.; Mansell, K.A.; Harper, A.E. What makes free-range broiler chickens range? In Situ measurement of habitat preference. *Anim. Behav.* **2003**, *66*, 151–160. [CrossRef]
17. Kirkden, R.D.; Pajor, E.A. Using preference, motivation and aversion test to ask scientific questions about animals' feelings. *Appl. Anim. Behav. Sci.* **2006**, *100*, 29–47. [CrossRef]
18. Hopper, L.M.; Egelkamp, C.L.; Fidino, M.; Ross, S.R. An assessment of touchscreens for testing primate food preferences and valuations. *Behav. Res. Methods* **2019**, *51*, 639–650. [CrossRef] [PubMed]
19. Hovland, A.L.; Mason, G.J.; Kirkden, R.D.; Bakken, M. The nature and strength of social motivations in young farmed silver fox vixens (*Vulpes vulpes*). *Appl. Anim. Behav. Sci.* **2008**, *111*, 357–372. [CrossRef]
20. Jackson, R.E.; Waran, N.K.; Cockram, M.S. Methods for measuring feeding motivation in sheep. *Anim. Welf.* **1999**, *8*, 53–63. [CrossRef]
21. Bacon, E.S.; Burghardt, G.M. Food preference testing of captive black bears. *Bears Biol. Manag.* **1983**, *5*, 102–105. [CrossRef]
22. Clay, A.W.; Bloomsmith, M.A.; Marr, M.J.; Maple, T.L. Systematic investigation of the stability of food preferences in captive orangutans: Implications for positive reinforcement training. *J. Appl. Anim. Welf. Sci.* **2009**, *12*, 306–313. [CrossRef]
23. Mehrkam, L.R.; Dorey, N.R. Is preference a predictor of enrichment efficacy in galapagos tortoises (*Chelonoidis nigra*)? *Zoo Biol.* **2014**, *33*, 275–284. [CrossRef]
24. Mehrkam, L.R.; Dorey, N.R. Preference assessments in the zoo: Keeper and staff predictions of enrichment preferences across species. *Zoo Biol.* **2015**, *34*, 418–430. [CrossRef]
25. Woods, J.M.; Lane, E.K.; Miller, L.J. Preference assessments as a tool to evaluate environmental enrichment. *Zoo Biol.* **2020**, *39*, 382–390. [CrossRef] [PubMed]
26. Truax, J.; Vonk, J. Silence is golden: Auditory preferences in zoo-housed gorillas. *J. Appl. Anim. Welf. Sci.* 2021, *in press*. [CrossRef] [PubMed]
27. Likert, R.A. Technique for the measurement of attitudes. *Arch. Psychol.* **1932**, *140*, 1–55.
28. Ho, G.W. Examining perceptions and attitudes: A review of Likert-type scales versus Q-methodology. *West. J. Nurs. Res.* **2017**, *39*, 674–689. [CrossRef] [PubMed]
29. Vonk, J. What's not to like about Likert? Developing a nonverbal animal preference scale (NAPS). *Am. J. Primatol.* **2022**, *84*, e23364. [CrossRef]
30. Alan, Ü.; Kabasakal, K.A. Effect of number of response options on the psychometric properties of Likert-type scales used with children. *Stud. Educ. Eval.* **2020**, *66*, 100895. [CrossRef]
31. Mellor, D.; Moore, K.A. The use of Likert scales with children. *J. Pediatr. Psychol.* **2014**, *39*, 369–379. [CrossRef]
32. Marmo, L.; Fowler, S. Pain assessment tool in the critically ill post–open heart surgery patient population. *Pain Manag. Nurs.* **2010**, *11*, 134–140. [CrossRef]
33. Vanhamme, J.; Chiu, C.K. Measuring different emotions in children with a pictorial scale: A self-reported nonverbal tool measures the emotions children experience when exposed to ads. *J. Advert. Res.* **2019**, *59*, 370–380. [CrossRef]
34. Soetenga, D.; Frank, J.; Pellino, T.A. Assessment of the validity and reliability of the University of Wisconsin Children's Hospital pain scale for preverbal and nonverbal children. *Pediatr. Nurs.* **1999**, *25*, 670.
35. Wong, D.; Baker, C. Pain in children: Comparison of assessment scales. *Pediatr. Nurs.* **1988**, *14*, 9–17.
36. Hall, L.; Hume, C.; Tazzyman, S. Five degrees of happiness: Effective smiley face likert scales for evaluating with children. In Proceedings of the 15th International Conference, Zakopane, Poland, 12–16 June 2016; pp. 311–321.

37. Coombes, L.; Bristowe, K.; Ellis-Smith, C.; Aworinde, J.; Fraser, L.K.; Downing, J.; Bluebond-Langner, M.; Chambers, L.; Murtagh, F.E.M.; Harding, R. Enhancing validity, reliability and participation in self-reported health outcome measurement for children and young people: A systematic review of recall period, response scale format, and administration modality. *Qual. Life Res.* **2021**, *30*, 1803–1832. [CrossRef] [PubMed]
38. Premack, D. Concordant preferences as a precondition for affective but not for symbolic communication (or how to do experimental anthropology). *Cognition* **1972**, *1*, 251–264. [CrossRef]
39. Fagot, J. *Picture Perception in Animals*; Psychology Press: London, UK, 2000.
40. Aust, U.; Huber, L. Picture-object recognition in pigeons: Evidence of representational insight in a visual categorization task using a complementary information procedure. *J. Exp. Psychol. Anim. Behav. Process.* **2006**, *32*, 190–195. [CrossRef] [PubMed]
41. Wein, A.; Gajdon, G.K.; Schwing, R. Picture–object recognition in kea (*Nestor notabilis*). *Ethology* **2015**, *121*, 1059–1070. [CrossRef]
42. Judge, P.G.; Kurdziel, L.B.; Wright, R.M.; Bohrman, J.A. Picture recognition of food by macaques (*Macaca silenus*). *Anim. Cogn.* **2012**, *15*, 313–325. [CrossRef]
43. Tabellario, S.; Babitz, M.A.; Bauer, E.B.; Brown-Palsgrove, M. Picture recognition of food by sloth bears (*Melursus ursinus*). *Anim. Cogn.* **2020**, *23*, 227–231. [CrossRef]
44. Vonk, J.; Truax, J.; McGuire, M. A Food for all seasons: Stability of food preferences in gorillas across testing methods and seasons. *Animals* **2022**, *12*, 685. [CrossRef]
45. Huskisson, S.M.; Jacobson, S.L.; Egelkamp, C.L.; Ross, S.R.; Hopper, L.M. Using a touchscreen paradigm to evaluate food preferences and response to novel photographic stimuli of food in three primate species (*Gorilla gorilla gorilla, Pan troglodytes, and Macaca fuscata*). *Int. J. Primatol.* **2020**, *41*, 5–23. [CrossRef]
46. Vonk, J.; Beran, M.J. Bears "count" too: Quantity estimation and comparison in black bears (*Ursus americanus*). *Anim. Behav.* **2012**, *84*, 231–238. [CrossRef]
47. Lyn, H.; Franks, B.; Savage-Rumbaugh, E.S. Precursors of morality in the use of the symbols "good" and "bad" in two bonobos (*Pan paniscus*) and a chimpanzee (*Pan troglodytes*). *Lang. Commun.* **2008**, *28*, 213–224. [CrossRef]
48. Hayes, K.J.; Nissen, C.H. Higher mental functions of a home-raised chim panzee. In *Behavior of Nonhuman Primates*; Schrier, A.M., Stoll Nitz, F., Eds.; Academic Press: Cambridge, MA, USA, 1971; Volume 4.
49. McGuire, M.; Vonk, J.; Fuller, G.; Allard, S. Using an ambiguous cue paradigm to assess cognitive bias in gorillas (*Gorilla gorilla gorilla*) during a browse manipulation. *Anim. Behav. Cogn.* **2017**, *4*, 91–104. [CrossRef]
50. McGuire, M.C.; Vonk, J. Gorillas (*Gorilla gorilla gorilla*) fail to learn abstract cues of differential outcomes in a novel cognitive bias test. *Anim. Behav. Cogn.* **2018**, *5*, 103–117. [CrossRef]
51. Patterson, F.G.P.; Cohn, R.H. Language acquisition by a lowland gorilla: Koko's first ten years of vocabulary development. *Word* **1990**, *41*, 97–143. [CrossRef]
52. Rossi, A.P.; Ades, C. A dog at the keyboard: Using arbitrary signs to communicate requests. *Anim. Cogn.* **2008**, *11*, 329–338. [CrossRef] [PubMed]
53. Herman, L.M.; Kuczaj II, S.A.; Holder, M.D. Responses to anomalous gestural sequences by a language-trained dolphin: Evidence for processing of semantic relations and syntactic information. *J. Exp. Psychol. Gen.* **1993**, *122*, 184–194. [CrossRef]
54. Pepperberg, I.M. Symbolic communication in nonhuman animals. In *APA Handbooks in Psychology: Basic Concepts, Methods, Neural Substrate, and Behavior*; Call, J., Burghardt, G.M., Pepperberg, I.M., Snowdon, C.T., Zentall, T., Eds.; American Psychological Association: Worcester, MA, USA, 2017; pp. 663–679. [CrossRef]
55. Castro, L.; Kennedy, P.L.; Wasserman, E.A. Conditional same-different discrimination by pigeons: Acquisition and generalization to novel and few-item displays. *J. Exp. Psychol. Anim. Behav. Process.* **2010**, *36*, 23. [CrossRef]
56. Brown, K.L.; Pagani, J.H.; Stanton, M.E. Spatial conditional discrimination learning in developing rats. *Dev. Psychobiol.* **2005**, *46*, 97–110. [CrossRef]
57. Hvorecny, L.M.; Grudowski, J.L.; Blakeslee, C.J.; Simmons, T.L.; Roy, P.R.; Brooks, J.A.; Boal, J.G. Octopuses (*Octopus bimaculoides*) and cuttlefishes (*Sepia pharaonis, S. officinalis*) can conditionally discriminate. *Anim. Cogn.* **2007**, *10*, 449–459. [CrossRef]
58. Thomas, R.K.; Kerr, R.S. Conceptual conditional discrimination in *Saimiri sciureus*. *Anim. Learn. Behav.* **1976**, *4*, 333–336. [CrossRef]
59. Dugdale, N.; Lowe, C.F. Testing for symmetry in the conditional discriminations of language-trained chimpanzees. *J. Exp. Anal. Behav.* **2000**, *73*, 5–22. [CrossRef] [PubMed]
60. Mahr, J.B.; Fischer, B. Internally triggered experiences of hedonic valence in nonhuman animals: Cognitive and welfare considerations. *Perspect. Psychol. Sci.* **2022**, 17456916221120425. [CrossRef] [PubMed]
61. Vonk, J.; Rastogi, G. The "sh-ape bias" in non-linguistic categorization: Comparisons between children and other apes. *J. Cogn. Dev.* **2019**, *20*, 380–398. [CrossRef]
62. Brosnan, S.F.; de Waal, F. Socially learned preferences for differentially rewarded tokens in the brown capuchin monkey (*Cebus apella*). *J. Comp. Psychol.* **2004**, *118*, 133. [CrossRef] [PubMed]
63. DeLeon, I.G.; Frank, M.A.; Gregory, M.K.; Allman, M.J. On the correspondence between preference assessment outcomes and progressive-ratio schedule assessments of stimulus value. *J. Appl. Behav. Anal.* **2009**, *42*, 729–733. [CrossRef] [PubMed]

64. Roane, H.S. On the applied use of progressive-ratio schedules of reinforcement. *J. Appl. Behav. Anal.* **2008**, *41*, 155. [CrossRef]
65. Brino, A.L.F.; Galvão, O.F.; Picanço, C.R.; Barros, R.S.; Souza, C.B.; Goulart, P.R.; McIlvane, W.J. Generalized identity matching-to-sample after multiple-exemplar training in capuchin monkeys. *Psychol. Rec.* **2014**, *64*, 693–702. [CrossRef]

animals

MDPI

Article

MDA-DETR: Enhancing Offending Animal Detection with Multi-Channel Attention and Multi-Scale Feature Aggregation

Haiyan Zhang [1,2], Huiqi Li [1,2], Guodong Sun [1,2] and Feng Yang [1,2,*]

[1] School of Information Science and Technology, Beijing Forestry University, Beijing 100083, China; zhyzml@bjfu.edu.cn (H.Z.); shangguanli21@bjfu.edu.cn (H.L.); sungd@bjfu.edu.cn (G.S.)

[2] Engineering Research Center for Forestry-Oriented Intelligent Information Processing of National Forestry and Grassland Administration, Beijing 100083, China

* Correspondence: fengyang@bjfu.edu.cn

Simple Summary: As agriculture and human settlements expand, conflicts between humans and animals become more frequent, resulting in resource loss and safety risks. Therefore, accurately identifying and locating offending animals is essential. This study presents a novel method for automatically detecting offending animals, especially in situations where they are obscured or the images are unclear. This research focused on six types of offending animals commonly found in northeastern China. By employing improved image processing techniques, this study enhanced detection accuracy in complex environments. Experimental results demonstrate that the proposed method outperforms existing techniques on the dataset used. This new approach helps improve the accuracy of intelligent monitoring systems, providing better technical support to minimize conflicts between humans and animals, thus protecting agriculture and ensuring the safety of both humans and animals.

Abstract: Conflicts between humans and animals in agricultural and settlement areas have recently increased, resulting in significant resource loss and risks to human and animal lives. This growing issue presents a global challenge. This paper addresses the detection and identification of offending animals, particularly in obscured or blurry nighttime images. This article introduces Multi-Channel Coordinated Attention and Multi-Dimension Feature Aggregation (MDA-DETR). It integrates multi-scale features for enhanced detection accuracy, employing a Multi-Channel Coordinated Attention (MCCA) mechanism to incorporate location, semantic, and long-range dependency information and a Multi-Dimension Feature Aggregation Module (DFAM) for cross-scale feature aggregation. Additionally, the VariFocal Loss function is utilized to assign pixel weights, enhancing detail focus and maintaining accuracy. In the dataset section, this article uses a dataset from the Northeast China Tiger and Leopard National Park, which includes images of six common offending animal species. In the comprehensive experiments on the dataset, the mAP_{50} index of MDA-DETR was 1.3%, 0.6%, 0.3%, 3%, 1.1%, and 0.5% higher than RT-DETR-r18, yolov8n, yolov9-C, DETR, Deformable-detr, and DCA-yolov8, respectively, indicating that MDA-DETR is superior to other advanced methods.

Keywords: object detection; RT-DETR; transformer; computer vision; attention mechanism

1. Introduction

With the rapid advancement of urbanization, human activities are increasingly encroaching on wildlife habitats. This has led to a rise in harmful wildlife incidents, despite efforts to establish national parks and protected areas. Such incidents not only pose a threat to human health and livelihoods but also highlight the urgent need to address wildlife–human conflicts [1]. As a result, mitigating these conflicts has become a global priority for governments and a focus of academic research.

Traditional methods for managing wildlife–human conflicts, such as erecting physical barriers like barbed wire or electric fences around settlements and agricultural areas, have notable limitations. These measures can induce stress responses or even fatalities in animals and pose potential risks to humans [2]. To overcome these challenges, there is a growing demand for intelligent surveillance systems capable of automatically detecting and identifying offending animals in real time.

Object detection, a fundamental yet challenging task in computer vision, involves assigning precise bounding boxes and classification labels to objects in images. Recent advances in deep learning, coupled with reduced hardware costs, have spurred the development of robust object detection algorithms. Two-stage detection frameworks such as Faster R-CNN [3,4] and R-CNN [5], as well as single-stage detectors like YOLO [6–9], have gained prominence. Moreover, the dominance of Convolutional Neural Networks (CNNs) in this field has recently been challenged by the rise of Transformers, which leverage attention mechanisms to boost model performance. DETR, the first Transformer-based object detection model [10], eliminates the need for manually designed anchor frames and non-maximal suppression (NMS). However, it faces limitations in processing speed. In 2023, Baidu's Flying Paddle team introduced RT-DETR, a breakthrough in object detection. This model transitions from "dense detection" to "sparse detection", eliminating threshold filtering and NMS. It delivers an end-to-end, real-time object detection solution [11], making it a promising tool for applications like offending animal detection.

Offending animals frequently appear near planting and breeding areas, causing damage to vegetation and livestock. These activities often obscure the animals within the damaged vegetation. Furthermore, many offending animals are nocturnal, leading to blurry images captured by trap cameras. This study emphasizes the crucial need for accurate detection and recognition of obscured or blurry images of offending animals. While the RT-DETR model exhibits strong robustness and generalization, it has limitations. The AIFI module, relying solely on a self-attention mechanism, provides rich semantic information but lacks precise positional information. Additionally, the CCFF module, despite integrating cross-scale information, struggles with detecting targets in blurry images.

To address these limitations, this paper proposes MDA-DETR, a novel method based on Multi-Channel Coordinated Attention and multi-scale feature aggregation, to enhance the detection of obscured and blurry images of offending animals. Inspired by channel attention mechanisms [12], we introduce a Multi-Channel Coordinated Attention (MCCA) mechanism. This mechanism extracts image features along three spatial directions, capturing long-range dependencies and providing precise positional information while preserving the rich semantic information from deep networks. This significantly improves the accuracy of detecting occluded targets.

Furthermore, inspired by the RepBiPAN module [13], we propose a Multi-Dimension Feature Aggregation Module (DFAM). This module effectively fuses positional information from shallow networks with semantic information from deeper networks, and aggregates features from multiple scales. This enhances the detection accuracy of blurry images. To reduce the overall parameter count, we incorporate the RepNCSPELAN4 module [14] into the backbone network.

MDA-DETR was rigorously evaluated on a subset of the Northeast China Tiger and Leopard National Park dataset [15]. Ablation experiments were conducted to assess the performance of each component, demonstrating the superior performance of MDA-DETR in detecting offending animals. The main contributions of this study are as follows:

(i) **Multi-Channel Coordinate Attention (MCCA) mechanism** This mechanism extracts image features along three spatial directions, integrating precise positional and long-range dependency information while preserving the rich semantic information from deep networks.
(ii) **Multi-Dimension Feature Aggregation Module (DFAM)** This module effectively fuses multi-scale feature maps, enabling the extraction of complementary and global information.
(iii) **Comparative experiments** demonstrate the superiority of MDA-DETR over state-of-the-art methods. Ablation experiments confirm the effectiveness of each key component.

2. Related Works

2.1. Transformer-Based Object Detection

Transformer is an encoder–decoder architecture based on a self-attention mechanism, first proposed by Google and applied to Natural Language Processing (NLP), achieving state-of-the-art (SOTA) results in various NLP tasks [16]. In 2020, Nicolas Carion et al. introduced Transformer to object detection with the DETR (DEtection TRansformer) model, simplifying the detection process by treating it as an ensemble prediction problem and eliminating the need for manually designing components. In addition, DETR converts the feature image output from the backbone network into a one-dimensional sequence, enabling the model to compute correlations between each pixel and others, thus achieving a wider receptive field than that of CNNs [17].

However, as a pioneering work, DETR has some limitations. Zhu et al. proposed Deformable-detr [18] to address the slow training speed of Transformer in computer vision. Its attentional module focuses only on the key sampling points, resulting in better performance on small targets and reduced training time. In 2022, Wang et al. introduced Anchor DETR [19], a new Transformer-based query mechanism for object detection. This model employs a new attentional variant to predict multiple objects in a region, thus effectively solving the problem of 'one region, multiple objects'.

Although the Transformer-based DETR series has challenged the monopoly of CNN in just two years, it is still no substitute for the CNN-based series of algorithms in terms of real-time performance in industrial applications. To address this, Baidu's team, composed of Lv et al., proposed RT-DETR (Real-Time DEtection Transformer) in 2023 [11]. It is a real-time end-to-end detector based on the DETR model and outperforms YOLO in real-time object detection. Compared with yolov8, RT-DETR requires a shorter training duration, fewer data enhancement strategies, and demonstrates stronger performance under the same test conditions.

Most state-of-the-art Transformer methods divide images into regular grids and represent each grid region with visual tokens. However, the fixed token distribution ignores the semantic information of different areas of the image, which leads to performance degradation. To address this issue, in 2024, Wang Zeng et al. proposed the Token Clustering Transformer (TCFormer) [20]. This method dynamically generates visual tokens based on semantic information, allowing regions with similar semantics to be represented by the same token, even if these regions are not adjacent. For areas containing important details, TCFormer uses fine-grained tokens to enhance the accuracy of image understanding. In the same year, Yansong Peng et al. redefined the bounding box regression task in the DETR model and proposed the D-FINE model [21]. This model introduces Fine-grained Distribution Refinement (FDR) and Global Optimal Localization Self-Distillation (GO-LSD) strategies, transforming the regression process from predicting fixed coordinates to iteratively refining probability distributions, thus achieving higher accuracy in object localization.

2.2. Wildlife Object Detection

Wildlife surveys are key to nature conservation. In 2021, Delplanque et al. evaluated the performance of three CNN algorithms: Faster-RCNN, Libra-RCNN [22], and RetinaNet [23]. They trained the models using an independent dataset, enabling them to detect and identify African mammal species based on high-resolution aerial imagery. The Libra-RCNN model [24] with the best detection results accurately detected animals in open and sparse grasslands, reducing detection time. In 2022, Cai Qianzhou et al. proposed a solution [25] for long-tailed data based on YOLOv4-Tiny. Their approach combined two-stage learning and reweighting. In the first stage, the model was trained without weighting. In the second stage, weights from the first stage were used in combination with the reweighting method, ultimately improving the model's accuracy for long-tailed data acquired by trap cameras. In 2023, Roy et al. proposed the WilDect-YOLO model [26] for the automatic detection of endangered wildlife. This model introduces residual blocks in the backbone network and integrates DenseNet blocks to extract and retain critical feature information. Additionally, Spatial Pyramid Pooling (SPP) and an improved Path Aggregation Network (PANet) are added to the feature fusion part, enhancing the model's perceptual field and preserving fine-grained local information. In 2024, Yang Wenhan et al. combined the Swin Transformer module with a CNN to propose a method [27] for detecting wildlife images captured by trap cameras based on YOLOV5s. The technique fuses the advantages of both networks in the feature extraction layer, expanding the receptive field of feature extraction and enabling the model to accurately detect animals despite severe occlusion, low contrast with the background, and other challenges.

2.3. Offending Animal Detection

Timely detection and identification of offending animals are crucial to reducing human–animal conflicts. In 2018, Ram et al. proposed an automatic unsupervised elephant image detection system [28] for human–elephant conflicts. This system acquires animal images and presence signals from cameras and sensors, respectively. Placed near conflict-prone areas in forest villages, the device generates a warning signal when an elephant is detected, and then sent to a specific location via a GSM module. In 2020, Ravoor et al. designed a cross-camera tracking system [29] for detecting jaguars, elephants, and other offending animals. The method uses the MobileNetv2-SSD [30] model to localize the animals and the Triplet Loss trained ResNet-50 [31] model for re-identification. The integrated model runs at 2–3 frames per second, enabling near real-time functionality. In 2022, Lee et al. proposed an extract–append data enhancement method [32] that extracts specific objects from a limited number of images through semantic segmentation and appends them to numerous images with arbitrary backgrounds. The technique generates images of offending animals with varied backgrounds, improving the model's detection performance by enriching the dataset. In 2023, Charles et al. developed a system [33] consisting of an Arduino, a PIR motion sensor, an LED flash, a speaker, and an acoustic cannon to monitor a field 24 × 7 h a day. When the PIR motion sensor detects an animal, the Arduino activates, the speaker emits a threatening animal sound, the LED light flashes, and the system sends a text message and a photo to the farmer within seconds of detection.

3. Materials and Methods

3.1. Dataset

This study uses the publicly available Northeast China Tiger and Leopard National Park dataset [15]. From this dataset, six common species of offending animals in north-eastern China—badger, black bear, leopard cat, red fox, weasel, and wild boar—were selected. The total dataset comprises 9641 images, including 5823 daytime images and

3818 nighttime images. The dataset was divided into training, validation, and test sets in a 7:2:1 ratio. Mosaic data augmentation was applied to the training set to enhance the dataset (see Table 1).

Table 1. Brief information about the dataset used in this article.

Animal	Total Number of Images	Number of Nighttime Images	Number of Daytime Images	Number of Instances
Badger	1309	921	388	1364
Black bear	1352	955	397	1413
Leopard cat	1974	1021	953	1982
Red fox	1862	983	879	1873
Weasel	1269	728	541	1296
Wild boar	1875	1215	660	2007
Total	9641	5823	3818	9935

3.2. *Overall Architecture*

The overall structure of MDA-DETR proposed in this paper is shown in Figure 1, which is a Transformer-based encoding–decoding structure. First, the improved ResNet-18 backbone network is used for encoding and extracting image features. This paper acquires three different scales of features with the resolutions: $S4[M, N]$, $S5[\frac{M}{2}, \frac{N}{2}]$, $S7[\frac{M}{4}, \frac{N}{4}]$. Then, the $S7$ deep feature images are inputted into MCCA to obtain accurate position information. In addition, the AIFI self-attention mechanism in the RT-DETR model is retained to extract rich semantic information. Next, the feature maps of three different scales—shallow, intermediate, and deep—are further fused by DFAM to obtain a new feature image with global information. Finally, the Variable Focal Loss function (VariFocal Loss) [34], GIoU loss function [35], and L1-loss [36] are used to measure the error between the model's prediction results and the actual labels and to adjust the weight parameters accordingly to optimize the overall model further.

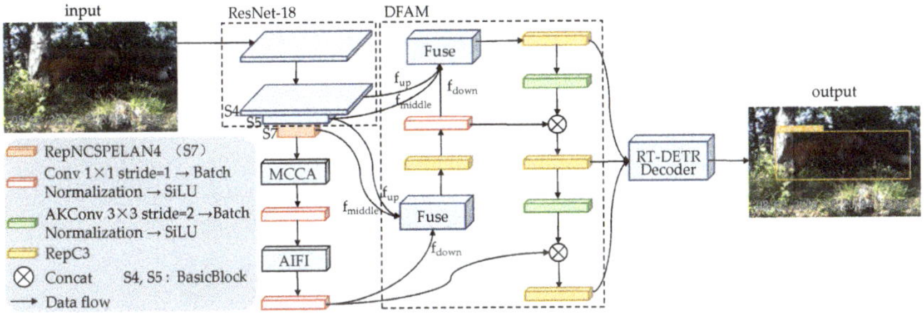

Figure 1. The overall structure of MDA-DETR.

3.3. *Multi-Channel Coordinate Attention (MCCA)*

Generally, channel attention mechanisms play a crucial role in improving model performance. However, traditional channel attention mechanisms focus primarily on extracting semantic information from feature maps, often neglecting the importance of target position [37]. In the field of detecting offending animals, obstructions often occur, making accurate positional information critical for enhancing model predictive performance. Inspired by channel attention mechanisms, this paper proposes the Multi-Channel Coordinated Attention Mechanism (MCCA). It effectively extracts semantic and positional information from feature maps through three spatial directions and captures long-range dependencies (as shown in Figure 2), ultimately improving the accuracy of detecting obscured offending animals.

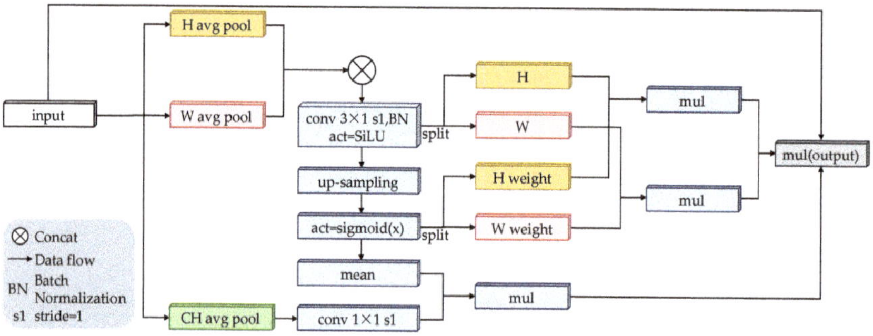

Figure 2. The structure diagram of MCCA.

Specifically, the input image of size $M \times N$ is first decomposed into three one-dimensional feature encoding processes. The three outputs can be expressed as follows:

$$\begin{cases} Z(h) = \frac{1}{W}\sum_{j=1}^{W} x(h,j) \\ Z(w) = \frac{1}{H}\sum_{i=1}^{H} x(i,w) \\ Z(ch) = \sum_{i=1}^{H}\sum_{j=1}^{W} x(i,j) \end{cases} \quad (1)$$

The features of H and W are then combined to obtain $Z_{concat} = [Z(h), Z(w)]$. These combined features are then fed into a 3×1 convolution kernel for convolution, followed by batch normalization and SiLU activation [38]. This process can be described as follows:

$$F_1 = B_{conv}(Z_{concat}) \quad (2)$$

where $B_{conv}(\cdot)$ represents a series of operations, including 3×1 convolution, batch normalization, and SiLU activation functions.

Subsequently, F_1 is decomposed into two separate tensors along the H and W directions to obtain F_1^h and F_1^w. Upsampling and $sigmoid(x)$ activation operations are then performed on F_1. The $sigmoid(x)$ activation function is used here to reduce the complexity and computational overhead of the module. This process can be described as follows:

$$F_2 = sigmoid(U(F_1)) \quad (3)$$

where $U(\cdot)$ denotes upsampling.

Decomposing F_2 into two separate attentional weights along the H and W directions yields F_2^h and F_2^w for subsequent refinement of the F_1^h and F_1^w tensors. Next, F_2 is batch normalized. Afterward, it is multiplied with $Z(ch)$, which has been processed through a 1×1 convolution, to obtain F_3. This process can be defined as follows:

$$F_3 = F_2 \times Conv(Z(ch)) \quad (4)$$

The resulting output F_{out} can be written as

$$F_{out} = F_1^h \times F_1^w \times F_2^h \times F_2^w \times F_3 \quad (5)$$

3.4. Multi-Dimension Feature Aggregation Module (DFAM)

When detecting offending animals, their rapid movements and nighttime appearances can result in blurry images and unclear boundaries when captured by trap cameras. Utilizing the MCCA module and the AIFI module, the model obtains feature images with precise localization and rich semantic information. However, recognition accuracy for such challenging conditions remains an issue. To leverage the complementarity and correlation between multi-scale features, this paper proposes a multi-scale feature aggregation module (DFAM), inspired by the RepBiPAN module, to fuse feature images from the backbone network with those processed through MCCA and AIFI operations. Figure 1 illustrates the overall structure of DFAM. The main contribution of DFAM is the fusion of multiple feature images at different scales, utilizing both element-wise addition and cascading operations. This approach enhances information at the same scale while benefiting from features at other scales. Additionally, this paper introduces an aggregation block (Fuse) that consolidates feature maps of varying sizes in the DFAM fusion path, followed by a convolution block (RepC3) composed of multiple 1×1 convolutional layers. These operations enable the model to integrate feature maps from each layer, yielding feature images with globally essential features and improving the detection accuracy of blurry animal images. Figure 3 shows the structure of Fuse and RepC3 in DFAM.

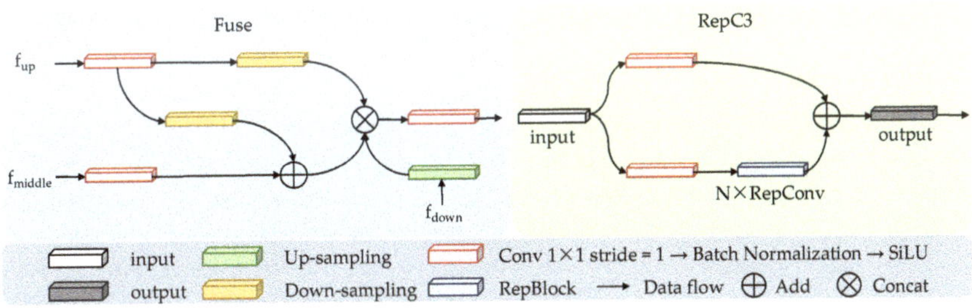

Figure 3. Structure of Fuse and RepC3 in DFAM.

Specifically, the Fuse module handles three inputs to the aggregation block: f_{up}, f_{middle}, and f_{down}. For the shallow feature map, f_{up} is first processed by a 1×1 convolution, followed by batch normalization and the SiLU activation function, and then undergoes downsampling to facilitate subsequent fusion operations. This process can be expressed as follows:

$$f_1 = D(B^1_{conv}(f_{up})) \quad (6)$$

where $B^1_{conv}(\cdot)$ is a series of sequential operations, including 1×1 convolution, batch normalization, and SiLU activation functions.

After that, the middle layer feature map f_{middle} is input into $B^1_{conv}(\cdot)$ for a convolution operation with a 1×1 kernel, and the deep layer feature map f_{down} is upsampled. This process can be represented as follows:

$$\begin{cases} f_2 = B^1_{conv}(f_{middle}) \\ f_3 = U(f_{down}) \end{cases} \quad (7)$$

where $U(\cdot)$ denotes upsampling.

Then, f_1 and f_2 are fused using element-wise addition, and the fused feature maps are cascaded with f_1 and f_3, and finally input into $B^1_{conv}(\cdot)$ for a convolution operation. This process can be defined as follows:

$$f_{out} = B^1_{conv}(concat((f_1 + f_2), f_1, f_2)) \tag{8}$$

In the RepC3 convolutional block, the input features first undergo sequential operations to adjust the number of channels. After this, feature extraction is performed on one of the branches using a RepBlock consisting of N RepConvs [39]. Finally, the outputs of the dual paths are fused using element-wise addition. By employing DFAM, the model can fuse features at three different scales and utilize cross-level features to acquire more global and complementary information. This results in a more expressive fused feature image that accurately detects the presence of offending animals in blurred images.

3.5. Backbone Network Improvement Strategy

Although the images processed by MCCA and DFAM retain accurate positional information and rich semantic information, the overall number of model parameters is high. To address this, we use the RepNCSPELAN4 structure to replace the last layer of the ResNet-18 backbone network. The RepNCSPELAN4 structure combines CSPNet with gradient path planning and ELAN, extending ELAN to support GELAN for any computational block. This design is focused on achieving lightweight inference speed and accuracy. This optimization strategy enables the model to reduce the number of parameters while maintaining the original accuracy.

3.6. Loss Function

The Variable Focal Loss function (VariFocal Loss) uses the IoU-aware Classification Score (IACS) to represent the probability of target object ranking of candidate detection frames. The loss function employs an asymmetric weighting approach to improve the model's classification performance by reducing the weight of negative samples and increasing the weight of positive samples. This allows the model to better focus on training with high-quality positive samples. The functional representation of VariFocal Loss is as follows:

$$L_{VFL} = \begin{cases} -y^{(i)}(y^{(i)}\log(\hat{y}^{(i)}) + (1 - y^{(i)})\log(1 - \hat{y}^{(i)})) & y^{(i)} > 0 \\ -\alpha\hat{y}^{(i)\gamma}\log(1 - \hat{y}^{(i)}) & y^{(i)} = 0 \end{cases} \tag{9}$$

where $y^{(i)}$ denotes the label of the sample i with positive class 1 and negative class 0. $\hat{y}^{(i)}$ denotes the predicted IACS for the sample i. α and $\hat{y}^{(i)\gamma}$ represent the scalability coefficients of the moderating loss.

Traditional object detection algorithms typically use the IoU loss function to measure the degree of overlap between the prediction frame and the ground truth. However, when the overlap is high, the IoU loss function suffers from the vanishing gradient problem, making further model optimization difficult [40]. To better localize the target and reduce the absolute error between the prediction and ground truth, the GIoU loss function and L1-loss are introduced. Specifically, the GIoU loss function introduces a penalty term on top of the IoU loss function to measure the non-overlap between the prediction and ground truth frames, addressing the vanishing gradient problem and enabling faster convergence. The L1-loss, or mean absolute error (MAE), is the average of the absolute errors between the model's predicted and true values. Using both the GIoU loss function and L1-loss provides better supervision during model training and ensures the accuracy of its results.

The expressions for the loss functions are as follows:

$$L_{GIoU} = 1 - IoU + \frac{|A_C - U|}{|A_C|} \tag{10}$$

$$L_1 = \frac{1}{N}\sum_{i=1}^{N}|y_i - y_i^{'}| \tag{11}$$

where IoU denotes the intersection ratio between the predicted and ground truth boxes, A_c represents the minimum bounding box containing both the predicted and ground truth boxes, and U represents the union of predicted and real bounding boxes. N is the number of samples, y_i is the true value of sample i, and $y_i^{'}$ is the predicted value of sample i.

Finally, the overall loss function for the proposed model is as follows:

$$L = L_{VFL} + L_{GIoU} + L_1 \tag{12}$$

4. Results

4.1. Evaluation Metrics

To evaluate the model proposed in this study, six widely used evaluation metrics in object detection tasks were adopted: Precision, Recall, mAP_{50} (mean Average Precision), $mAP_{50\text{–}95}$, parameters, and FPS (Frames Per Second). In object detection tasks, the Intersection over Union (IoU) between the predicted bounding box and the ground truth box is generally used to assess their overlap, thereby evaluating the model's detection accuracy on a specific dataset. A higher IoU value indicates greater similarity between the two boxes. If the IoU exceeds a certain threshold (commonly set at 0.5, also used in this paper), the model is considered to have successfully detected the object, and the result is classified as a positive example (P); otherwise, it is treated as a negative example (N). For a positive example, if the model's classification result is correct, it is considered a true positive (TP); otherwise, it is regarded as a false positive (FP). If a ground truth box is not detected, it is labeled as a false negative (FN). Based on these, Precision and Recall can be calculated using the following formulas:

$$Precision = \frac{TP}{TP + FP} \tag{13}$$

$$Recall = \frac{TP}{TP + FN} \tag{14}$$

However, Precision and Recall do not fully reflect the performance of an object detection model across different thresholds. When Precision increases, Recall often decreases; conversely, when Recall increases, Precision tends to decrease. This is because the stricter the criteria for predicting a sample as a positive example, the more likely the model is to miss some true positive examples. On the other hand, the more lenient the criteria, the more likely the model is to increase the number of false positives. Therefore, this study includes two additional evaluation metrics: mAP_{50} and $mAP_{50\text{–}95}$. mAP_{50} represents the average precision (AP) across multiple categories at an IoU threshold of 0.5. $mAP_{50\text{–}95}$ represents the average of mAP results at 10 IoU thresholds ranging from 0.5 to 0.95, with a step size of 0.05. The higher the mAP value, the better the model's detection performance. The formulas for calculating AP and mAP are as follows:

$$AP = \sum_{k=1}^{N} P(s) \cdot \Delta R(s) \tag{15}$$

$$mAP = \frac{1}{c}\sum_{i=1}^{c} AP(i) \tag{16}$$

where P represents Precision, R represents Recall, s represents the classification confidence of the detection box, N represents the number of discrete points, and c represents the number of categories.

For model complexity and detection speed, this study uses two metrics: Frames Per Second (FPS) and parameters. FPS indicates the number of frames the object detection model can process per second. The higher the FPS value, the faster the detection speed and the better the real-time performance of the model. Parameters reflect the spatial complexity of the model. The expression for FPS is as follows:

$$FPS = \frac{1000\,\text{ms}}{preprocess + inference + NMS} \tag{17}$$

where 1000 ms refers to 1000 milliseconds, $preprocess$ refers to the time required for the preprocessing stage, $inference$ refers to the time required for the model's inference, and NMS refers to the post-processing time required when the model applies Non-Maximum Suppression.

4.2. Parameter Settings

The overall model is implemented using the Pytorch framework and trained on a Titan RTX3090 GPU. During training, the pre-trained weights of the ResNet-18 [31] on ImageNet-1k [41] were used and trained on the public dataset used in this paper. The model used the AdamW optimizer with a weight decay of 0.0001, an initial learning rate set to 0.0001, and a mosaic data enhancement probability of 0.5.

4.3. Comparative Experiments on Backbone Network Improvement Strategies

This paper compares the model with only the backbone network part to the one with the RT-DETR decoder part. The results show that the improved model reduces the number of parameters by 46.75% compared with the original model using the ResNet-18 backbone network. Additionally, it achieves some improvement in accuracy, as shown in Table 2.

Table 2. Comparison of the improved backbone network model with the ResNet-18 backbone network model on the dataset.

Model	Parameter	mAP_{50} %	$mAP_{50–95}$ %
ResNet-18 backbone network (RT-DETR decoder)	15.4 M	0.949	0.772
Improved backbone network (RT-DETR decoder)	8.2 M	0.959	0.78

Both the ResNet-18 backbone network and the improved backbone network do not include MCCA and DFAM.

4.4. Ablation Study

To evaluate the effectiveness of each key component in MDA-DETR, this paper conducted comprehensive ablation experiments focusing on (1) the effectiveness of backbone network improvement strategies, (2) the effectiveness of MCCA, and (3) the importance of DFAM. During these experiments, two challenging scenarios—occlusion and blurring—were selected to observe and record the performance of these key components under different conditions, demonstrating their effectiveness in various situations.

(1) Effectiveness of Backbone Network Improvement Strategies: The backbone network employs the RepNCSPELAN4 structure, initially enhancing the backbone's performance and shows improvement across four indicators compared with the model with only the ResNet-18 backbone network (as shown in the second row of Table 3).

However, when animals are occluded, the detection accuracy is low (as shown in the second row and fourth column of Figure 4).

(2) Effectiveness of MCCA: This study retained only the improved backbone network and MCCA for training and testing to demonstrate its effectiveness. The results indicate that MCCA significantly improves all four evaluation metrics and further enhances the accuracy of detecting occluded offending animals (as shown in the third row and fourth column of Figure 4).

(3) The effectiveness of DFAM. DFAM adjusts the number of channels in the feature map generated by the AIFI module and further fuses it with the feature map generated by the improved backbone network. To demonstrate DFAM's effectiveness, this study retained only DFAM and the improved backbone network for testing. Table 3 shows that adding DFAM leads to a slight decrease in recall. However, the 0.3% decrease in recall is accompanied by a 1.2% increase in precision, making the slight drop in recall acceptable. Furthermore, Figure 4 shows that using this module has enhanced the accuracy of generating the minimum bounding rectangle. Particularly when the offending animal in the image is blurry, it can more accurately locate and identify the offending animal. Finally, by combining the three proposed improvement methods, the model achieved a 2.9% increase in accuracy, a 3.3% increase in recall, a 2.9% improvement in mAP_{50}, and a 4.7% improvement in $mAP_{50\text{–}95}$ compared with the original baseline. The detection performance was exceptional, proving the effectiveness of the improvement strategy proposed in this article.

Figure 4. Comparison of ablation experiments. (The orange animal in the upper right corner of the third column is the dataset icon.)

Table 3. Ablation experiments of the three key components on the dataset constructed in this paper.

Improved Backbone Network	MCCA	DFAM	P %	R %	mAP_{50} %	mAP_{50-95} %
ResNet-18 backbone network (RT-DETR decoder)			0.951	0.931	0.949	0.772
✓			0.957	0.94	0.959	0.78
✓	✓		0.964	0.95	0.967	0.794
✓		✓	0.969	0.937	0.956	0.795
✓	✓	✓	0.98	0.964	0.978	0.819

4.5. Comparison with Other Models

To comprehensively evaluate the performance of MDA-DETR, this paper compares it with six state-of-the-art object detection models: RT-DETR-r18 [11], yolov8n [42], yolov9-C [14], DETR [10], Deformable-detr [18], and DCA-yolov8 [43]. The results for all models were generated using the official code or based on published papers.

Table 4 shows the results of comparing MDA-DETR with six state-of-the-art methods, where the red font and blue font indicate the best and second best performance. From the results in Table 4, it can be seen that the proposed model achieves 97.8% mAP_{50} metrics on the dataset, outperforming the original RT-DETR-r18 model by 1.3%, and it is 0.6% better than yolov8n, 0.3% better than yolov9-C, 3% better than DETR, 1.1% better than Deformable-detr, and 0.5% better than DCA-yolov8. Overall, the model proposed in this article outperforms other advanced methods in the detection of offending animals on the part of the publicly available Northeast China Tiger and Leopard National Park dataset and can better detect images of accident-prone animals.

Table 4. Comparison of MDA-DETR with 6 state-of-the-art methods.

Model	P %	R %	mAP_{50} %	mAP_{50-95} %	Parameter	FPS
RT-DETR-r18 [11]	0.958	0.934	0.965	0.806	21 M	52.6
yolov8n [42]	0.943	0.926	0.972	0.798	3.2 M	76.9
yolov9-C [14]	0.971	0.958	0.975	0.809	25.5 M	71.5
DETR [10]	0.935	0.921	0.948	0.741	41 M	57.4
Deformable-detr [18]	0.957	0.927	0.967	0.757	40 M	59.8
DCA-yolov8 [43]	0.947	0.938	0.973	0.801	11.5 M	102
MDA-DETR (ours)	0.98	0.964	0.978	0.819	18.4 M	54.5

Figures 5 and 6 show the detection results of MDA-DETR and advanced methods in different scenarios. The images selected in this article include three scenarios: animals being occluded, only having partial animal features, and animals being blurry at night. Among them, DETR and Deformable-detr have better performance in detecting images with only partial animal features, but there are false detections when detecting occluded animal images (leopard cats), and DETR also has false detections when detecting occluded wild boars. Yolov8n and yolov9-C can effectively handle scenes with only partial animal features and animal occlusion, but when detecting blurry animal images at night, yolov8n has lower detection accuracy, while yolov9-C lacks comprehensive detection (badgers). This article selects the DCA-yolov8 model, which can also detect occluded scenes, as the comparison model. From the results, it can be seen that the MDA-DETR proposed in this article has higher detection accuracy than DCA-ylov8 in several complex tasks. Overall, the method proposed in this article outperforms other methods on the publicly available dataset used. MDA-DETR not only accurately classifies the offending animal but also accurately locates and detects the offending animal in scenes where the animal is occluded and blurred at night.

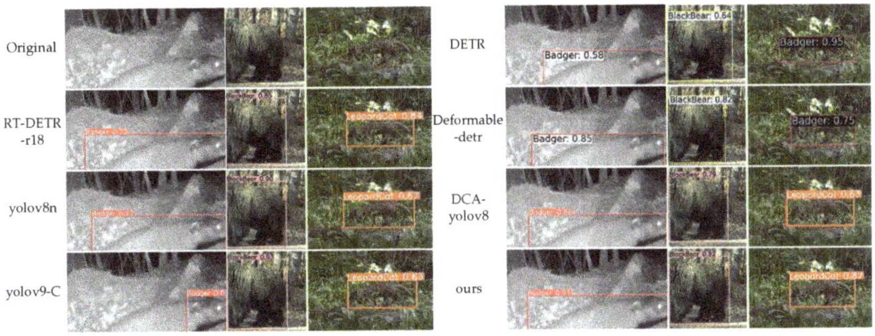

Figure 5. Comparison between ours and advanced methods (badger, black bear, leopard cat).

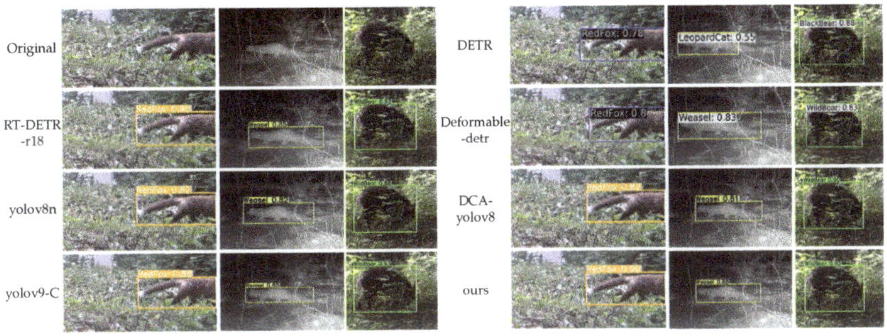

Figure 6. Comparison between ours and advanced methods (red fox, yellow weasel, wild boar).

5. Discussion

This study proposed MDA-DETR, a novel object detection model, to address the challenges of detecting offending animals, particularly in obscured and blurry nighttime images. Comprehensive evaluations on the Northeast China Tiger and Leopard National Park dataset demonstrated that MDA-DETR significantly outperforms six state-of-the-art models, achieving superior results in mAP50 (+1.3% to +3%) and mAP50-95 (+0.5% to +4.7%). These results validate the effectiveness of the Multi-Channel Coordinated Attention (MCCA) mechanism and the Multi-Dimension Feature Aggregation Module (DFAM) in improving detection accuracy under challenging conditions.

The main findings of this research align with the initial hypothesis that incorporating multi-scale feature aggregation and enhanced attention mechanisms would address limitations in existing models. The observed improvements can be attributed to the following factors: The MCCA mechanism effectively integrates spatial and long-range dependencies, enhancing the accuracy of detecting occluded animals. The DFAM module fuses features across multiple scales, providing global and complementary information to improve the detection of blurry images. The use of VariFocal Loss further enhances classification accuracy by emphasizing high-quality positive samples.

Compared with the YOLO series, DETR, and its variants, MDA-DETR demonstrates significant advantages in handling occlusion and nighttime detection on the dataset used in this study. While the existing YOLO and DETR series methods perform well in most conventional object detection tasks, YOLO models tend to lose crucial information when processing blurred images of animals at night, leading to inaccurate target localization. Although DETR and its variants have improved target detection localization, they often fail to correctly classify objects in complex backgrounds, especially in occlusion or cluttered backgrounds. In contrast, MDA-DETR effectively overcomes these challenges by combining MCCA and DFAM. By incorporating multi-scale feature fusion and integrating the image's positional information, semantic information, and long-range dependencies, MDA-DETR avoids detection failures caused by the loss of local features. Additionally, this multi-level information integration enhances the model's ability to detect occluded animals and blurred nighttime animal images, reducing false negatives and false positives, and providing stronger robustness.

Despite the significant advantages demonstrated by MDA-DETR, this study still has certain limitations. First, the dataset used in this study contains only six specific species of offending animals, which limits the model's applicability to other species and broader scenarios. Second, the dataset does not cover more complex environmental factors, such as changing lighting conditions or motion blur caused by rapid movement, which can often affect the model's performance in detection tasks. Furthermore, although MDA-DETR has achieved significant accuracy improvements, its computational complexity may pose challenges for real-time deployment on edge devices.

To address these limitations, future research will focus on the following areas: Expanding the dataset: Plans are in place to create a dataset that includes more animal species and more complex environmental factors. Additionally, the use of generative models or trap cameras will be considered to acquire the relevant images for further evaluating the model's performance in various scenarios, thereby enhancing its generalization ability and applicability to a broader range of species. Optimizing the model architecture: considering the computational complexity involved in real-time deployment, future work will explore lighter model architectures, such as integrating more efficient attention mechanisms or designing low-computation overhead feature extraction modules, to better suit resource-constrained edge devices. Expanding the application scope: Future studies will further explore the potential of MDA-DETR in a wider range of application scenarios, including general wildlife detection and endangered species protection. By expanding its application scope, this study wishes MDA-DETR to play a greater role in protecting ecological environments and biodiversity.

6. Conclusions

We propose a novel model, MDA-DETR, for detecting offending animals. To better handle scenes where animals are occluded, the MCCA module is designed to extract semantic information from feature maps, target location information, and long-distance dependencies. Subsequently, the DFAM module aggregates features at three different scales to obtain global features, enhancing the accuracy of detecting offending animals in blurry images. The data used in this study come from the Northeast Tiger and Leopard National Park dataset. Ablation experiments on this dataset demonstrate the effectiveness of the key components in MDA-DETR. Additionally, comprehensive comparative experimental results on this dataset confirm the robustness and effectiveness of the MDA-DETR model.

Author Contributions: All authors have made significant contributions to this research. Conceptualization, H.L. and H.Z.; methodology, H.L. and H.Z.; validation, H.L.; writing, H.L., H.Z., G.S., and F.Y.; supervision, H.Z., G.S., and F.Y. All authors have read and agreed to the published version of the manuscript.

Funding: This research was funded by National Key R&D Program of China (Grant number 2022YFF1302700), The Emergency Open Competition Project of National Forestry and Grassland Administration (Grant number 202303), and Outstanding Youth Team Project of Central Universities (Grant number QNTD202308).

Institutional Review Board Statement: Ethical review and approval were waived for this study because the research involves the development and evaluation of a computer vision model for detecting offending animals. The model was trained and tested on existing datasets of images and videos, such as the NTLNP dataset (https://github.com/myyyyw/NTLNP, (accessed on 9 March 2023)), which does not involve any direct interaction with or harm to animals.

Informed Consent Statement: Not applicable.

Data Availability Statement: This study uses the publicly available Northeast China Tiger and Leopard National Park dataset, which can be downloaded at https://github.com/myyyyw/NTLNP (accessed on 9 March 2023).

Acknowledgments: We appreciate the hard work of every author, the valuable suggestions of every reviewer, and the patience and meticulousness of academic editors.

Conflicts of Interest: The authors declare no conflicts of interest.

References

1. Huang, Y.; Yang, J.; Zhang, H.; Su, K.W.; Wen, Y.L. A Comparative Study of Human-wildlife Conflicts Management in Nature Reserves at Home and Abroad. *World For. Res.* **2021**, *34*, 27–32.
2. Mamat, N.; Othman, M.F.; Yakub, F. Animal intrusion detection in farming area using YOLOv5 approach. In Proceedings of the 2022 22nd International Conference on Control, Automation and Systems (ICCAS), Jeju, Republic of Korea, 27 November–1 December 2022; pp. 1–5.
3. Girshick, R. Fast R-CNN. In Proceedings of the 2015 IEEE International Conference on Computer Vision (ICCV), Santiago, Chile, 7–13 December 2015; pp. 1440-1448.
4. Ren, S.; He, K.; Girshick, R.; Sun, J. Faster R-CNN: Towards real-time object detection with region proposal networks. *IEEE Trans. Pattern Anal. Mach. Intell.* **2016**, *39*, 1137–1149. [CrossRef] [PubMed]
5. Girshick, R.; Donahue, J.; Darrell, T.; Malik, J. Rich feature hierarchies for accurate object detection and semantic segmentation. In Proceedings of the IEEE Conference on Computer Vision and Pattern Recognition, Columbus, OH, USA, 23–28 June 2014; pp. 580–587.
6. Redmon, J.; Divvala, S.; Girshick, R.; Farhadi, A. You Only Look Once: Unified, Real-Time Object Detection. In Proceedings of the 2016 IEEE Conference on Computer Vision and Pattern Recognition (CVPR), Las Vegas, NV, USA, 27–30 June 2016; pp. 779–788.
7. Redmon, J. Yolov3: An incremental improvement. *arXiv* **2018**, arXiv:1804.02767.
8. Li, C.; Li, L.; Jiang, H.; Weng, K.; Geng, Y.; Li, L.; Ke, Z.; Li, Q.; Cheng, M.; Nie, W.; et al. YOLOv6: A single-stage object detection framework for industrial applications. *arXiv* **2022**, arXiv:2209.02976.
9. Wang, C.-Y.; Bochkovskiy, A.; Liao, H.-Y.M. YOLOv7: Trainable bag-of-freebies sets new state-of-the-art for real-time object detectors. In Proceedings of the IEEE/CVF Conference on Computer Vision and Pattern Recognition, Vancouver, BC, Canada, 17–24 June 2023; pp. 7464–7475.
10. Carion, N.; Massa, F.; Synnaeve, G.; Usunier, N.; Kirillov, A.; Zagoruyko, S. End-to-end object detection with transformers. In Proceedings of the European Conference on Computer Vision, Glasgow, UK, 23 August 2020; pp. 213–229.
11. Zhao, Y.; Lv, W.; Xu, S.; Wei, J.; Wang, G.; Dang, Q.; Liu, Y.; Chen, J. Detrs beat yolos on real-time object detection. In Proceedings of the IEEE/CVF Conference on Computer Vision and Pattern Recognition, Seattle, WA, USA, 16–22 June 2024; pp. 16965–16974.
12. Guo, M.-H.; Xu, T.-X.; Liu, J.-J.; Liu, Z.-N.; Jiang, P.-T.; Mu, T.-J.; Zhang, S.-H.; Martin, R.R.; Cheng, M.-M.; Hu, S.-M. Attention mechanisms in computer vision: A survey. *Comput. Vis. Media* **2022**, *8*, 331–368. [CrossRef]
13. Li, C.; Li, L.; Geng, Y.; Jiang, H.; Cheng, M.; Zhang, B.; Ke, Z.; Xu, X.; Chu, X. Yolov6 v3.0: A full-scale reloading. *arXiv* **2023**, arXiv:2301.05586.
14. Wang, C.-Y.; Yeh, I.-H.; Liao, H.-Y.M. Yolov9: Learning what you want to learn using programmable gradient information. *arXiv* **2024**, arXiv:2402.13616.
15. GitHub Repository. 2023. Available online: https://github.com/myyyyw/NTLNP (accessed on 2 January 2025).
16. Ashish, V. Attention is all you need. In Proceedings of the Advances in Neural Information Processing Systems, Long Beach, CA, USA, 4–9 December 2017.
17. Lecun, Y.; Bottou, L.; Bengio, Y.; Haffner, P. Gradient-based learning applied to document recognition. *Proc. IEEE* **1998**, *86*, 2278–2324. [CrossRef]
18. Zhu, X.; Su, W.; Lu, L.; Li, B.; Wang, X.; Dai, J. Deformable detr: Deformable transformers for end-to-end object detection. *arXiv* **2020**, arXiv:2010.04159.
19. Wang, Y.; Zhang, X.; Yang, T.; Sun, J. Anchor detr: Query design for transformer-based detector. In Proceedings of the AAAI Conference on Artificial Intelligence, Online, 22 February–1 March 2022; pp. 2567–2575.
20. Zeng, W.; Jin, S.; Xu, L.; Liu, W.; Qian, C.; Ouyang, W.; Luo, P.; Wang, X. TCFormer: Visual Recognition via Token Clustering Transformer. *IEEE Trans. Pattern Anal. Mach. Intell.* **2024**, *46*, 9521–9535. [CrossRef] [PubMed]

21. Peng, Y.; Li, H.; Wu, P.; Zhang, Y.; Sun, X.; Wu, F. D-FINE: Redefine Regression Task in DETRs as Fine-grained Distribution Refinement. *arXiv* **2024**, arXiv:2410.13842.
22. Pang, J.; Chen, K.; Shi, J.; Feng, H.; Ouyang, W.; Lin, D. Libra r-cnn: Towards balanced learning for object detection. In Proceedings of the IEEE/CVF Conference on Computer Vision and Pattern Recognition, Long Beach, CA, USA, 15–20 June 2019; pp. 821–830.
23. Lin, T.-Y.; Goyal, P.; Girshick, R.; He, K.; Dollár, P. Focal loss for dense object detection. In Proceedings of the IEEE Conference on Computer Vision and Pattern Recognition, Venice, Italy, 22–29 October 2017; pp. 2980–2988.
24. Delplanque, A.; Foucher, S.; Lejeune, P.; Linchant, J.; Théau, J. Multispecies detection and identification of African mammals in aerial imagery using convolutional neural networks. *Remote Sens. Ecol. Conserv.* **2022**, *8*, 166–179. [CrossRef]
25. Cai, Q.; Zheng, B.; Zeng, X.; Hou, J. Wildlife object detection combined with solving method of long-tail data. *J. Comput. Appl.* **2022**, *42*, 1284–1291.
26. Roy, A.M.; Bhaduri, J.; Kumar, T.; Raj, K. WilDect-YOLO: An efficient and robust computer vision-based accurate object localization model for automated endangered wildlife detection. *Ecol. Inform.* **2023**, *75*, 101919. [CrossRef]
27. Yang, W.; Liu, T.; Zhou, J.; Hu, W.; Jiang, P. CNN-Swin Transformer Detection Algorithm of Forest Wildlife Images Based on Improved YOLOv5s. *Sci. Silvae Sin.* **2024**, *60*, 121–130.
28. Ram, A.V.; Prakash, A.S.; Ahamed, A.I.; Anirudh, K.; Arvindh, M.M.; Nithyavathy, N. A self induced warning system for wild animal trespassing using machine vision system. In Proceedings of the 2018 International Conference on Intelligent Computing and Communication for Smart World (I2C2SW), Erode, India, 14–15 December 2018; pp. 349–353.
29. Ravoor, P.C.; Sudarshan, T.S.B.; Rangarajan, K. Digital Borders: Design of an Animal Intrusion Detection System Based on Deep Learning. In Proceedings of the International Conference on Computer Vision and Image Processing, Prayagraj, India, 4–6 December 2020; pp. 186–200.
30. Chiu, Y.-C.; Tsai, C.-Y.; Ruan, M.-D.; Shen, G.-Y.; Lee, T.-T. Mobilenet-SSDv2: An improved object detection model for embedded systems. In Proceedings of the 2020 International Conference on System Science and Engineering (ICSSE), Kagawa, Japan, 31 August–3 September 2020; pp. 1–5.
31. He, K.; Zhang, X.; Ren, S.; Sun, J. Deep residual learning for image recognition. In Proceedings of the IEEE Conference on Computer Vision and Pattern Recognition, Las Vegas, NV, USA, 27–30 June 2016; pp. 770–778.
32. Lee, J.; Lim, K.; Cho, J. Improved Monitoring of Wildlife Invasion through Data Augmentation by Extract—Append of a Segmented Entity. *Sensors* **2022**, *22*, 7383. [CrossRef] [PubMed]
33. Charles, P.S.; Dharmalingam, S.; Baskaran, M.; Manoharan, P.S.; Hemanth, G.R. Intelligence Harmless Animal Avoider in Agriculture Farm. In Proceedings of the 2023 International Conference on Computer Communication and Informatics (ICCCI), Coimbatore, India, 23–25 January 2023; pp. 1–4.
34. Zhang, H.; Wang, Y.; Dayoub, F.; Sunderhauf, N. Varifocalnet: An iou-aware dense object detector. In Proceedings of the IEEE/CVF Conference on Computer Vision and Pattern Recognition, Nashville, TN, USA, 20–25 June 2021; pp. 8514–8523.
35. Rezatofighi, H.; Tsoi, N.; Gwak, J.; Sadeghian, A.; Reid, I.; Savarese, S. Generalized intersection over union: A metric and a loss for bounding box regression. In Proceedings of the IEEE/CVF Conference on Computer Vision and Pattern Recognition, Long Beach, CA, USA, 15–20 June 2019; pp. 658–666.
36. Barron, J.T. A general and adaptive robust loss function. In Proceedings of the IEEE/CVF Conference on Computer Vision and Pattern Recognition, Long Beach, CA, USA, 15–20 June 2019; pp. 4331–4339.
37. Hou, Q.; Zhou, D.; Feng, J. Coordinate Attention for Efficient Mobile Network Design. In Proceedings of the IEEE/CVF Conference on Computer Vision and Pattern Recognition, Nashville, TN, USA, 20–25 June 2021; pp. 13713–13722.
38. Elfwing, S.; Uchibe, E.; Doya, K. Sigmoid-weighted linear units for neural network function approximation in reinforcement learning. *Neural Netw.* **2018**, *107*, 3–11. [CrossRef] [PubMed]
39. Ding, X.; Zhang, X.; Ma, N.; Han, J.; Ding, G.; Sun, J. Repvgg: Making vgg-style convnets great again. In Proceedings of the IEEE/CVF Conference on Computer Vision and Pattern Recognition, Nashville, TN, USA, 20–25 June 2021; pp. 13733–13742.
40. Yu, J.; Jiang, Y.; Wang, Z.; Cao, Z.; Huang, T. Unitbox: An advanced object detection network. In Proceedings of the 24th ACM International Conference on Multimedia, Amsterdam, The Netherlands, 15–19 October 2016; pp. 516–520.
41. Russakovsky, O.; Deng, J.; Su, H.; Krause, J.; Satheesh, S.; Ma, S.; Huang, Z.; Karpathy, A.; Khosla, A.; Bernstein, M.; et al. ImageNet Large Scale Visual Recognition Challenge. *Int. J. Comput. Vis.* **2015**, *115*, 1405–1573. [CrossRef]
42. Jocher, G.; Qiu, J.; Chaurasia, A. Ultralytics YOLO. GitHub Repository. 2023. Available online: https://github.com/ultralytics/ultralytics (accessed on 2 January 2025).
43. Yang, W.; Wu, J.; Zhang, J.; Gao, K.; Du, R.; Wu, Z.; Firkat, E.; Li, D. Deformable convolution and coordinate attention for fast cattle detection. *Comput. Electron. Agric.* **2023**, *211*, 108006. [CrossRef]

animals

MDPI

Article

Performance Comparison of Genomic Best Linear Unbiased Prediction and Four Machine Learning Models for Estimating Genomic Breeding Values in Working Dogs

Joseph A. Thorsrud [1], Katy M. Evans [2,3], Kyle C. Quigley [2], Krishnamoorthy Srikanth [1] and Heather J. Huson [1,*]

1 Department of Animal Sciences, College of Agriculture and Life Sciences, Cornell University, 201 Morrison Hall, 507 Tower Road, Ithaca, NY 14853, USA; jat325@cornell.edu (J.A.T.); hjh3@cornell.edu (H.J.H.)
2 The Seeing Eye Inc., 1 Seeing Eye Wy, Morristown, NJ 07960, USA; kevans@seeingeye.org (K.M.E.)
3 School of Veterinary Medicine and Science, University of Nottingham, Sutton Bonington, Loughborough LE12 5RD, UK
* Correspondence: hjh3@cornell.edu; Tel.: +1-(607)-255-2289

Simple Summary: This study aims to improve the breeding of guide dogs by using genetic information to predict important health and behavior traits. Guide dogs need to be healthy and attentive to effectively assist people with visual impairments. This study compares several methods for predicting whether a dog might develop certain health issues, such as anodontia (missing teeth), distichiasis (extra eyelashes that can irritate the eyes), or oral papillomatosis (oral tumors caused by a virus), as well as behaviors like high distractibility, based on their genetic makeup. Data from German Shepherds, Golden Retrievers, Labrador Retrievers, and their crosses were analyzed to see which prediction methods work best given different models and data parameters. The results show that all the tested methods were similarly effective in predicting these traits. Notably, simpler and less time-intensive methods and data collection processes perform just as well as more complex ones. This means that dog breeders can use these genetic prediction tools without investing in expensive technology or genetic testing. By applying these methods, breeders can make better informed decisions when selecting dogs for breeding, focusing on those more likely to be healthy and exhibit desirable behaviors. Ultimately, this approach can lead to the development of healthier and more capable guide dogs, benefiting individuals who rely on them and contributing to the overall well-being of the dog population.

Abstract: This study investigates the efficacy of various genomic prediction models—Genomic Best Linear Unbiased Prediction (GBLUP), Random Forest (RF), Support Vector Machine (SVM), Extreme Gradient Boosting (XGB), and Multilayer Perceptron (MLP)—in predicting genomic breeding values (gEBVs). The phenotypic data include three binary health traits (anodontia, distichiasis, oral papillomatosis) and one behavioral trait (distraction) in a population of guide dogs. These traits impact the potential for success in guide dogs and are therefore routinely characterized but were chosen based on differences in heritability and case counts specifically to assess gEBV model performance. Utilizing a dataset from The Seeing Eye organization, which includes German Shepherds (n = 482), Golden Retrievers (n = 239), Labrador Retrievers (n = 1188), and Labrador and Golden Retriever crosses (n = 111), we assessed model performance within and across different breeds, trait heritability, case counts, and SNP marker densities. Our results indicate that no significant differences were found in model performance across varying heritabilities, case counts, or SNP densities, with all models performing similarly. Given its lack of need for parameter optimization, GBLUP was the most efficient model. Distichiasis showed the highest overall predictive performance, likely due to its higher heritability, while anodontia and distraction exhibited moderate accuracy, and oral papillomatosis had the lowest accuracy, correlating with its low heritability. These findings underscore that lower density SNP datasets can effectively construct gEBVs, suggesting that high-cost, high-density genotyping may not always be necessary. Additionally, the similar performance of all models indicates that simpler models like GBLUP, which requires less fine tuning, may be sufficient for genomic prediction in canine breeding programs. The research highlights the importance of standardized phenotypic assessments and carefully constructed reference populations to optimize the utility of genomic selection in canine breeding programs.

Keywords: genomic prediction; breeding values; machine learning models; dog breeding; genetic selection; genomic best linear unbiased prediction; random forest; support vector machine; extreme gradient boosting; multilayer perceptron

1. Introduction

Modern dog breeding took shape in the mid-19th century with a focus on enhancing phenotypic attributes and aligning them with esthetic ideals and functional roles established by breed standards [1]. The use of phenotypes and pedigree data to track lineages allowed breeders to direct selection toward their desired attributes. Modern updates to pedigree-based selection allow for the quantification of estimated breeding values (EBVs). EBVs predict the genetic potential of an individual to pass on traits to offspring by combining pedigree and phenotypic data. Unlike tests that simply identify if an animal is a carrier of a specific gene, EBVs provide a comprehensive assessment of an animal's genetic merit for quantitative traits, enabling breeders to make more informed selection decisions and improve traits across generations. Notable successes of this technique include the estimation of hip and elbow dysplasia breeding values in dogs, leading to improved joint health [2–5]. However, compared to agricultural species, the implementation of EBVs in canine breeding programs has lagged due to challenges such as smaller reference populations, less standardized selection criteria, and limited collaboration among breeders. This gap highlights the need for research focusing on optimizing genetic prediction methods for dogs, which is the aim of the present study.

As these methods are refined, it is crucial to ensure that breeding decisions prioritize the welfare of individual animals, preserve genetic diversity, and mitigate the perpetuation of harmful traits. Overemphasis on narrowly defined breed-specific characteristics can exacerbate existing hereditary problems, raising ethical concerns around breeding practices that may compromise long-term health and well-being [6]. Researchers and breeders must remain vigilant in balancing performance or esthetic goals with robust welfare standards, including the responsible use of EBVs. By employing breeding values to identify and minimize hereditary disorders and maintaining balanced trait selection, breeders can harness these tools ethically and effectively to select for or against traits [7]. Though systemic problems arise with dog breeding, through a concerted focus on health and behavior traits, responsible breeding can avoid pitfalls that have plagued the field [8]. In this way, the primary focus remains on safeguarding and enhancing the quality of life of dogs while improving genomic tool development.

The introduction of genomic data-driven selection marked a transformative shift in breeding practices for many agricultural species [9]. Genomic selection utilizes DNA marker data to provide a more precise assessment of an individual's genetic merit compared to traditional pedigree-based methods. This advancement took off with the usage of single-nucleotide polymorphisms (SNPs) for genotyping. By analyzing SNP markers and using them to infer regions in linkage, breeders gain deeper insights into the genetic composition of individuals, thus improving the accuracy and efficiency of breeding decisions [10]. The success of genomic selection has been particularly pronounced in agricultural species like dairy cattle, where it has led to significant advancements in milk production, health, and fertility [9,11]. One of the key benefits of genomic selection is the ability to identify an individual's genetic merit at birth and differentiate between littermates, reducing the generation interval and accelerating the rate of genetic gain through selection compared to pedigree-based methods. These agricultural examples highlight the importance of large reference populations and consistently agreed-upon phenotypes among breeders for selection.

Several advanced genomic prediction models, including machine learning techniques, have been developed to leverage genomic data effectively. Machine learning models are particularly appealing due to their ability to model complex, nonlinear relationships in large datasets, which is advantageous in genomic prediction. Among traditional methods, Genomic Best Linear Unbiased Prediction (GBLUP) extends the traditional Best Linear Unbiased Prediction (BLUP) model by replacing the pedigree-based relationship matrix with a genomic relationship matrix derived from SNP markers [12]. This approach captures genetic relationships with greater accuracy, especially in scenarios where pedigree data are incomplete or unavailable. GBLUP is valued for its robustness and its ability to handle extensive datasets with numerous genetic markers, making it a widely adopted tool in modern breeding programs.

The first of the five machine learning models tested is Random Forest (RF), which represents a powerful model that has been previously used in genomic prediction [13–15]. This ensemble learning method constructs multiple decision trees during training and aggregates their predictions to enhance accuracy and mitigate overfitting. RF excels in managing complex interactions among multiple genes and demonstrates robustness against overfitting. The strengths of RF may make it particularly beneficial for predicting traits influenced by complex genetic factors. Support Vector Machine (SVM) is a machine learning technique designed for classification tasks that has been used in genomic selection before [14,16,17]. It identifies the hyperplane that best separates different classes in a high-dimensional space, making it well suited for genomic prediction where data dimensionality is high and traits are binary. Extreme Gradient Boosting (XGB) is an advanced implementation of gradient boosting algorithms. XGB builds models sequentially, with each new model aiming to correct the errors of its predecessors. This approach is highly efficient and effective at managing large, complex datasets, capturing intricate genetic patterns and potentially improving the accuracy of breeding value predictions. XGB's capability to handle complex trait architectures makes it an already utilized tool in genomic prediction [17,18]. Multilayer Perceptron (MLP) comprises neural network models inspired by the structure and function of the human brain. MLPs consist of multiple layers of interconnected nodes that can model complex, nonlinear relationships within the data. Although MLPs require substantial computational resources and large datasets for training, they have shown promise in capturing complex genetic interactions and providing accurate predictions for traits with intricate genetic architectures [19,20].

The effectiveness of these genomic prediction models is closely tied to the quality and quantity of the genomic data available. SNP markers are critical data points used in these models, offering detailed insights into genetic variation across the genome. The density of SNP markers—referring to the number of markers analyzed—can potentially impact the accuracy of predictive models [21–23]. Generally, higher marker density improves prediction accuracy by capturing a greater portion of the genetic variation associated with traits. However, an excessive number of markers can lead to overfitting, where the model becomes too specialized to the training data and performs poorly on new data during validation, can increase computation time, or is cost-prohibitive [24]. Therefore, finding an optimal marker density is crucial for balancing predictive accuracy with practical considerations such as cost and computational capacity.

The focus of this study is on specific traits that vary in their epidemiology, heritability, and case count: anodontia, distichiasis, oral papillomatosis, and distraction. These traits were selected to provide insight into how heritability estimates and the number of cases can influence model performance. They also represent health and behavioral concerns with tangible ramifications for both dogs and breeders. Preventing painful ocular irritations (distichiasis) and health complications (anodontia and papillomatosis) helps maintain dogs' well-being. Ensuring reliable working performance (distraction) can also minimize the number of dogs that need to be rehomed due to behavioral failures. Anodontia, a congenital absence of teeth, can lead to severe oral health issues and difficulties with feeding and has variability in presentation, cause, and severity [25]. Distichiasis, characterized by the presence of extra eyelashes that may grow towards the eye, can cause corneal irritation and impaired vision [26]. In other breeds, heritability has been estimated to be high, ranging from 0.276 to 0.720, indicating a potentially strong genetic component to the disease [27,28]. Oral papillomatosis, caused by the canine papillomavirus, results in benign oral tumors that, while generally non-life-threatening, can cause significant discomfort and interfere with eating [29,30]. As it is a viral infection, genetic factors may influence susceptibility, but environmental exposure plays a significant role, suggesting a lower heritability [29]. Distraction, a behavioral trait affecting a dog's focus, is crucial for guide dogs whose effectiveness depends on their attentiveness [31]. Behavioral traits often have moderate heritability and can be more subjective in assessment but selection for focus and non-distracted behavior has been identified as an important aspect of guide dog breeding [32].

This study aims to compare the performance of genomic prediction models—GBLUP, RF, SVM, XGB, and MLP—in predicting breeding values for three binary health traits and one behavioral trait in a population of guide dogs. By evaluating these models across different breeds—including Labrador Retrievers, Golden Retrievers, German Shepherds, and Labrador and Golden Retriever crosses—the research aims to illuminate model performance under various population strategies. This includes both within-breed and across-breed analyses and takes into account different reference population sizes. Additionally, this study assesses model performance across traits with different heritabilities and a behavioral trait, as well as the impact of SNP marker density on predictive accuracy, offering a deeper understanding of how different models handle breed-specific genetic variations and trait characteristics.

The implications of this study extend beyond guide dogs, potentially influencing breeding programs for other populations of animals with smaller reference populations and unique characteristics that differentiate them from research conducted primarily on agricultural species. Improved predictive accuracy can lead to healthier, more capable working dogs and contribute to more effective and informed breeding decisions. By addressing the challenges unique to canine breeding, such as smaller population sizes and less standardized phenotyping, this study contributes valuable knowledge toward closing the gap in the application of genomic selection in dogs compared to agricultural species.

2. Materials and Methods

2.1. Population

This study utilized a dataset provided by The Seeing Eye organization encompassing the 26-year period of 1998–2024. The Seeing Eye breeds and trains guide dogs, including

German Shepherds, Golden Retrievers, Labrador Retrievers, and Labrador and Golden Retriever crosses. The crosses are subsequently bred back to Labrador Retrievers. The crosses between Labradors and Golden Retrievers were originally introduced to combine the well-known temperament and working abilities of these breeds. Subsequent backcrossing to Labradors was employed to maintain a consistent appearance aligned with The Seeing Eye's working requirements while also aiming to enrich genetic diversity. Phenotypic data were collected during the puppy and young dog raising and training phases before their placement with a visually impaired individual or selection for breeding. Phenotypic data were collected by The Seeing Eye from birth until approximately 4.5 years of age. The dataset encompasses both successful guide dogs and those identified as failures. Disease traits were recorded by an in-house veterinary team, while behavioral traits, specifically distractibility, were assessed by trainers. Models were run on the all-breed data, then within the breeds of Labrador Retrievers (LRs), Golden Retrievers (GRs), and German Shepherds (GSs) and combining the more similar breeds and crosses of Labrador Retrievers, Golden Retrievers, and their crosses (LR/GR). Animals with incomplete phenotypic or genotypic data were excluded from the analysis. The sample size included 2020 dogs for the health traits and 1290 dogs for the behavioral trait.

2.2. The Seeing Eye Phenotypic Dataset

Health phenotypes were diagnosed through physical examinations conducted at specific developmental stages. These included a puppy physical around 5 weeks of age (or slightly older if the puppy was purchased), a pre-training physical (PTP) at approximately 14–16 months, a pre-breeder physical about 3 months after the PTP, and a pre-class physical roughly 4 months after the PTP. The Seeing Eye uses a coding system with defined diagnostic criteria for each trait to standardize health trait characterization. There are typically five veterinarians on staff with new veterinarians trained to use the coding system. Difficult cases are discussed amongst the team of veterinarians with an agreed upon final diagnostic code.

Distraction was assessed through evaluations during the training period. There are currently 31 dog trainers who are similarly trained to use a coding system with defined criteria for standardized trait characterization. Trainer identification is recorded along with the scoring of individual dog performance at each time point. Dogs return from puppy raisers and start specialized training at 14–16 months of age. The mid-term blindfold test typically occurs 6–8 weeks after the start of training, while the final blindfold test is conducted 12–14 weeks into training. Distractibility was initially rated on a categorical scale but was subsequently converted to a binary classification of 'high' and 'low' for this study due to ninety-three percent of animals being categorized into only two groups. This avoids overfitting the models to skewed ordinal data and allows for a more robust comparison to be made between all binary traits. One hundred and fifteen instructors scored dogs during the 26-year period.

For health traits, the sample consisted of 1188 Labrador Retrievers, 482 German Shepherds, 239 Golden Retrievers, and 111 Labrador and Golden Retriever crosses. For distraction, there were 847 Labrador Retrievers, 257 German Shepherds, 125 Golden Retrievers, and 61 Labrador and Golden Retriever crosses in the dataset. The case counts for each trait are listed below (Table 1).

Table 1. Counts of cases for the behavior and health traits by breed.

Trait	Total Cases	Labrador Retrievers	German Shepherds	Golden Retrievers	Labrador/Golden Retriever Crosses
Distichiasis	116	18	0	93	5
Anodontia	272	193	46	17	16
Oral Papillomatosis	216	125	43	28	20
High Distractibility	588	364	126	71	27

2.3. Genotypic Data

The Seeing Eye routinely collects whole-blood samples from all their dogs for DNA extraction and storage in their own biobank. They perform DNA extraction in house and create genotype subsets of dogs annually. Additional archived blood samples were extracted for this study and similarly genotyped. DNA extraction protocols in both laboratories followed general Qiagen (Qiagen, Germantown, MD, USA) PureGene Extraction protocols with in-house buffers. Quality control and quantification procedures ensured the integrity of the extracted DNA and compliance with genotyping standards. Genotyping was performed using three different single-nucleotide polymorphism (SNP) chips: the EMBARK (Embark Veterinary Inc., Boston, MA, USA) panel, a 220k Illumina (Illumina Inc., San Diego, CA, USA) microarray chip, and a 173k Illumina chip. The markers in the panel were mapped to CanFam 3.1 [33].

These datasets were merged, resulting in a comprehensive dataset of 239,478 SNPs across 2176 samples, with a genotyping rate of 91.4%. The data were filtered for genotype call rate and sample call rate genotypes (<90%), Hardy Weinberg equilibrium ($p < 1 \times 10^{-5}$), and minor allele frequency (<0.01) using PLINK ver 1.9 [34]. Post quality control, 166,463 SNPs and 2176 dogs were available for imputation. Imputation was performed following the method previously described in [35]. Briefly, the genotypes were phased and imputed using a reference panel mapped to CanFam 3.1 [33] comprising 660 dogs across 157 modern breeds and village dogs, including 15 German Shepherds, 20 Golden Retrievers, and 23 Labrador Retrievers [36]. We used Beagle ver 5.2 for both phasing and imputation [37,38]. Imputation was performed on a per-chromosome basis using mostly default settings; however, the effective population size (*ne*) was set to 200 based on previous work [39]. The imputed dataset included 10 dogs for which whole-genome sequence data were available. The accuracy of imputation was assessed based on genotype concordance and imputation quality scores [40]. Variants with a Beagle Dosage R-squared (DR2) $\geq$ 0.6 were retained for downstream analyses. The average genotype concordance rate between the imputed and true genotypes post DR2 filtering was 96.2%. The dataset was further pruned for LD, with PLINK ver 1.9 [34], using the -indep-pairwise flag (window size of 50 SNPs, step size of 5, and r2 threshold of 0.7), resulting in a final dataset of 1,219,623 SNPs. The dataset was then parsed to the 220 k SNPs present on the Illumina chip and then a random 50 k further subsample. The SNP density datasets, consisting of 1.2 million, 220 K, and 50 K SNPs, were selected to represent varying levels of marker densities commonly used in livestock genomic prediction. The 1.2 million SNP dataset reflects the potential of emerging, cost-effective, low-pass sequencing technologies to capture much higher density data compared to traditional SNP chips. Both smaller datasets correspond to existing work related to livestock, exploring similar densities' effects on model performance.

2.4. Genomic Prediction Models

Genomic Best Linear Unbiased Prediction (GBLUP): The genomic relationship matrix was constructed using SNP data to capture genetic relationships among individuals. Analyses and matrix construction were performed using SNP & Variation Suite v8.9.1 (Golden Helix, Inc., Bozeman, MT, USA, www.goldenhelix.com, accessed on 6 December 2024). Pseudo-heritability was calculated from a full-population GBLUP analysis of each trait. The restricted maximum likelihood (REML) algorithm used was Efficient Mixed-Model Association (EMMA).

All machine learning models were evaluated using the area under the receiver operating characteristic curve (AUROC) to optimize hyperparameters, as it allows for assessment even with potential low sensitivity issues in overtrained data. Hyperparameters were optimized with a random search followed by a grid search of the characteristics of the top-performing versions of the models.

Random Forest (RF): Various hyperparameters, including n estimators, max features, max depth, min samples split, min samples leaf, bootstrap, criterion, and class weight, were optimized. This analysis was performed using the scikit-learn v1.5.1 Python package [32].

Support Vector Machine (SVM): Various kernel functions (linear, polynomial, radial basis function) were tested. Model training involved optimizing hyperparameters such as C, kernel, gamma, degree, coef0, and shrinking probability. This analysis was performed using the scikit-learn v1.5.1 Python package [32].

Extreme Gradient Boosting (XGB): The XGB model underwent iterative training with hyperparameter tuning focusing on n estimators, max depth, learning rate, subsample, col sample by tree, gamma, lambda, alpha, scale pos weight, objective, and eval metric. This analysis was performed using the XGBoost v2.1.1 Python package [41].

Multilayer Perceptron (MLP): Neural network hyperparameters, including hidden layer sizes, activation, solver, alpha, batch size, learning rate, learning rate init, max iter, tol, momentum, and early stopping, were optimized. This analysis was performed using the scikit-learn v1.5.1 Python package [32].

2.5. Model Validation and Evaluation

Models were evaluated using 5-fold cross-validation, which involved partitioning the data into five subsets, training on four subsets and validation on the remaining subset. This ensured a comprehensive evaluation and allowed each individual to be in the validation dataset once. Performance metrics were calculated by comparing predicted phenotypes to true values from the run when a given individual was in the validation dataset.

AUROC Calculation: The area under the receiver operating characteristic curve (AUROC) measures the ability of a model to distinguish between classes. It is calculated by plotting the true positive rate (sensitivity) against the false positive rate (1—specificity) at various threshold settings. The AUROC score ranges from 0 (completely incorrect) to 0.5 (no discrimination) to 1 (perfect discrimination).

MCC Calculation: The Matthews correlation coefficient (MCC) is used to assess the quality of binary classifications, especially in the presence of imbalanced datasets. It considers true and false positives and negatives and is calculated using Equation (1).

$$MCC = (TP \times TN - FP \times FN)/\sqrt{((TP + FP) \times (TP + FN) \times (TN + FP) \times (TN + FN))} \quad (1)$$

TP is the number of true positives, TN is the number of true negatives, FP is the number of false positives, and FN is the number of false negatives. The values for the MCC range from −1 to 1, with 0 representing a naïve model.

3. Results

The traits varied in heritability among each other and when calculated from the different breed groups and SNP densities (Figure 1). Distichiasis had the highest pseudo-heritability calculated from the Genomic Best Linear Unbiased Prediction (GBLUP) analysis in the all-breed group (0.33–0.34) and Labrador/Golden Retriever (LR/GR) group (0.30–0.31). Within breeds, Golden Retrievers (GRs) had a higher heritability (0.18–0.19) than Labrador Retrievers (LRs) (0.09–0.11), and German Shepherds (GS) had no cases present. Anodontia was similarly heritable in the all-breed, GS, LR, and LR/GR breed groups (0.22–0.27) but had a heritability estimate of zero in the GR group, possibly due to low case numbers. Oral papillomatosis had a low heritability of 0.04–0.06 across all groups. Distraction was slightly more heritable among all groups (0.13–0.16) except the GR group, which had a lower heritability. There was no difference in the pseudo-heritabilities across the single-nucleotide polymorphism (SNP) datasets for any of the traits.

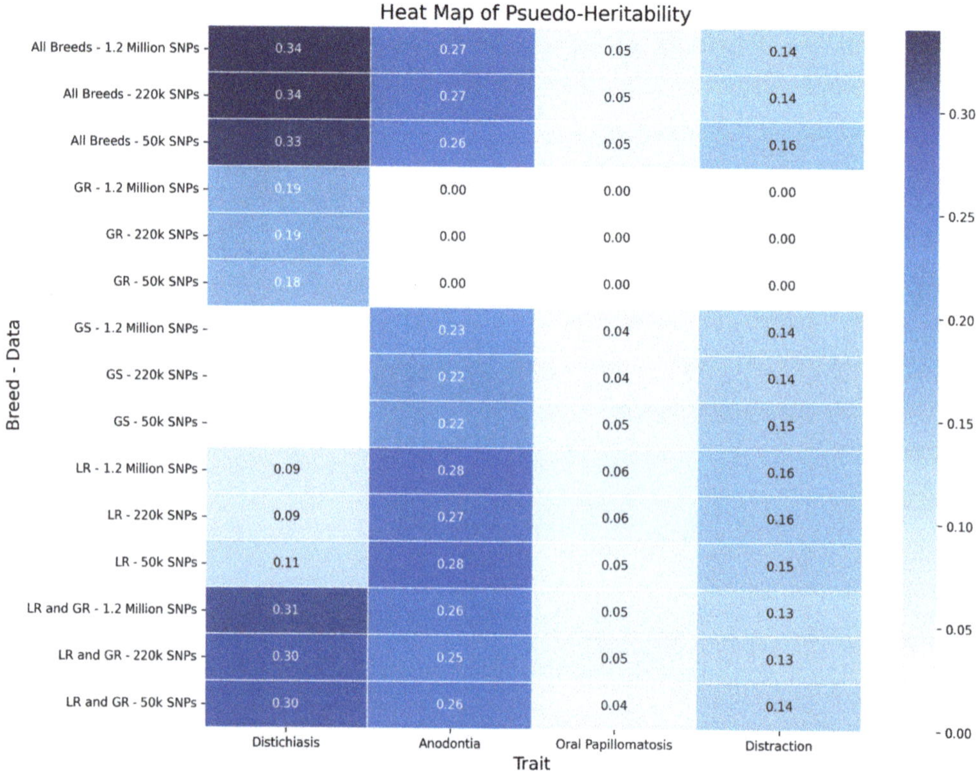

Figure 1. Heat Map of pseudo-heritability of traits for different breed datasets (all breeds (includes all dogs regardless of breed), GR = Golden Retriever, GS = German Shepherd, LR = Labrador Retriever, and LR/GR (includes all LRs, GRs, and LR/GR crosses)) and SNP density datasets. Heritabilities were calculated from GBLUP analysis of each trait organized by breed and dataset.

The trait with the highest overall performance was distichiasis (Figure 2), which had the highest heritability among the four traits assessed, ranging from 0.33–0.34 in the all-breed group to lower values within individual breed groups (GR: 0.18–0.19; LR: 0.09–0.11). The total number of cases included 116 dogs: 93 GR, 18 LR, 5 LR/GR, and 0 GS cases. There was no significant difference in model performance within each breed group/SNP dataset, but Extreme Gradient Boosting (XGB) had the highest average Matthews correlation coefficient (MCC) (0.25), while GBLUP and Random Forest (RF) had the highest area under the receiver operating characteristic curve (AUROC) across the dataset (0.81). Comparing across breeds, the all-breed and LR/GR groups had the highest performance, with an average MCC of 0.37 and 0.33 and AUROCs of 0.90 and 0.89, respectively. The GR group was next with an MCC of 0.16 but a lower AUROC of 0.63. No model was able to perform highly among the LR group, and there were no cases among the GS group. There was a slight decline in the average MCC as the SNP density decreased but no change in the average AUROC.

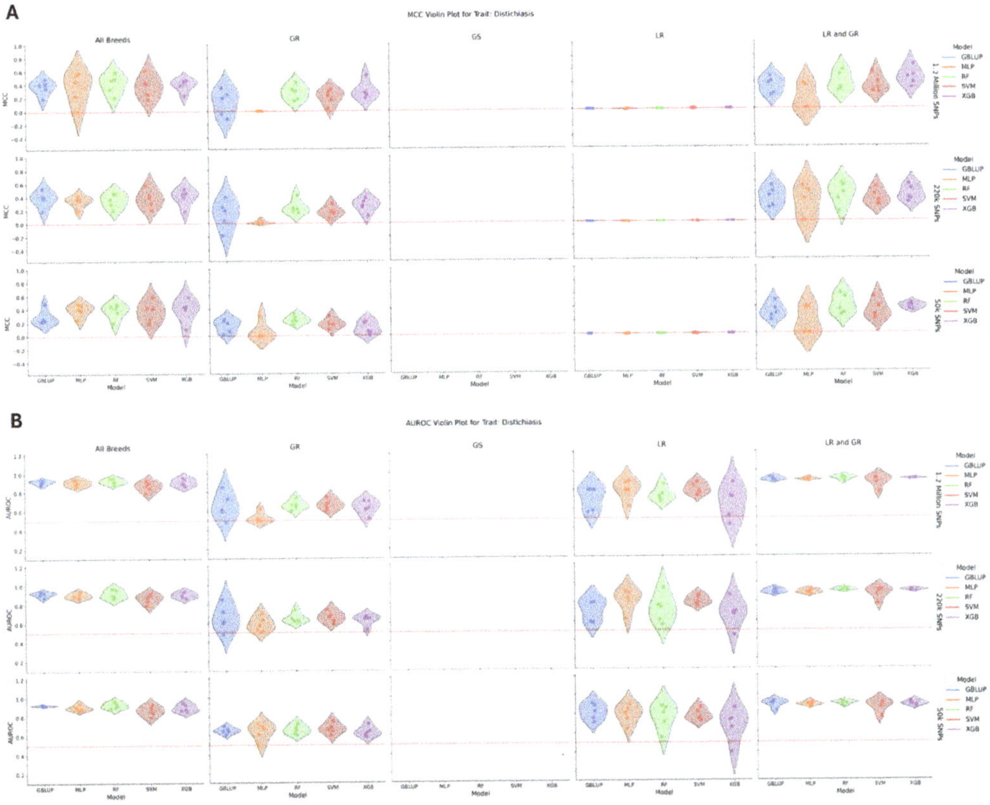

Figure 2. (**A**) A violin plot of model performance by MCC for distichiasis. The data are partitioned by SNP dataset in the rows and breed group in the columns with the models colored as follows: GBLUP—blue, MLP—orange, RF—green, SVM—red, and XGB—purple. The red dotted line represents the expected performance of a totally naïve model with an MCC of 0. There was no significant difference in model performance within each breed group/SNP dataset. (**B**) A violin plot of model performance by AUROC for distichiasis. The data are partitioned by SNP dataset in the rows and breed group in the columns with the models colored as follows: GBLUP—blue, MLP—orange, RF—green, SVM—red, and XGB—purple. The red dotted line represents the expected performance of a totally naïve model with an AUROC of 0.5. There was no significant difference in model performance within each breed group/SNP dataset.

In the case of anodontia, which had a heritability ranging from 0.22 to 0.27 in the all-breed, GS, LR, and LR/GR breed groups but had a heritability estimate of zero in the GR group, there was no significant difference in model performance within each breed group/SNP dataset (Figure 3). The total number of cases of anodontia included 272 dogs, with 17 GR, 193 LR, 16 LR/GR, and 46 GS cases. Support Vector Machine (SVM) outperformed all other models with an average MCC of 0.13 across the anodontia groups. The lowest performing models were RF and GBLUP with average MCCs of 0.07. However, Multilayer Perceptron (MLP) and RF had the lowest average AUROCs of 0.62 and 0.65, while GBLUP had the highest average AUROC of 0.68. The breed group with the highest average MCC was LR at 0.17, followed by LR/GR at 0.15 and the all-breed group at 0.13. Neither the GS nor GR group had high-performing models by either metric, with the GR group having a heritability estimate of zero and the GS group having a heritability similar to other breeds despite lower performance. There was no change in the MCC or AUROC across the SNP data types.

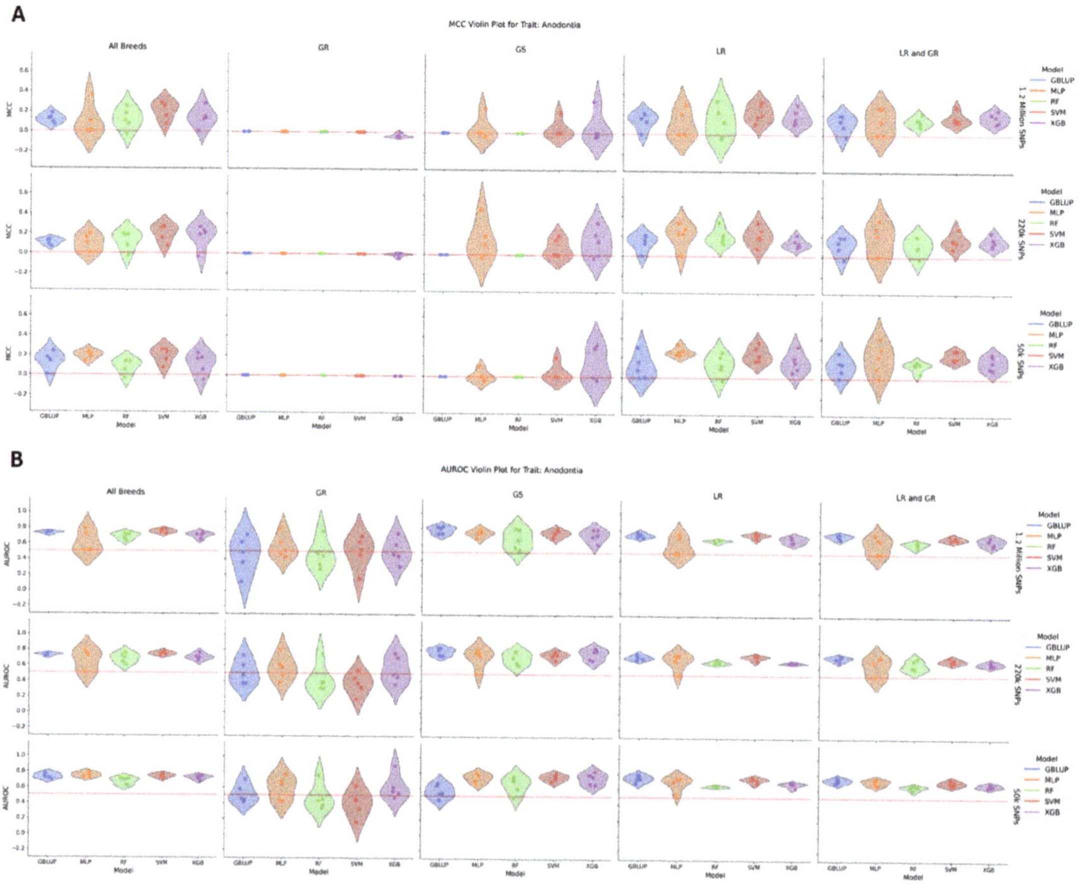

Figure 3. (**A**) A violin plot of model performance by MCC for anodontia. The data are partitioned by SNP dataset in the rows and breed group in the columns with the models colored as follows: GBLUP—blue, MLP—orange, RF—green, SVM—red, and XGB—purple. The red dotted line represents the expected performance of a totally naïve model with an MCC of 0. There was no significant difference in model performance within each breed group/SNP dataset. (**B**) A violin plot of model performance by AUROC for anodontia. The data are partitioned by SNP dataset in the rows and breed group in the columns with the models colored as follows: GBLUP—blue, MLP—orange, RF—green, SVM—red, and XGB—purple. The red dotted line represents the expected performance of a totally naïve model with an AUROC of 0.5. There was no significant difference in model performance within each breed group/SNP dataset.

For oral papillomatosis, which had a low heritability of 0.04–0.06 across all groups, there was no significant difference in model performance within each breed group/SNP dataset (Figure 4). The total number of cases included 216 dogs, with 193 LR, 43 GS, 28 GR, and 20 LR/GR cases. This trait exhibited the lowest average performance across all traits, with no standout model among any breed group by MCC or AUROC. Additionally, there was no increase or decrease in performance between the SNP density groups.

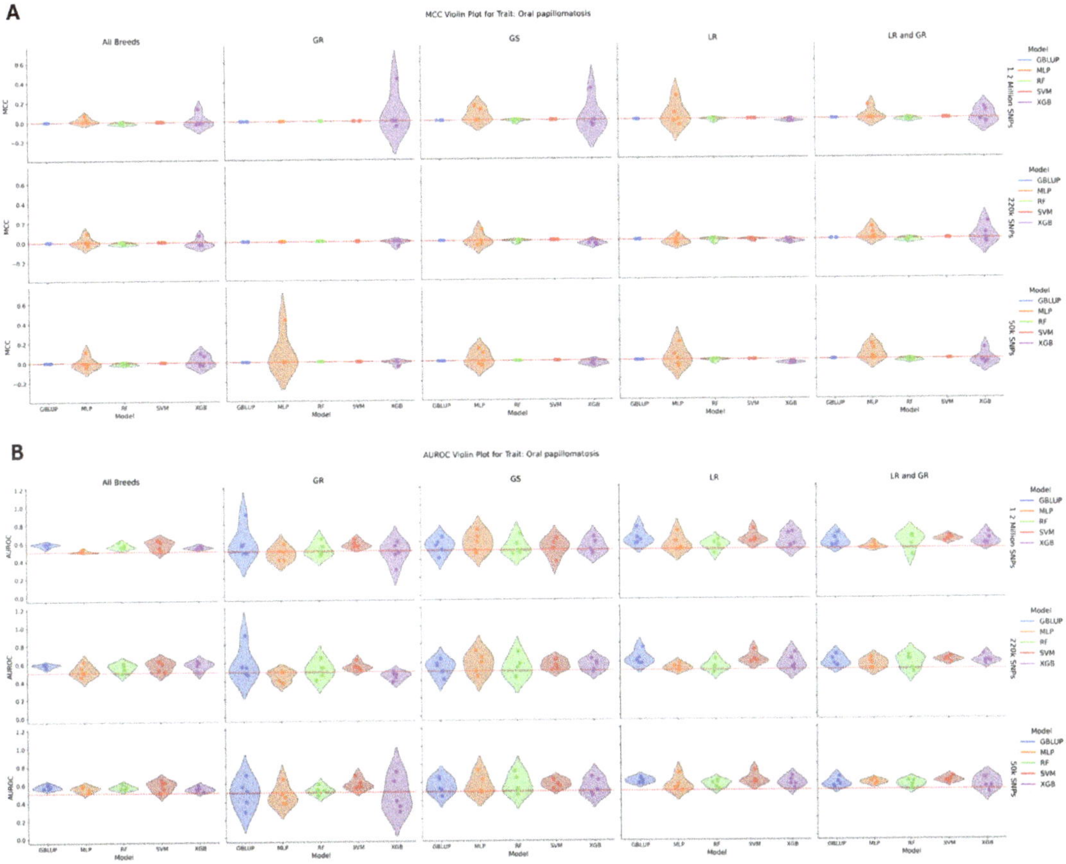

Figure 4. (**A**) A violin plot of model performance by MCC for oral papillomatosis. The data are partitioned by SNP dataset in the rows and breed group in the columns with the models colored as follows: GBLUP—blue, MLP—orange, RF—green, SVM—red, and XGB—purple. The red dotted line represents the expected performance of a totally naïve model with an MCC of 0. There was no significant difference in model performance within each breed group/SNP dataset. (**B**) A violin plot of model performance by AUROC for oral papillomatosis. The data are partitioned by SNP dataset in the rows and breed group in the columns with the models colored as follows: GBLUP—blue, MLP—orange, RF—green, SVM—red, and XGB—purple. The red dotted line represents the expected performance of a totally naïve model with an AUROC of 0.5. There was no significant difference in model performance within each breed group/SNP dataset.

For the behavioral trait, distraction, which had a heritability ranging from 0.13 to 0.16 across most groups but lower in the GR group, there was no significant difference in model performance within each breed group/SNP dataset (Figure 5). The total number of high-distraction cases included 588, with 364 LR, 126 GS, 71 GR, and 27 LR/GR cases. MLP had the lowest average MCC of 0.03, while RF, XGB, and GBLUP averaged at 0.12. All models, apart from MLP, had similar AUROC scores as well. The models performed best by MCC for the LR (0.15) and all-breed (0.13) datasets, followed by the LR/GR group (0.11), and struggled in the GR (0.05) and GS (0.04) groups. The AUROC scores followed the same trend, with the top three having AUROCs of 0.58–0.60, while the lower ones ranged between 0.52 and 0.53. There was a slight dip in the MCC from 1.2 million SNPs to 220k SNPs, with averages decreasing from 0.11 to 0.08, but the average MCC increased to 0.09 at 50k SNPs. The AUROC also remained consistent between 0.56 and 0.57.

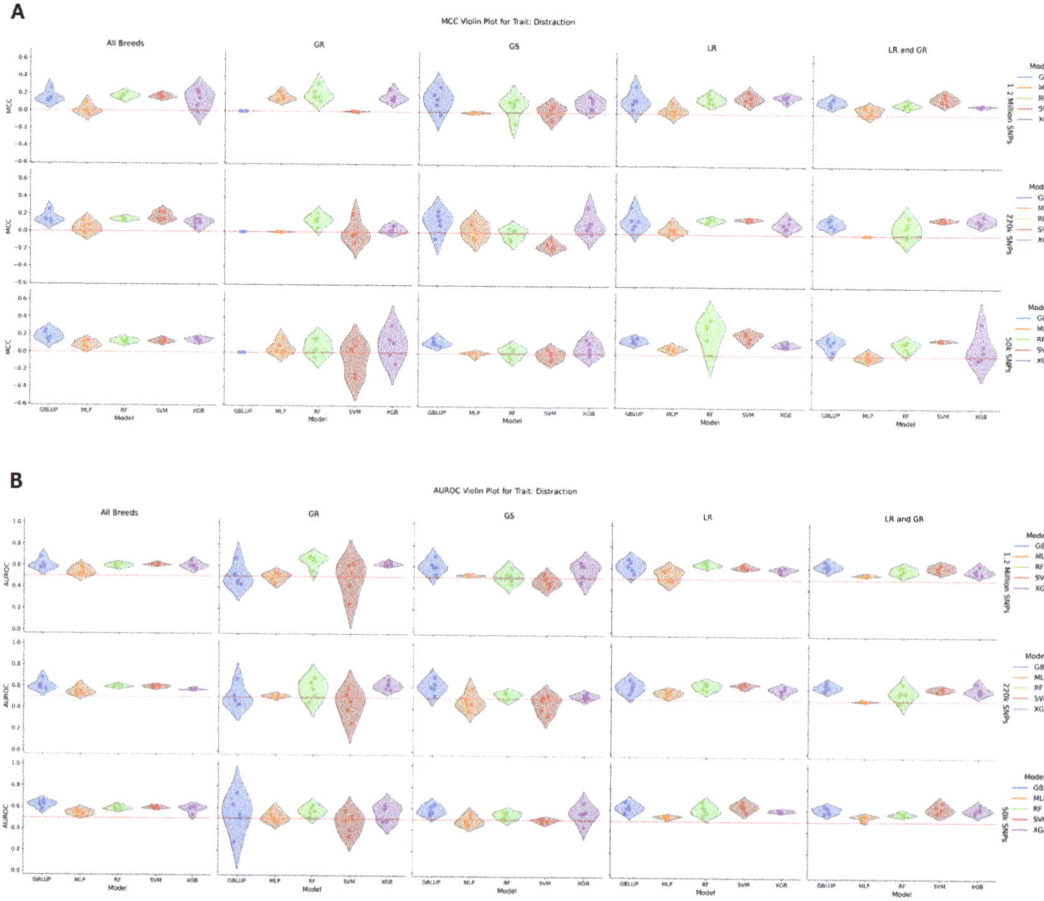

Figure 5. (**A**) A violin plot of model performance by MCC for distraction. The data are partitioned by SNP dataset in the rows and breed group in the columns with the models colored as follows: GBLUP—blue, MLP—orange, RF—green, SVM—red, and XGB—purple. The red dotted line represents the expected performance of a totally naïve model with an MCC of 0. There was no significant difference in model performance within each breed group/SNP dataset. (**B**) A violin plot of model performance by AUROC for distraction. The data are partitioned by SNP dataset in the rows and breed group in the columns with the models colored as follows: GBLUP—blue, MLP—orange, RF—green, SVM—red, and XGB—purple. The red dotted line represents the expected performance of a totally naïve model with an AUROC of 0.5. There was no significant difference in model performance within each breed group/SNP dataset.

4. Discussion

This study compared the performance of Genomic Best Linear Unbiased Prediction (GBLUP) with several machine learning (ML) models—Random Forest (RF), Support Vector Machine (SVM), Extreme Gradient Boosting (XGB), and Multilayer Perceptron (MLP)—in predicting binary health traits and a behavioral trait in guide dogs. The traits analyzed included distichiasis, anodontia, oral papillomatosis, and distractibility, demonstrating varying degrees of predictive success across the different models and datasets. The performance metrics used, Matthews correlation coefficient (MCC) and area under the receiver operating characteristic curve (AUROC), offer different perspectives on model effectiveness. While the AUROC measures a model's ability to distinguish between classes across all threshold levels, the MCC provides a balanced evaluation that accounts for true and false positives and negatives, making it particularly informative for imbalanced datasets. The differences observed between the MCC and AUROC across traits and breed groups indicate that the MCC may be more sensitive to class imbalance and prevalence rates, affecting the interpretation of predictive success in different scenarios. This underscores the importance of considering multiple performance metrics when evaluating model performance for various traits and populations.

Machine learning approaches are continuing to be investigated with various levels of success but continual improvement in efficiency, model architectures, and increased research offer a promising future. For their usage in breeding values, ML approaches have been found to perform comparably or even outperform other methods such as GBLUP depending on specific usage and datasets [14,16,17,42]. Both an advantage and drawback for the adoption of ML algorithms in various breeding programs is the documented diversity in model performance [43]. With the variety of models and their different strengths and weaknesses, finding a single robust model that performs well across many different phenotypes and reference populations remains a challenge. Another avenue of ML adoption includes deep learning approaches, although the "black box" nature of deep learning algorithms can make interpreting the results in genomic datasets more difficult [44].

Additionally, environmental factors such as training, diet, and general environmental exposure can influence health and behavior traits. The data included in this study mitigated environmental influence on model performance by using only dogs from The Seeing Eye which had very similar breeding, nutrition, training, and health evaluations. However, there was still variation across the puppy raiser households which reflects the variables affecting pet dogs. The goal of this study was to demonstrate the feasibility and value of genomic prediction in a population with reliable phenotypes yet common environmental exposure inside and outside of households similar to pet dogs. Identifying and incorporating environmental factors influencing traits, especially on a broader and more diversified dataset of dogs, is an important avenue for future research and application of gEBVs.

Across the models, no single approach consistently outperformed others for any specific data type, as the models largely overlapped in the ranges of their run scores for both MCC and AUROC. This finding aligns with previous research that utilized GBLUP as a benchmark, confirming its robustness [45]. Considering the lack of parameter optimization, which reduces steps in breeding value creation, along with its high performance, GBLUP appears to be well suited for estimating binary trait breeding values in small, closely related populations of working dogs, as present in this study.

Additionally, there was no discernible trend of performance increase or decrease in either the MCC or AUROC with different single-nucleotide polymorphism (SNP) densities. This observation supports previous studies demonstrating that lower density SNP datasets can still effectively construct breeding values in poultry and cattle [22–24]. For canine breeding values, this suggests that lower cost SNP chips may suffice, negating the necessity for imputation to larger datasets for binary health traits.

Distichiasis exhibited the highest overall predictive performance and pseudo-heritability. Both GBLUP and RF models achieved the highest AUROC of 0.81, while Extreme Gradient Boosting (XGB) obtained the highest average MCC of 0.25 across the breed groups. Predictive accuracy varied among breeds, with the all-breed and Labrador Retriever/Golden Retriever groups achieving AUROCs of 0.90 and 0.89, respectively. This is consistent with prior research highlighting a significant genetic component for distichiasis, particularly in breeds such as the English Cocker Spaniel and Havanese [27,28]. The high pseudo-heritability and model performance within the breeds in the present study may be attributed to the close lineage of individuals exhibiting the trait. Interestingly, while Labrador Retrievers had a lower pseudo-heritability compared to Golden Retrievers, the combination of Labradors and Golden Retrievers outperformed both individual groups based on the MCC and AUROC. Additionally, the all-breed group had the highest pseudo-heritability and correspondingly higher scores for the MCC and AUROC. This highlights the importance of reference population construction. The increased performance among the mixed-breed groupings may be due in part to breed-specific markers being used to exclude dogs from the breeds with low to no cases, which can artificially increase the model's performance. As such, understanding the breeds and prevalence rates can help in identifying possible inflation in reported model performance. In smaller and highly homogeneous breed groups, exploring breed-specific variants could further refine predictions, but it also raises the risk of overfitting if the variants are limited to very few breeding lines. While adding dogs, particularly increasing the number of cases, can increase the external validity, introducing dogs without the uniform phenotypic and genotypic data in the present population introduces additional variation into the theoretical dataset and may alter internal validity.

Anodontia and distractibility displayed higher predictive performance in some breed groupings compared to others, with Golden Retrievers showing the lowest pseudo-heritability, MCC, and AUROC for both traits. This may be due to the fact that the Golden Retriever group was the smallest, comprising 125 individuals in the distractibility dataset and 239 in the health traits dataset. This follows the trend present in distichiasis, where the Golden Retriever group had similarly lower predictive accuracy. Across all other groups, the pseudo-heritabilities and performance metrics were similar within each trait. This is despite the variation in group size and case counts between the Labrador Retrievers and German Shepherds. This result may suggest that Golden Retrievers may have either breed-specific variants that do not allow for as accurate predictions or that a threshold of population size is needed for the model to reach a plateau in performance. Possible explanations that may also be at play include the breed-specific variants having more complex polygenic interactions and only being identifiable with sufficient power among the specific population, in this case Golden Retrievers.

Conversely, oral papillomatosis offered an opportunity to investigate the effects of model and dataset on performance for low-heritability traits. However, as none of the models or datasets exhibited significant differences, this suggests that model selection may have less impact on predictive performance for traits with low heritability. Consequently, a standardized modeling approach may be sufficient for estimating breeding values in such cases, regardless of the specific heritability of the binary health trait.

The variability in breed composition, genetic diversity, and trait expression within the reference population illustrates the challenges inherent in constructing breeding values for functional applications. As genomic breeding values are developed for dogs, careful consideration must be given to the groups on which the models are trained. Working dogs raised in a single location offer consistency and standardized phenotypes; however, this may lead to diminished external validity due to the homogeneity of these closed populations compared to the broader breed population.

Future directions for breeding values in dogs must address unique challenges, including the lack of uniform selection criteria across different breeds and breeders. Standardization suggestions have been proposed, and working dog colonies may serve as ideal testing grounds for these approaches [46]. Additionally, the fragmented nature of dog breeding, with many breeders working with small numbers of animals with unique selective goals, may pose a challenge. The lack of a large overseeing authority akin to the Council on Dairy Cattle Breeding (CDCB) in dairy cattle to standardize breeding goals and phenotypes and collect both phenotypic and genotypic datasets presents another significant hurdle to standardization attempts. Although both the American Kennel Club (AKC) and the International Working Dog Registry have the potential to fill this gap, they currently lack large reference populations with genotypic and phenotypic data. Finally, as evidenced by the variation in the results across breed in the present study, reference populations must be carefully constructed with target populations in mind when attempting to construct a functional breeding value.

Our findings demonstrate that GBLUP, along with ML models, can effectively predict both health and behavioral traits in guide dogs, suggesting that genomic estimated breeding values (gEBVs) are valuable tools in selection programs. By enabling breeders to make informed decisions based on genetic potential, gEBVs can accelerate genetic progress and improve the overall health and performance of dog populations. Furthermore, our comparative analysis revealed that no single model consistently outperformed others across different traits and breeds, highlighting the robustness of GBLUP due to the absence of hyperparameter optimization along with the potential flexibility in model choice depending on specific breeding objectives. These results emphasize the practicality of incorporating gEBVs into canine breeding strategies, especially when considering traits of varying heritabilities and prevalence. The results also provide a proof of concept for the creation of multi-trait indices to capture the genomic influence in a series of related traits. Despite some challenges, there remains substantial potential for improvement in the current landscape. Advancements in validated health trait breeding values and the development of selection indices incorporating multiple traits could significantly enhance the utility of breeding values in dogs.

5. Conclusions

In conclusion, this study demonstrates that genomic prediction models—including Genomic Best Linear Unbiased Prediction (GBLUP) and machine learning approaches like Random Forest (RF), Support Vector Machine (SVM), Extreme Gradient Boosting (XGB), and Multilayer Perceptron (MLP)—are effective tools for predicting breeding values for both health and behavioral traits in guide dogs. By evaluating these models across different breeds, traits with varying heritabilities, and SNP marker densities, we found that all models performed similarly, with no single model consistently outperforming the others. Notably, GBLUP emerged as the most logistically efficient model for breeders to quickly master and implement (due to the lack of hyperparameter optimization), making it a practical choice for canine breeding programs.

These findings suggest that lower density SNP datasets are sufficient for constructing accurate genomic estimated breeding values (gEBVs), potentially reducing the costs associated with high-density genotyping. This is particularly significant for breeding programs with limited resources. By enabling breeders to make more informed selection decisions based on genetic potential, the incorporation of genomic prediction models can accelerate genetic progress and improve the overall health and performance of dog populations. Future research should focus on standardizing phenotypic assessments and expanding reference populations to enhance the utility and applicability of genomic selection in canine breeding, bridging the gap between dogs and agricultural species in genetic breeding practices.

Author Contributions: Conceptualization, K.M.E. and H.J.H.; methodology, K.S. and J.A.T.; software, K.S. and J.A.T.; validation, K.S. and J.A.T.; formal analysis, J.A.T.; investigation, J.A.T.; resources, K.M.E., K.C.Q. and H.J.H.; data curation, K.M.E., K.C.Q. and H.J.H.; writing—original draft preparation, J.A.T.; writing—review and editing, H.J.H., K.M.E., K.C.Q. and K.S.; visualization, J.A.T.; supervision, H.J.H. and K.M.E.; project administration, H.J.H.; funding acquisition, H.J.H., K.M.E. and K.S. All authors have read and agreed to the published version of the manuscript.

Funding: This research was funded by the AKC (American Kennel Club) Canine Health Foundation, grant number 03069.

Institutional Review Board Statement: The animal study protocol was approved by the Institutional Animal Care and Use Committee (IACUC) of Cornell University (2014-0121; approved 21 January 2021).

Informed Consent Statement: Not applicable.

Data Availability Statement: The datasets presented in this article are not readily available as they are the private ownership of The Seeing Eye. Author, Katy M. Evans, may be contacted to inquire about the datasets. The script for the machine learning models is available on google colab: https://colab.research.google.com/drive/1CTYO6ly20ZpTwLsJGXace4KYbBXZL7gP?usp=sharing (accessed on 6 December 2024).

Acknowledgments: The authors thank Patricia Tardive from The Seeing Eye and are grateful for her work in the DNA laboratory. The authors thank Toni Reverter-Gomez and Laercio Porto Neto for their expertise and training on genomic breeding values.

Conflicts of Interest: Authors K.M.E. and K.C.Q. were employed by The Seeing Eye. The remaining authors declare that the research was conducted in the absence of any commercial or financial relationships that could be construed as a potential conflict of interest. The authors declare that this study received funding from the AKC Canine Health Foundation. The funder was not involved in the study design, collection, analysis, interpretation of data, the writing of this article or the decision to submit it for publication.

References

1. Worboys, M.; Strange, J.-M.; Pemberton, N. *The Invention of the Modern Dog*; Johns Hopkins University Press: Baltimore, MD, USA, 2022; ISBN 978-1-4214-2659-4.
2. Hedhammar, Å. Swedish Experiences From 60 Years of Screening and Breeding Programs for Hip Dysplasia—Research, Success, and Challenges. *Front. Vet. Sci.* **2020**, *7*, 228. [CrossRef]
3. James, H.K.; McDonnell, F.; Lewis, T.W. Effectiveness of Canine Hip Dysplasia and Elbow Dysplasia Improvement Programs in Six UK Pedigree Breeds. *Front. Vet. Sci.* **2020**, *6*, 490. [CrossRef]
4. Soo, M.; Worth, A. Canine Hip Dysplasia: Phenotypic Scoring and the Role of Estimated Breeding Value Analysis. *N. Z. Vet. J.* **2015**, *63*, 69–78. [CrossRef]
5. Leighton, E.A.; Holle, D.; Biery, D.N.; Gregor, T.P.; McDonald-Lynch, M.B.; Wallace, M.L.; Reagan, J.K.; Smith, G.K. Genetic Improvement of Hip-Extended Scores in 3 Breeds of Guide Dogs Using Estimated Breeding Values: Notable Progress but More Improvement Is Needed. *PLoS ONE* **2019**, *14*, e0212544. [CrossRef] [PubMed]
6. Farstad, W. Ethics in Animal Breeding. *Reprod. Domest. Anim.* **2018**, *53*, 4–13. [CrossRef]
7. Tsairidou, S.; Anacleto, O.; Woolliams, J.A.; Doeschl-Wilson, A. Enhancing Genetic Disease Control by Selecting for Lower Host Infectivity and Susceptibility. *Heredity* **2019**, *122*, 742–758. [CrossRef]
8. Menor-Campos, D.J. Ethical Concerns about Fashionable Dog Breeding. *Animals* **2024**, *14*, 756. [CrossRef] [PubMed]
9. Wiggans, G.R.; Carrillo, J.A. Genomic Selection in United States Dairy Cattle. *Front. Genet.* **2022**, *13*, 994466. [CrossRef]
10. Calus, M.P.L. Genomic Breeding Value Prediction: Methods and Procedures. *Animal* **2010**, *4*, 157–164. [CrossRef] [PubMed]
11. Scholtens, M.; Lopez-Villalobos, N.; Lehnert, K.; Snell, R.; Garrick, D.; Blair, H.T. Advantage of Including Genomic Information to Predict Breeding Values for Lactation Yields of Milk, Fat, and Protein or Somatic Cell Score in a New Zealand Dairy Goat Herd. *Animals* **2020**, *11*, 24. [CrossRef]
12. Clark, S.A.; Van Der Werf, J. Genomic Best Linear Unbiased Prediction (gBLUP) for the Estimation of Genomic Breeding Values. In *Genome-Wide Association Studies and Genomic Prediction*; Gondro, C., Van Der Werf, J., Hayes, B., Eds.; Methods in Molecular Biology; Humana Press: Totowa, NJ, USA, 2013; Volume 1019, pp. 321–330, ISBN 978-1-62703-446-3.
13. Naderi, S.; Yin, T.; König, S. Random Forest Estimation of Genomic Breeding Values for Disease Susceptibility over Different Disease Incidences and Genomic Architectures in Simulated Cow Calibration Groups. *J. Dairy Sci.* **2016**, *99*, 7261–7273. [CrossRef]
14. Ogutu, J.O.; Piepho, H.-P.; Schulz-Streeck, T. A Comparison of Random Forests, Boosting and Support Vector Machines for Genomic Selection. *BMC Proc.* **2011**, *5*, S11. [CrossRef] [PubMed]

15. Sarkar, R.K.; Rao, A.R.; Meher, P.K.; Nepolean, T.; Mohapatra, T. Evaluation of Random Forest Regression for Prediction of Breeding Value from Genomewide SNPs. *J. Genet.* **2015**, *94*, 187–192. [CrossRef] [PubMed]
16. Srivastava, S.; Lopez, B.I.; Kumar, H.; Jang, M.; Chai, H.-H.; Park, W.; Park, J.-E.; Lim, D. Prediction of Hanwoo Cattle Phenotypes from Genotypes Using Machine Learning Methods. *Animals* **2021**, *11*, 2066. [CrossRef]
17. Hamadani, A.; Ganai, N.A.; Mudasir, S.; Shanaz, S.; Alam, S.; Hussain, I. Comparison of Artificial Intelligence Algorithms and Their Ranking for the Prediction of Genetic Merit in Sheep. *Sci. Rep.* **2022**, *12*, 18726. [CrossRef] [PubMed]
18. Li, B.; Zhang, N.; Wang, Y.-G.; George, A.W.; Reverter, A.; Li, Y. Genomic Prediction of Breeding Values Using a Subset of SNPs Identified by Three Machine Learning Methods. *Front. Genet.* **2018**, *9*, 237. [CrossRef] [PubMed]
19. Ghotbaldini, H.; Mohammadabadi, M.; Nezamabadi-pour, H.; Babenko, O.I.; Bushtruk, M.V.; Tkachenko, S.V. Predicting Breeding Value of Body Weight at 6-Month Age Using Artificial Neural Networks in Kermani Sheep Breed. *Acta Sci. Anim. Sci.* **2019**, *41*, 45282. [CrossRef]
20. Rosado, R.D.S.; Cruz, C.D.; Barili, L.D.; De Souza Carneiro, J.E.; Carneiro, P.C.S.; Carneiro, V.Q.; Da Silva, J.T.; Nascimento, M. Artificial Neural Networks in the Prediction of Genetic Merit to Flowering Traits in Bean Cultivars. *Agriculture* **2020**, *10*, 638. [CrossRef]
21. Lopez, B.I.M.; An, N.; Srikanth, K.; Lee, S.; Oh, J.-D.; Shin, D.-H.; Park, W.; Chai, H.-H.; Park, J.-E.; Lim, D. Genomic Prediction Based on SNP Functional Annotation Using Imputed Whole-Genome Sequence Data in Korean Hanwoo Cattle. *Front. Genet.* **2021**, *11*, 603822. [CrossRef] [PubMed]
22. Salvian, M.; Moreira, G.C.M.; Silveira, R.M.F.; Reis, Â.P.; Dias D'auria, B.; Pilonetto, F.; Gervásio, I.C.; Ledur, M.C.; Coutinho, L.L.; Spangler, M.L.; et al. Estimation of Breeding Values Using Different Densities of SNP to Inform Kinship in Broiler Chickens. *Livest. Sci.* **2023**, *267*, 105124. [CrossRef]
23. Solberg, T.R.; Sonesson, A.K.; Woolliams, J.A.; Meuwissen, T.H.E. Genomic Selection Using Different Marker Types and Densities. *J. Anim. Sci.* **2008**, *86*, 2447–2454. [CrossRef] [PubMed]
24. Reverter, A.; Hudson, N.J.; McWilliam, S.; Alexandre, P.A.; Li, Y.; Barlow, R.; Welti, N.; Daetwyler, H.; Porto-Neto, L.R.; Dominik, S. A Low-Density SNP Genotyping Panel for the Accurate Prediction of Cattle Breeds. *J. Anim. Sci.* **2020**, *98*, skaa337. [CrossRef] [PubMed]
25. Feuer, R.; Mulherin, B.L. Different Presentations of Unerupted Canine Teeth in Three Juvenile Dogs. *Vet. Rec. Case Rep.* **2023**, *11*, e652. [CrossRef]
26. Jondeau, C.; Gounon, M.; Bourguet, A.; Chahory, S. Epidemiology and Clinical Significance of Canine Distichiasis: A Retrospective Study of 291 Cases. *Vet. Ophthalmol.* **2023**, *26*, 339–346. [CrossRef]
27. Bellamy, K.K.L.; Lingaas, F.; Madsen, P. Heritability of Distichiasis in Havanese Dogs in Norway. *Canine Genet. Epidemiol.* **2021**, *8*, 11. [CrossRef] [PubMed]
28. Petersen, T.; Proschowsky, H.F.; Hardon, T.; Rasch, S.N.; Fredholm, M. Prevalence and Heritability of Distichiasis in the English Cocker Spaniel. *Canine Genet. Epidemiol.* **2015**, *2*, 11. [CrossRef] [PubMed]
29. Yhee, J.-Y.; Kwon, B.-J.; Kim, J.-H.; Yu, C.-H.; Im, K.-S.; Lee, S.-S.; Lyoo, Y.-S.; Chang, B.-J.; Sur, J.-H. Characterization of Canine Oral Papillomavirus by Histopathological and Genetic Analysis in Korea. *J. Vet. Sci.* **2010**, *11*, 21. [CrossRef] [PubMed]
30. Thaiwong, T.; Sledge, D.G.; Wise, A.G.; Olstad, K.; Maes, R.K.; Kiupel, M. Malignant Transformation of Canine Oral Papillomavirus (CPV1)-Associated Papillomas in Dogs: An Emerging Concern? *Papillomavirus Res.* **2018**, *6*, 83–89. [CrossRef] [PubMed]
31. Batt, L.S.; Batt, M.S.; Baguley, J.A.; McGreevy, P.D. Factors Associated with Success in Guide Dog Training. *J. Vet. Behav.* **2008**, *3*, 143–151. [CrossRef]
32. Pedregosa, F.; Varoquaux, G.; Gramfort, A.; Michel, V.; Thirion, B.; Grisel, O.; Blondel, M.; Müller, A.; Nothman, J.; Louppe, G.; et al. Scikit-Learn: Machine Learning in Python. *J. Mach. Learn. Res.* **2012**, *12*, 2825–2830. [CrossRef]
33. Hoeppner, M.P.; Lundquist, A.; Pirun, M.; Meadows, J.R.S.; Zamani, N.; Johnson, J.; Sundström, G.; Cook, A.; FitzGerald, M.G.; Swofford, R.; et al. An Improved Canine Genome and a Comprehensive Catalogue of Coding Genes and Non-Coding Transcripts. *PLoS ONE* **2014**, *9*, e91172. [CrossRef]
34. Chang, C.C.; Chow, C.C.; Tellier, L.C.; Vattikuti, S.; Purcell, S.M.; Lee, J.J. Second-Generation PLINK: Rising to the Challenge of Larger and Richer Datasets. *Gigascience* **2015**, *4*, s13742-015. [CrossRef]
35. Srikanth, K.; Von Pfeil, D.J.F.; Stanley, B.J.; Griffitts, C.; Huson, H.J. Genome Wide Association Study with Imputed Whole Genome Sequence Data Identifies a 431 Kb Risk Haplotype on CFA18 for Congenital Laryngeal Paralysis in Alaskan Sled Dogs. *Genes* **2022**, *13*, 1808. [CrossRef] [PubMed]
36. Plassais, J.; Kim, J.; Davis, B.W.; Karyadi, D.M.; Hogan, A.N.; Harris, A.C.; Decker, B.; Parker, H.G.; Ostrander, E.A. Whole Genome Sequencing of Canids Reveals Genomic Regions under Selection and Variants Influencing Morphology. *Nat. Commun.* **2019**, *10*, 1489. [CrossRef] [PubMed]
37. Browning, B.L.; Tian, X.; Zhou, Y.; Browning, S.R. Fast Two-Stage Phasing of Large-Scale Sequence Data. *Am. J. Hum. Genet.* **2021**, *108*, 1880–1890. [CrossRef]

38. Browning, B.L.; Zhou, Y.; Browning, S.R. A One-Penny Imputed Genome from Next-Generation Reference Panels. *Am. J. Hum. Genet.* **2018**, *103*, 338–348. [CrossRef]
39. Friedenberg, S.G.; Meurs, K.M. Genotype Imputation in the Domestic Dog. *Mamm. Genome* **2016**, *27*, 485–494. [CrossRef]
40. Ramnarine, S.; Zhang, J.; Chen, L.-S.; Culverhouse, R.; Duan, W.; Hancock, D.B.; Hartz, S.M.; Johnson, E.O.; Olfson, E.; Schwantes-An, T.-H.; et al. When Does Choice of Accuracy Measure Alter Imputation Accuracy Assessments? *PLoS ONE* **2015**, *10*, e0137601. [CrossRef]
41. Chen, T.; Guestrin, C. XGBoost: A Scalable Tree Boosting System. In Proceedings of the 22nd ACM SIGKDD International Conference on Knowledge Discovery and Data Mining, San Francisco, CA, USA, 13–17 August 2016; ACM: New York, NY, USA, 2016; pp. 785–794.
42. Liang, M.; Miao, J.; Wang, X.; Chang, T.; An, B.; Duan, X.; Xu, L.; Gao, X.; Zhang, L.; Li, J.; et al. Application of Ensemble Learning to Genomic Selection in Chinese Simmental Beef Cattle. *J. Anim. Breed. Genet.* **2021**, *138*, 291–299. [CrossRef] [PubMed]
43. Chafai, N.; Hayah, I.; Houaga, I.; Badaoui, B. A Review of Machine Learning Models Applied to Genomic Prediction in Animal Breeding. *Front. Genet.* **2023**, *14*, 1150596. [CrossRef]
44. Xu, C.; Jackson, S.A. Machine Learning and Complex Biological Data. *Genome Biol.* **2019**, *20*, 76. [CrossRef] [PubMed]
45. Hay, E.H. Machine Learning for the Genomic Prediction of Growth Traits in a Composite Beef Cattle Population. *Animals* **2024**, *14*, 3014. [CrossRef] [PubMed]
46. Chen, F.L.; Zimmermann, M.; Hekman, J.P.; Lord, K.A.; Logan, B.; Russenberger, J.; Leighton, E.A.; Karlsson, E.K. Advancing Genetic Selection and Behavioral Genomics of Working Dogs Through Collaborative Science. *Front. Vet. Sci.* **2021**, *8*, 662429. [CrossRef] [PubMed]

MDPI AG
Grosspeteranlage 5
4052 Basel
Switzerland
Tel.: +41 61 683 77 34

Animals Editorial Office
E-mail: animals@mdpi.com
www.mdpi.com/journal/animals

www.ingramcontent.com/pod-product-compliance
Lightning Source LLC
Chambersburg PA
CBHW061450210926
51850CB00008B/1672
* 9 7 8 3 7 2 5 8 5 1 8 1 2 *